Collins

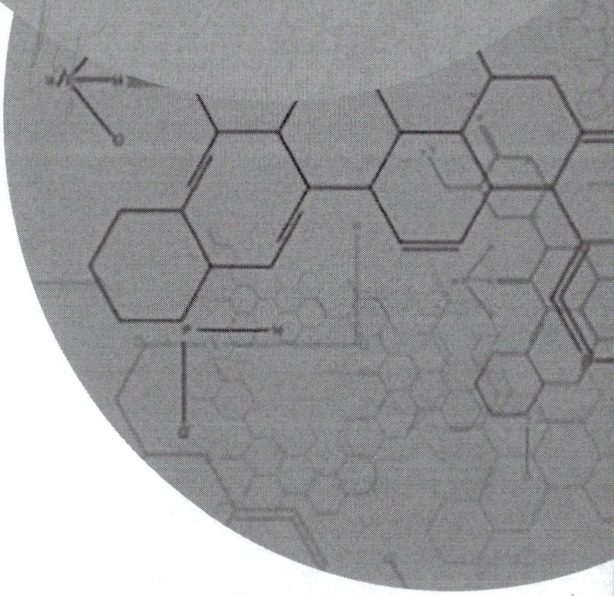

AQA GCSE (9-1) Biology
Teacher Pack

Sunetra Berry
Louise Smiles
Series editor: Ed Walsh

William Collins' dream of knowledge for all began with the publication of his first book in 1819. A self-educated mill worker, he not only enriched millions of lives, but also founded a flourishing publishing house. Today, staying true to this spirit, Collins books are packed with inspiration, innovation and practical expertise. They place you at the centre of a world of possibility and give you exactly what you need to explore it.

Collins. Freedom to teach

HarperCollins Publishers
1 London Bridge Street
London SE1 9GF

Browse the complete Collins catalogue at
www.collins.co.uk

First edition 2016

10 9 8 7 6 5 4 3 2 1

ISBN 978-0-00-815879-8

www.collins.co.uk

A catalogue record for this book is available from the British Library

Commissioned by Lucy Rowland and Lizzie Catford
Edited by Hamish Baxter
Project managed by Elektra Media Ltd
Development and copy edited by Rebecca Ramsden
Proofread by Amanda and Mike Harman
Illustrated by Ann Paganuzzi
Typeset by Jouve and Ken Vail Graphic Design
Cover design by We are Laura
Printed and bound by Martins the Printers
Cover images © Shutterstock/Komsan Loonprom, Shutterstock/Everett Historical

ACKNOWLEDGEMENTS

The Publishers gratefully acknowledge the permissions to reproduce copyright material in this book. Every effort has been made to contact the holders of copyright material, but if any have been inadvertently overlooked, the publisher will be pleased to make necessary arrangements at the first opportunity.

Worksheets

Worksheet 1.3 AMMRF, UNIVERSITY OF SYDNEY/SCIENCE PHOTO LIBRARY, CNRI/SCIENCE PHOTO LIBRARY, DR JEREMY BURGESS/SCIENCE PHOTO LIBRARY; Worksheet 1.4.1, 1.4.2 CNK02/Shutterstock, Jose Luis Calvo/Shutterstock, Pan Xunbin/Shutterstock, CNK02/Shutterstock; Worksheet 1.5.1, 1.5.2 Lukiyanova Natalia/frenta/Shutterstock, Dimarion/Shutterstock; Worksheet 1.6 RAJ CREATIONZS/Shutterstock, royaltystockphoto.com/Shutterstock, Jose Luis Calvo/Shutterstock, Sebastian Kaulitzki/Shutterstock; Worksheet 1.10 Designua/Shutterstock, ellepigrafica/Shutterstock; Worksheet 1.11 Martial Red/Shutterstock, RAJ CREATIONZS/Shutterstock, royaltystockphoto.com/Shutterstock, Sebastian Kaulitzki/Shutterstock, Jose Luis Calvo/Shutterstock; Worksheet 1.13 Satirus/Shutterstock, KuLouKu/Shutterstock, Satirus/Shutterstock; Worksheet 5.1.1, 5.1.2 SpeedKingz/Shutterstock, Andrey Armyagov/Shutterstock; Worksheet 5.25.2, 5.25.3, 5.25.3 Anest/Shutterstock, Pat_Hastings/Shutterstock, Studio Barcelona/Shutterstock, Marco Uliana/Shutterstock; Worksheet 5.28 sasaken/Shutterstock, ChameleonsEye/Shutterstock, Jamroen Jaiman/Shutterstock, Elena Elisseeva/Shutterstock, Christophe Baudot/Shutterstock, Quang Ho/Shutterstock, Nattika/Shutterstock, alexsvirid/Shutterstock, Studio Barcelona/Shutterstock; Worksheet 7.2.1, 7.2.2, 7.2.3 Steve McWilliam/Shutterstock; Worksheet 7.4.1, 7.4.2 MarcelClemens/Shutterstock, Rattana/Shutterstock; Worksheet 7.5.1, 7.5.2, 7.5.3 Greg Amptman/Shutterstock; Worksheet 7.6.1, 7.6.2, 7.6.3 kaschibo/Shutterstock, Dirk Ercken/Shutterstock, Matt Jeppson/Shutterstock, Matt Jeppson/Shutterstock; Worksheet 7.7.1, 7.7.2, 7.7.3 Bildagentur Zoonar GmbH/Shutterstock, nattanan726/Shutterstock, visceralimage/Shutterstock, Mikkel Bigandt/Shutterstock, Coprid/Shutterstock, Practical sheet 7.8.1 Alex Staroseltsev/Shutterstock; Practical sheet 7.8.2 IlzeTheBeast/Shutterstock, Worksheet 7.12 JONATHAN PLEDGER/Shutterstock; Worksheet 7.19 Norma Cornes/Shutterstock; David Carbo/Shutterstock, Gilmanshin/Shutterstock; Worksheet 8.2.1, 8.2.2 Vaclav Volrab/Shutterstock; Worksheet 8.22 Johan Larson/Shutterstock.

PowerPoints

58798_biol_ch8_lesson3 PAUL D STEWART/SCIENCE PHOTO LIBRARY; 58798_biol_ch8_lesson3 VOLKER STEGER/SCIENCE PHOTO LIBRARY;
58798_biol_ch8_lesson3 NATURAL HISTORY MUSEUM, LONDON/SCIENCE PHOTO LIBRARY;
58798_biol_ch8_lesson5 LAURIE O'KEEFE/SCIENCE PHOTO LIBRARY;
58798_biol_ch8_lesson5 JIM AMOS/SCIENCE PHOTO LIBRARY;
58798_biol_ch8_lesson5 NATIONAL SCIENCE FOUNDATION/SCIENCE PHOTO LIBRARY;
58798_biol_ch8_lesson6 AMERICAN PHILOSOPHICAL SOCIETY/SCIENCE PHOTO LIBRARY;
58798_biol_ch8_lesson7 RANS LANTING, MINT IMAGES / SCIENCE PHOTO LIBRARY;
58798_biol_ch8_lesson7 PAUL D STEWART/SCIENCE PHOTO LIBRARY;
58798_biol_ch8_lesson10 SCIENCE PHOTO LIBRARY;
58798_biol_ch8_lesson12 CORDELIA MOLLOY/SCIENCE PHOTO LIBRARY;
58798_biol_ch8_lesson12 TONY CAMACHO/SCIENCE PHOTO LIBRARY;
58798_biol_ch8_lesson13 CORDELIA MOLLOY/SCIENCE PHOTO LIBRARY;
58798_biol_ch8_lesson13 KLAUS GULDBRANDSEN/SCIENCE PHOTO LIBRARY;
58798_biol_ch8_lesson13 PHILIPPE PSAILA/SCIENCE PHOTO LIBRARY;
58798_biol_ch8_lesson14 MAKOTO IWAFUJI/EURELIOS/SCIENCE PHOTO LIBRARY;
58798_biol_ch8_lesson18 JEROME WEXLER/SCIENCE PHOTO LIBRARY;
58798_biol_ch8_lesson18 SCIENCE PHOTO LIBRARY; all other images © Shutterstock.

Contents

8 Ecology in action

Introduction to the course

Welcome to the Collins GCSE Science course. These materials have been developed by a team of experts drawing on many years' experience in teaching, curriculum development and educational publishing. They have been produced to support you, the professional in the classroom, teach more effectively and your students to make good progress.

The aim is to both develop students' interest in science and enable them to attain grades that reflect a mastery in the subject. The course is not intended to be prescriptive or restrictive, shoe horning teachers into a very particular way of teaching, but to be used in a range of ways. On the other hand, it doesn't offer a bewildering array of options which require a lot of filtering and selection of activities.

We are aware that specifications sometimes seem to change at an alarming rate. We have tried to reflect both the key features of the new specifications and the classic elements of high quality teaching.

Key changes for the courses introduced in 2016

Non-modular

The new specifications have a non-unitised structure. Content can be covered in any order deemed appropriate. This doesn't mean that the order doesn't matter. There are a number of principles that can be used when selecting a 'running order', including:

- deciding which topics are more accessible for students at an earlier stage in their GCSE studies
- identifying which topics need to be covered to give access to which other concepts
- using topics that have possibly greater appeal, due to the role of practical work or engaging contexts, to sustain interest over the whole course.

However, what is also important is to build in assessment, tracking and intervention. You need to know if students are developing an understanding that at least corresponds to their ability and to be able to respond.

The Collins course offers a particular route. The exams are designed to assess one part of the content in Paper 1 and the other part in Paper 2. One approach is to teach the 'Paper 1 content' first and then the 'Paper 2 content'. This means that a past Paper 1 could be used as a 'half way assessment point'.

However, the materials don't have to be used in this order. It may be decided that the needs of students are not best served by concentrating all the material on one topic in one place, but rather by revisiting it later on.

Maths

Science has always had a strong relationship with maths and this continues. What has been strengthened in this set of courses is the specificity of the location of the skills and the role of the skills in exams. It has been made very clear in the specifications which skills relate to which topics and the exams will disadvantage candidates who can't demonstrate mastery of the relevant skills.

In the course we have taken a 'two pronged approach'. Mathematical skills are used in context as key concepts are covered. They are reflected in questions that are set and students are shown how they are drawn upon at particular points. However, we know that for some students this won't be sufficient. Therefore, we have developed other spreads that will focus on a particular set of mathematical skills and explore how they develop. These always place the skills into a scientific context, so it is clear why they are relevant, but the text is led by the developing mathematical skill.

Practical skills

Practical work has a strong role in supporting students' developing understanding of science, and many people involved in the British system are passionate about ensuring this role is maintained. In recent years much of the debate is about how it should be assessed, and successive specifications have used different approaches.

In this series there is no direct assessment of practical skills. Instead the approach is being taken that enquiry will be put in a better place if students are assessed on their mastery of skills and processes in the final exams. A minimum of 15% of marks will be allocated to questions relating to the stipulated practical investigations. However, these won't necessarily be AO1 questions; as well as understanding and recalling the procedures, candidates will also be expected to apply the skills to other contexts and to interpret and evaluate evidence.

As well as covering the effective running of the investigations themselves, the Collins course also offers guidance on placing these in a wider context, so that students can see how their skills and understanding can be used and assessed.

Synoptic questions

Another of the key changes is the introduction of synoptic questions, drawing upon more than one set of ideas. The justification behind this is that scientists often need to draw upon ideas and processes from different areas of science. Again, this is reflected in the assessment approach developed.

Extended written responses

Unlike the previous set of specifications, QWC (Quality of Written Communication) is no longer being assessed in science. However, there is a requirement for candidates to be able to display an ability to develop and sustain a longer response. This is reflected in the questions in the end of topic tests as well as in the Collins Connect digital assessment materials.

Assessment

Grades

The grading system is being changed to a numbered system, in which 1 is the lowest grade awarded and 9 the highest. There is no one-to-one correlation between grades in the incoming and outgoing systems, but the following features are being deployed:

- The new grade 4 is equivalent to the old grade C.
- The new grade 7 is equivalent to the old grade A.
- The threshold for performance indicators will be the new grade 5.

Endorsed publications are no longer permitted to carry specific reference to how certain grades may be attained. Nevertheless, although the calibration has altered, what has not changed is what constitutes a better quality of response. Each of the spreads in the Student Book is designed to present students with an increasing challenge and provide access to higher grades. However, we have also been mindful when developing the resources that a wide range of students will be using them, and so there is a focus upon accessibility as well.

Attribution grids

There has long been an important role attached to the use of end of topic tests. When students have been taught something, we (and they) want to know how well they've understood it and so we've included tests. However, we have developed these further, both in terms of structure and use.

In order to be useful, we think that as far as possible tests should be authentic to the form and function of assessment in GCSE. In other words, if students do well in the tests, this should correspond to doing well in the final exams. This is important, both to improve the validity of any data generated and also in terms of communicating to students what it feels like to answer the kind of questions that they will face in the exam. These aren't mock exams, but they should have a similar balance of components.

Each of the end of topic tests has been written to an attribution grid, so that the basis is the same. The tests are marked out of 40 and have ten marks at each of four different levels of demand. Broadly speaking, Foundation Tier students should do the first two bands and Higher Tier students the final three bands. If they do so, each group will be answering questions which have a 2:2:1 ratio of assessment objectives 1, 2 and 3, which mirror the make-up of the final exams. This is important. AO1 questions focus on knowledge and understanding and are sometimes over-represented in internal tests. As many marks go on AO2 (application). Another 20% go on AO3 (interpretation and evaluation); students need to be aware of this. The constant proportions mean that it is possible to analyse student performance against AO and modify subsequent teaching accordingly.

Similarly, the tests also reflect a mix of response types (objective test questions, short written responses and longer written responses), and the presence of questions relating to mathematical skills and to the stipulated practicals.

Intervention

The composition of the tests is designed to inform the process of intervention. They will obviously provide a global score which can be used to draw conclusions about attainment in that topic. However, intervention is then somewhat problematical; if the group has moved on to another topic, the only option might be to set up additional sessions elsewhere in the school day. These may be successful but it is useful to be able to move analysis beyond the perspective of looking for which areas of content cause problems. Some do, but this isn't the only source of underperformance.

With the end of chapter tests (and the Collins Connect tests), the attribution grids mean that student performance can be analysed in other ways (see above), including against:

- assessment objectives
- style of response
- command of mathematical skills
- mastery of enquiry skills.

As these are all generic components of science education, whichever topic is being addressed next, there will be opportunities to focus upon improving performance.

It will then be possible in the next module test to see if the analysis of marks indicates that this has been addressed.

Command words

Awarding organisations use certain command words as the stems of questions. The Collins resources have been written to reflect these words and to provide students with exposure to them. This is important as it then provides a basis to gauge the extent to which students understand what is expected of them – whether they are, for example, explaining a difference as opposed to just describing it. These are used throughout the book, in the in text questions as well as the end of topic questions and the Collins Connect resources.

AQA list the following command words and their meanings for GCSE Science. Command words marked * are new for teaching from 2016.

Calculate Students should use numbers given in the question to work out the answer.

Choose* Select from a range of alternatives.

Compare This requires the student to describe the similarities and/or differences between things, not just write about one.

Complete Answers should be written in the space provided, for example, on a diagram, in spaces in a sentence or in a table.

Define* Specify the meaning of something.

Describe Students may be asked to recall some facts, events or process in an accurate way.

Design* Set out how something will be done.

Determine* Use given data or information to obtain an answer.

Draw To produce, or add to, a diagram.

Estimate Assign an approximate value.

Evaluate Students should use the information supplied as well as their knowledge and understanding to consider evidence for and against.

Explain Students should make something clear, or state the reasons for something happening.

Give Only a short answer is required, not an explanation or a description.

Identify* Name or otherwise characterise.

Justify Use evidence from the information supplied to support an answer.

Label Provide appropriate names on a diagram.

Measure* Find an item of data for a given quantity.

Name Only a short answer is required, not an explanation or a description. Often it can be answered with a single word, phrase or sentence.

Plan* Write a method.

Plot* Mark on a graph using data given.

Predict* Give a plausible outcome.

Show* Provide structured evidence to reach a conclusion.

Sketch* Draw approximately.

Suggest Apply knowledge and understanding to a new situation.

Use The answer must be based on the information given in the question. Unless the information given in the question is used, no marks can be given. In some cases, students might be asked to use their own knowledge and understanding.

Work out* Students should use numbers given in the question to work out the answer.

Write Only a short answer is required, not an explanation or a description.

How to use these resources

Flexibility

The first thing to say is that the resources are designed to be used flexibly. There is no assumption that all students will do all of the activities suggested in the order indicated or that no other resources or learning activities will be used. The materials have an open and transparent structure, which clearly indicates key components and supports the teacher to select activities that are right for their students.

Features

Student Book

Topic introduction: every chapter starts with the page on the left showing students ideas they have met previously that they will be making use of, and a page on the right indicates why the ideas in the forthcoming chapter are both interesting and important. The purpose of this is to engage students, indicate that they already know something about it but persuade them that it is worth engaging with.

Main content spread: each spread includes material that is ramped in challenge. There is no set grade equivalence for the start and end but the spread always starts with more accessible material. Each of the sections includes questions to embed understanding of the ideas, learning objectives, key words and a 'did you know?' feature to engage students further.

Key concept spread: in every topic we have identified a key concept and revisited this in a slightly different way. These concepts have a kind of 'gatekeeper' function: if students grasp these it will give them access to several other areas of content and so they are worth focusing upon. The highlighting of the spread means they can easily be referred back to, either from further on in the same chapter or subsequent ones.

Required practical spread: the specification indicates particular investigations that students should have undertaken. Questions relating to these will feature in terminal examinations; however, these will be not only AO1 (knowledge and understanding) questions but also AO2 (application) and AO3 (interpretation and evaluation) questions. These spreads have been written to support students to take ideas explored in these experiments and place them in a wider context, thus reflecting the range of skills that examiners will be assessing.

Maths skill spread: mathematical skills have long been a key part of GCSE Science courses and exams. However, the proportion and challenge of these questions is being increased. We have developed spreads to reflect this. Maths is still embedded in the main content, with skills being drawn upon as necessary. However, some students will need more than this, and so these additional features have been written. Each takes a particular set of maths skills and shows how they are used in a scientific context, taken from that topic. As the spread becomes more challenging, it shows how the mathematics can be used to answer more complex questions.

Check your progress spread: at the end of each chapter the main ideas are set out as progressive outcomes. In essence they show what it looks like to be better in the use of a particular idea. They can be used in various ways but students self-assessing is one of the more usual ones.

Worked examples: these provide a commentary on typical student responses to GCSE-style questions, showing what has been done well and how further improvement might be achieved.

End of topic questions: these are written to an attribution grid as described above. The questions provide a broad coverage of ideas within the topic, items with different levels of demand (both in terms of cognitive and conceptual complexity) and needing different types of response. This means that they provide varying degrees of challenge and in various ways; they should indicate both what students can do and also what the next steps in their learning need to be.

Teacher Pack
The support for individual lessons includes features such as learning objectives, outcomes, working scientifically focus, maths skills focus, key words and resources needed. The outline for the lesson structure is based around a five-stage learning cycle; this proposes that in every lesson there should be these components:

- Engage: in every lesson students need to be engaged, so that they become involved in the lesson and see it as being of interest and relevance.
- Challenge and develop: students will already have a certain level of understanding of a topic. The idea of this phase is to challenge this and move their thinking on. This is what will drive progress; students should see that their current mode of thinking can be developed so that they can make sense of more complex ideas or a wider range of contexts. This might be a demonstration, question to discuss or video clip.
- Explain: this is where the teacher uses some means of offering an explanation. It could be formal explanation, a demonstration or commentary on a diagram; it might use targeted questioning. It should offer a way of making sense of a problem or phenomenon. The Student Book often plays a crucial role in this phase.
- Consolidate and apply: this is a crucial phase in which students embed the ideas. It should consist of opportunities to 'take on board' new ideas and own them. Written questions and discussion may support this. The worksheets often have a significant role here.
- Extend: students may be ready to extend their learning to something broader or more challenging.

In the ideas for learning activities there is a choice and the options relate to planning provision for students working at different levels.

Teacher Pack Resources

Each lesson is supported by worksheets, which provide activities for students to apply their understanding. They have a ramped structure, offering different levels of challenge.

Practical worksheets support students with planning, carrying out and writing up practical and investigation work.

Technicians' notes provide equipment lists and set-up instructions for practicals.

Collins Connect

Collins Connect makes Collins GCSE Science available at home and at school online and can be used as a front-of-class teaching tool, as well as a way to set homework and tests.

AQA GCSE (9–1) Science on Collins Connect contains:

- Bookview – an interactive digital version of the Student Books. It is ideal for whiteboard use and gives total fluidity between digital and print, with a page-for-page match.
- Quick starters – activities that pose an interesting question to hook students in at the start of a lesson. You can click to show a hint and then click to show the answer.
- Videos and animations – to support you to make real world links and model abstract or challenging concepts.
- Slideshow presentations – available in PowerPoint so you can edit and adapt them to suit the needs of your class.
- This Teacher Pack – so you can access the lesson plans and supporting resources online.
- Homework activities for every lesson – you can assign automarked homework quizzes to your class. Students can log in from home to complete the activities and get immediate feedback. For every lesson there are also 'creative' homework sheets with more open-ended activities.

You can also assess and track students' progress in GCSE Science using the full suite of digital resources on Collins Connect. Collins Connect provides regular and timely assessments to help you analyse student performance.

- The digital tests come in two versions: 'fully automarked' and 'automarked plus teacher response'. The fully automarked tests enable you to get a very quick gauge of students' knowledge and understanding. The 'automarked plus teacher response' tests include longer written response questions, giving students practice of the full range of question styles they will encounter in the final exams. All tests come with a mark scheme.
- End of chapter tests help you to review students' progress on a topic-by-topic basis.
- End of teaching block tests (at the end of every two chapters) provide additional common review checkpoints to help you review progress through the linear course and check students' retention of earlier learning.

- End of year and end of course tests help students to prepare and practice for the final exams.

Differentiation

Differentiation is a key feature of the student books, the lesson plans and the worksheets.

- The student books support differentiation with their ramped structure in each spread. The early section offers an accessible 'entry point' and later text is more challenging. This is reflected in the questions, both in the spreads and at the end of each topic. The progress checking spreads also support an understanding of progressing to higher levels of understanding.
- The teacher packs support differentiation by offering learning activities at different levels of challenge for different phases of the lesson, thus making it easier for teachers to tailor provision to a particular class.
- The worksheets also have a ramped structure.

Four stage plan

One of the ways of running the GCSE course is to use a four-stage plan. The idea behind this is that:

- some topics need to be taught before others and so should be placed earlier in the course
- with some topics there is no preferred running order and departments might want to avoid all groups in the same year doing exactly the same topic at the same time
- having what might be termed 'common assessment points' enables periodic assessment and reporting of progress to take place at points at which all students have covered the same ground – the same assessment can be used and students switched between groups if necessary.
- To provide additional common assessment points, Collins AQA Biology can be divided into four teaching blocks:

    - Teaching block 1: chapters 1 and 2 (Paper 1)
    - Teaching block 2: chapters 3 and 4 (Paper 1)
    - Teaching block 3: chapters 5 and 6 (Paper 2)
    - Teaching block 4: chapters 7 and 8 (Paper 2)

End of teaching block tests are provided on Collins Connect and can be used as common assessment points.

The order of the teaching blocks can be altered, if you prefer to teach content from Paper 1 and 2 over the duration of the course. The four-stage plan can be represented diagrammatically as follows:

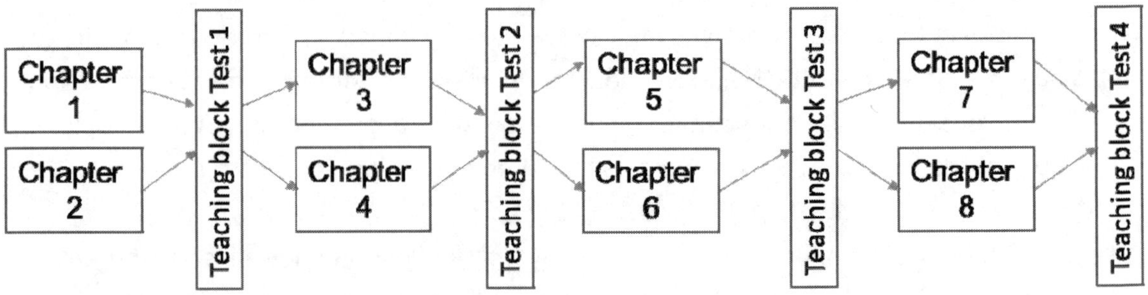

Cell Biology: Introduction

When and how to use these pages

This unit builds on the idea that the cell is the building block of life. Cells can be specialised and may be part of multicellular organisms or unicellular organisms (like bacteria). This chapter links to all chapters in which the structure and functions of different systems are considered, for example, Photosynthesis, and Health, and to chapters in which meiosis is explained.

Overview of the unit

In this unit, students will learn about the structure of plant, animal, prokaryotic and eukaryotic organisms, and the functions of major structures. They will compare the level of detail revealed by light and electron microscopes, calculating magnifications. Students will describe how cells divide by mitosis and the application of this in cloning techniques. The use of stem cells in producing new tissues and organs will be evaluated in terms of ethical and moral considerations. Students will consider the differences between aerobic and anaerobic respiration, and learn about the uses of anaerobic respiration in baking and brewing. They will learn how to grow cultures of bacteria safely and investigate the effectiveness of different disinfectants on bacterial growth.

This unit offers a number of opportunities for the students to use mathematics to carry out magnification calculations, plan and carry out investigations into the use of anaerobic respiration in baking, evaluate the effectiveness of disinfectants by preparing bacterial cultures and use graphical skills to analyse data in a number of different contexts. Students will debate the use of embryonic stem cells in terms of moral and ethical considerations.

Obstacles to learning

Students may need extra guidance with the following terms and concepts:

- Cells and related topics use abstract concepts and are hard to visualise. The use of cell models may help some students to make connections between different types of cells. Students often believe that cells are inactive, two-dimensional structures and the use of videos and electron micrographs will enable them to see this is not the case.
- Mitosis is also a difficult concept to follow, and the use of models will help students to see the process more clearly and follow the stages more easily. Students often think that cells grow by simply becoming bigger, rather than by cell division.
- The idea that a stem cell can become a nerve cell or a muscle cell, for example, can be difficult for students to understand.
- Students may believe that bacterial cells are the same as animal cells.
- Respiration is often confused with breathing (ventilation), and needs to be linked to the mitochondria within cells so its role within each cell can be emphasised.

Practicals in this unit

In this unit students will do the following practical work:

- (Required Practical) prepare plant and animal slides and observe them using a light microscope
- investigate the amount of energy in food
- prepare uncontaminated cultures of bacteria
- (Required Practical) investigating disinfectants.

	Lesson title	Overarching objectives
1	Looking at cells	Describe the structure of eukaryotic cells and functions of subcellular components.
2	The light microscope	How to observe animal and plant cells using the light microscope and its limitations.
3	Looking at cells in more detail	Compare the light microscope with the electron microscope, explaining how the development of the electron microscope has increased our understanding of cells.
4	Required practical: Using a light microscope to observe and record animal and plant cells	How to look at everyday material and cells using a light microscope. Understand the difference between low and high power; draw and describe images at high and low magnification.
5	Primitive cells	Describe the differences between prokaryotic and eukaryotic cells, and how they might have evolved over time.
6	Cell division	Describe the process of mitosis using models.
7	Cell differentiation	Explain the importance of cell differentiation and link specialised cells to their tissues, organs and body systems. Describe the organisation within a multicellular organism.
8	Cancer	Describe what cancer is and the factors that can trigger cells to become cancerous.
9	Stem cells	Describe the functions and uses of stem cells. Compare the use of embryonic and adult stem cells and the ethical implications of their use.
10	Stem cell banks	Explain the uses and risks of using stem cells in medicine, evaluating their benefits and disadvantages.
11	Key concept: Cell development	Revise ideas about cell structure, division and stem cells. Describe how plant meristems can be used in cloning.
12	Cells at work	Explain the process of aerobic respiration.
13	Life without oxygen	Describe the process of anaerobic respiration and compare it to aerobic respiration. Plan an investigation into factors affecting anaerobic respiration in dough-making.
14	Growing microorganisms	Describe how to prepare uncontaminated cultures of microorganisms and how bacteria reproduce.
15	Testing new antibiotics	Describe the disc-diffusion technique and how to use techniques to ensure samples are representative.
16	Required practical: Investigating disinfectants	Investigate and evaluate the effectiveness of disinfectants on the growth of bacteria.
17	Maths skills: Size and number	Making estimates, ratio and proportion, standard and decimal form.

Lesson 1: Looking at cells

Lesson overview

AQA Specification reference

AQA 4.1.1.1; AQA 4.1.1.2

Learning objectives

- Describe the structure of eukaryotic cells.
- Recognise the order of magnitude of cells.
- Explain how the main sub-cellular structures are related to their functions.

Learning outcomes

- Name the parts in a eukaryotic cell. [O1]
- Relate the size of a cell to other objects. [O2]
- Explain the function and reasons for sub-cellular structures. [O3]

Skills development

- WS 4.1 Use scientific vocabulary, terminology and definitions.
- WS 4.2 Recognise the importance of scientific quantities and understand how they are determined.
- WS 4.5 Interconvert units.

Maths focus

Recognise and use expressions in standard form.

Resources needed Worksheets 1.1.1, 1.1.2 and 1.1.3

Digital resources PowerPoint

Key vocabulary DNA, chloroplast, chlorophyll, chromosome, eukaryotic, order of magnitude

Teaching and learning

Engage

- Ask students to **write down** 10 things they already know about cells. Show students slide 1 on the PowerPoint and discuss students' responses as to what the images have in common. Elicit what a cell is and what features they have. [O1]
- Use slide 2 to introduce the term 'eukaryotic' to describe a cell with a true nucleus. Ask students to **compare** the images and **identify** what they have in common. [O1]
- Ask students to **imagine** how big a cell might be. **Discuss** what they would compare the size to. [O2]

Challenge and develop

- Show students the simulation which will help them to **identify** the order of magnitude of cells in relation to other objects. This can be found using the search terms 'cell size' and 'scale' at http://learn.genetics.utah.edu. In addition, students could look at page 15 of the Student Book. [O2]
- Discuss the different units of size. Ensure students understand the relationships between mm, μm, nm and m. Ensure they are familiar with standard form. Use page 15 from the Student Book to help or use PowerPoint slide 3 to demonstrate how to compare orders of magnitude.
- Higher demand, standard demand and lower demand students to carry out the appropriate card sort: lower demand, Worksheet 1.1.1; standard demand, Worksheet 1.1.2; higher demand, Worksheet 1.1.3.

  Provide students with the card sort from the appropriate worksheet. They should **group** the cards into sizes of the same dimension. For example, 1 m could be grouped with 1000 mm and 100 cm. This card sort is differentiated for different learners. [O2]

Show students worksheet task 2. Ask students to **imagine** the diameter of a human hair scaled up to 70 mm. Ask higher and standard demand students to **work out** the relative lengths of the other cells in the table. Ask all students to **draw** these lengths on their graph and **compare** the diameter of a hair as given on the worksheet. Ask students to **draw conclusions** about the sizes of cells. [O2]

Explain

- Ask students to **read** page 14 of the Student Book to remind themselves of the different parts of animal and plant cells. Students should answer all the questions from the Student Book. [O2, O3]

Consolidate and apply

- Ask students to **draw and summarise** the differences between animal and plant cells in the table given on the worksheet. [O2, O3]
- Students could work in small groups and **make** a three-minute presentation about cells to the class. They should include ideas about size, as well as explaining what a cell is and explaining the differences between animal and plant cells. [O1, O2, O3]

Extend

Ask students who are able to progress further to:

- **research** the history of the development of cells and **identify** on the timeline when eukaryotic cells first evolved. [O1, O2, O3]
- **compare** the structure of the 'cyanobacteria' with that of a typical plant cell. [O1, O2, O3]

Plenary suggestions

- Use the PowerPoint slide showing plant and animal cells with many mistakes on it. Ask students to **correct** the mistakes.
- Provide students with a list of structures from the PowerPoint. Ask them to **put them in order** of magnitude.

Answers to Worksheets 1.1.1, 1.1.2 and 1.1.3

1.

1 m	1 metre		100 cm	10^3 mm	1 000 000 µm	
1 cm	1 centimetre	10^{-2} m		10 mm	10^4 µm	
1 mm	1 millimetre	10^{-3} m			1000 µm	1 000 000 nm
1 µm	1 micrometre	10^{-6} m		10^{-3} mm		1000 (10^3) nm
1 nm	1 nanometre	10^{-9} m	10^{-7} cm	10^{-6} mm	10^{-3} µm	

2.

Object	Original size (µm)	New size (mm)
Human hair diameter	70	70
Bacterium	1	1
Red blood cell	7	7
Leaf cell	70	70
HIV	0.1	0.1

All lengths are drawn to the new size and graphs are labelled

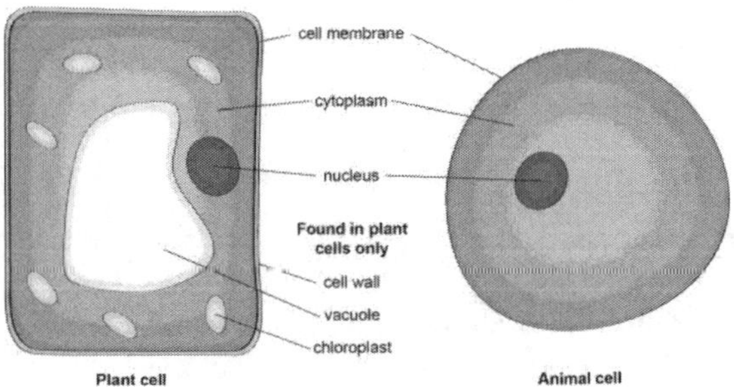

3.

Animal cell	Plant cell	Function of structure in the cell
nucleus	nucleus	controls reactions in the cell; contains DNA; controls reproduction of the cell
cell membrane	cell membrane	controls which substances enter and leave the cell
cytoplasm	cytoplasm	where all the chemical reactions in the cell take place
	vacuole	contains cell sap to provide internal strength to the cell
	chloroplasts	absorb light energy for photosynthesis
	cell wall	made from cellulose; provides strength and protection for the cell

Lesson 2: The light microscope

Lesson overview

AQA Specification reference

AQA 4.1.1.5

Learning objectives

- Describe how to use a microscope.
- Observe plant and animal cells with a light microscope.
- Understand the limitations of light microscopy.

Learning outcomes

- Describe how to prepare a microscope slide of animal and cheek cells. [O1]
- Make observations of cheek and animal cells. [O2]
- Calculate the size of objects using ideas about magnification. [O3]

Skills development

- WS 1.2 Use a variety of models such as representational, spatial, descriptive, computational and mathematical to solve problems, make predictions and to develop scientific explanations and understanding of familiar and unfamiliar facts.
- WS 2.6 Make and record observations and measurements using a range of apparatus and methods.
- WS 4.2 Recognise the importance of scientific quantities and understand how they are determined.

Maths focus

Make order of magnitude calculations.

Resources needed Worksheet 1.2; Practical sheet 1.2; Technician's notes 1.2

Digital resources PowerPoint

Key vocabulary iodine, magnification, methylene blue, resolving power

Teaching and learning

Engage

- Show students some images of plant or animal cells from the PowerPoint slide. Ask students to **identify** the plant cells and the animal cells. They should justify their responses. Ask students how scientists came to obtain such images. [O2]
- Use the PowerPoint slide to demonstrate how the light microscope works. Students should **read** page 16–17 of the Student Book. [O1]

Challenge and develop

- Demonstrate how to use a microscope, and how to work out the magnification of an object using the PowerPoint slide. Students should **answer** the questions in the Student Book to determine the magnification of different objects. [O2, O3]

  Provide students with Practical Sheet 1.2. Ask them to **follow** the procedures given to produce a slide of a cheek cell and an onion cell. Students should **observe** their cells under a light microscope and **record** their observations. [O1, O2]

Explain

- Ask students to **work out** the magnification of the images they recorded. [O3]
- Students should **explain** the differences in their images, using ideas about cell structure. [O2]

Consolidate and apply

- Students should **produce** a poster about the light microscope. They could **carry out** further research on this if required. [O1]

Higher demand:

- Students should **explain** how to work out the actual sizes of the objects from the microscope. [O3]

Extend

Ask students who are able to progress further to:

- **work out** the magnification, size of images and size of objects on the extension task from the worksheet. [O3]

Plenary suggestions

- Ask students to **write** 10 top tips for preparing plant and animals cell slides and observing them under a microscope. [O1, O2]
- Students should **summarise** all the limitations of using a microscope in three sentences. [O3]

Answers to Worksheet 1.2

Explaining differences

Plant cells should be highly ordered, with a much more distinctinctive cell wall. The cells may contain vacuoles, but these can be hard to observe.

Animal cells have a less distinctive shape and no vacuole.

Extension

Object	Measured size	Magnification
Ant	20 mm	20/3 = 6.7
Hair diameter	3000 μm	3000/100 = 30
Leaf cell	35 000 μm	35 000/70 = 500
Red blood cell	20 000 μm	20 000/7 = 2857
Bacterium	25 000 μm	25 000/1 = 25 000
HIV	30 000 000 nm	30 000 000/100 = 3 000 000
DNA	5 000 000 nm	5 000 000/2.5 = 2 000 000
Carbon atom	34 000 000 nm	34 000 000/0.34 = 100 000 000

Lesson 3: Looking at cells in more detail

Lesson overview

AQA Specification reference

AQA 4.1.1.5

Learning objectives

- Identify the differences in the magnification and resolving power of light and electron microscopes.
- Describe simply how electron microscopes work in comparison to light microscopes.
- Explain how electron microscopy has increased our understanding of sub-cellular structures.

Learning outcomes

- Describe what features can be seen with an electron microscope. [O1]
- Explain the differences in magnification and resolving power between a light and electron microscope. [O2]
- Explain the impact of the electron microscope on our understanding of cellular structures. [O3]

Skills development

- WS 1.1 Understand how scientific methods and theories develop over time.
- WS 1.4 Explain everyday and technological applications of science; evaluate associated personal, social, economic and environmental implications; and make decisions based on the evaluation of evidence and arguments.
- WS 4.1 Use scientific vocabulary, terminology and definitions.

Resources needed Worksheet 1.3

Digital resources PowerPoint

Key vocabulary mitochondria, ribosome, scanning electron microscope, transmission electron microscope

Teaching and learning

Engage

- Show students the images from an electron microscope and from a light microscope on the PowerPoint slide. Ask students to **answer** the following questions about the images: [O1, O2]

  Where do the images come from?

  What might they show?

  Why are there differences between the images?

  How might the images have been taken?

Challenge and develop

- Ask students to **read** the sections in the Student Book and/or on the worksheet on electron microscopes and light microscopes. Ask them to use the table on the worksheet to **compare and contrast** the electron microscope with the light microscope. [O1, O2]
- Show students the interactive simulation of internal structures of plant and animal cells found using the search term 'inside a cell' at http://learn.genetics.utah.edu. [O1]
- Introduce students to further structures found in the cytoplasm from studies using electron microscopes; ribosomes and mitochondria. Ask students to **answer** the differentiated questions in Worksheet 1.3.1 'What is inside a cell?' section (lower demand, Q1–3; standard demand, Q4; higher demand, Q5–7). [O1, O2]

Explain

- Students should **write** a short explanation about the impact the electron microscope has had on our understanding of cells. [O3]

Consolidate and apply

- Students could **answer** the questions from the Student Book.
- Students should use ideas about magnification to **explain** which of the examples on the worksheet can be viewed by an electron microscope and which can be viewed by a light microscope.

Extend

Ask students who are able to progress further to:

- **research** other structures that have been found using electron microscopes and **present** a case to the class for the further impact these inventions have had on the developments made in cell biology.

Plenary suggestions

Students could **imagine** that a biology research lab is in need of further optical equipment. Half of the class could represent a company selling light microscopes, the other half a company selling electron microscopes. Each half should **prepare a case** to sell their equipment.

Answers to Worksheet 1.3

1.

Optical microscope	Electron microscope
Used for hundreds of years	Very recently invented
Uses light rays	Uses electron beams of high energy
Specimen can be living	Specimen is dead
×1000 to ×2000 magnification	About ×2 000 000 magnification
Not possible to see internal structures inside the cytoplasm clearly	Internal structures inside the cytoplasm can be viewed
Quite cheap	Very expensive
Anyone can use this and observe images	Highly trained scientists needed to operate and analyse results
Not much space needed	Lots of space required
2D image only	3D image can be produced
Not possible to get better magnification with this technology	Technology can be improved over time

2.

1) Mitochondria
2) Ribosomes
3) Chloroplasts
4)
   a) Ribosomes – both animal and plant
   b) Mitochondria – both animal and plant
   c) Chloroplasts – plant only
5) Light microscope does not have a high enough magnification or resolution to view these structures.
6) The internal structures of these organelles; for example, the folds inside the mitochondria and the chloroplasts
7) How these organelles move, reproduce and function under normal conditions

3. Many answers are possible – here is an example:

If we still only had the light microscope, we would only be able to see the main structures of the cell, like the nucleus and cell membrane in animal cells, and vacuole and nucleus inside plant cells.

We would not understand that the cytoplasm is made up of many different types of structures. We would not be able to determine the function of different organelles inside the cytoplasm and could not make breakthroughs regarding health and the fight against viruses and cancers.

The electron microscope has enabled us to have a better understanding of how cells work and how to fight against diseases.

4.

1) Nucleus – visible under both microscopes
2) Cell wall – visible under both microscopes
3) Mitochondria – mainly visible by electron microscope; too small for optical
4) White blood cell ingesting a bacteria – only optical as the cells must be alive
5) Chloroplast – electron
6) Embryo dividing – only optical as cells must be alive
7) Sperm cell swimming – only optical as cell must be alive
8) Ribosomes – electron microscope
9) Detailed structure of the cytoplasm – electron microscopeFertilisation taking place – only optical as cells must be alive

Lesson 4: Required practical: Using a light microscope to observe and record animal and plant cells

Lesson overview

AQA Specification reference

AQA 4.1.1.2

Learning objectives

- Apply knowledge to select techniques, instruments, apparatus and materials to observe cells.
- Make and record observations and measurements.
- Present observations and other data using appropriate methods.

Learning outcomes

- Use a light microscope to observe a range of different objects. [O1]
- Record observations from a light microscope using simple diagrams of low and high power magnification. [O2]
- Use a scale bar to identify the size of images from microscope slides. [O3]

Skills development

- WS 2.6 Make and record observations and measurements using a range of apparatus and methods.
- WS 3.1 Present observations and other data using appropriate methods.
- WS 3.5 Interpret observations and other data (presented in verbal, diagrammatic, graphical, symbolic or numerical form), including identifying patterns and trends, making inferences and drawing conclusions.

Maths focus

- 2h Make order of magnitude calculations

Resources needed Range of materials to look at through the microscope, e.g. human hair, newspaper, string, prepared range of animal and plant cells. Worksheet 1.4.1 (low and standard demand); Worksheet 1.4.2 (high demand); Practical sheet 1.4.1 (low and standard demand); Practical sheet 1.4.2 (high demand) Technician's notes 1.4

Digital resources

Key vocabulary field of view, scale

Teaching and learning

Engage

- Provide each pair of students with a microscope. Ask them to **recap** and **discuss** the name and function of each part of the microscope and how it should be used. [O1]

Challenge and develop

- **Discuss** the importance of staining the slides and of why mounting animal cells in saline solution instead of water is important. [O1]
- Ask students to **look at** the images on Worksheet 1.4.1 (low and standard demand) or 1.4.2 (high demand). Ask them to **identify** if they were taken under low or high resolution. [O1, O2]
- Ask students to **observe** a range of everyday materials first through the microscope. Students should follow the instructions on Practical sheet 1.4.1 (low and standard demand) or 1.4.2 (high demand). They should **make annotated drawings** of their observations on both low power and then high power, as accurately as possible on a poster, which can then be later displayed. They should **explain** what they see in as much detail as possible. Where appropriate, students should **count** the numbers of cells they observe.

- High demand students should **discuss** the use of a scale bar using the section on recording images on page 21 of the Student Book. They should also **estimate** the actual size of individual cells on their worksheet by using the given scale bars. [O2]
- If possible, provide students with a range of different prepared animal and plant tissue. Ask students to **identify** if the cells are animal or plant, and to **draw** what they see, on low power first and then on high power, as accurately as possible. Students should **identify** different regions of the tissues in the slides as accurately as possible, where appropriate. [O2]

Explain

- Low and standard demand students should **write** a short explanation on their posters of the importance of viewing an object on a low magnification first and then on a higher magnification. [O1, O2]
- High demand students should **write** an explanation of how to determine the appropriate scale bar to use on their posters. They should **highlight** the importance of ensuring it is appropriate. [O3]

Consolidate and apply

- **Discuss** with students the success criteria for determining what a good image from a microscope looks like and write a class list for students to follow.
- Display all the posters on the wall. Ask students to **peer assess** their images, using post-its, providing one or two strengths and one area for development. [O1, O2, O3]
- Students should **read** pages 20–21 of the Student Book and **answer** questions 1–6.
- High demand students should also **answer** questions 7–8.

Extend

Ask students able to progress further to:

- **find out** how scale bars are determined for very small objects. [O3]

Plenary suggestion

Students could prepare a 1-minute talk on the importance of the light microscope, how to use it correctly, what information can be observed and how to record these observations appropriately. [O1, O2, O3]

Answers to Worksheet 1.4

1. Low resolution – A, B and D; high resolution – C
2. (High demand): A = ~2 μm; B = ~5 μm; C = ~2 μm; D = ~10 μm

Lesson 5: Primitive cells

Lesson overview

AQA Specification reference

AQA 4.1.1.1; AQA 4.1.1.2

Learning objectives

- Describe and explain the differences between prokaryotic cells and eukaryotic cells.
- Explain how prokaryotic and eukaryotic cells evolved over time.

Learning outcomes

- Recognise that cells without a nucleus are prokaryotic cells. [O1]
- Compare and contrast prokaryotic and eukaryotic cells. [O2]
- Explain the evolutionary link between prokaryotic and eukaryotic cells. [O3]

Skills development

- WS 1.1 Understand how scientific methods and theories develop over time.
- WS 3.6 Present reasoned explanations including relating data to hypotheses.
- WS 4.1 Use scientific vocabulary, terminology and definitions.

Resources needed Worksheet 1.5

Digital resources PowerPoint

Key vocabulary domain, genome, nucleic acid, plasmid, prokaryotic

Teaching and learning

Engage

- Show students the PowerPoint slide of the oldest fossil on Earth. Ask students to **generate five questions** regarding things they would like to know about this fossil. Take feedback and write a list of questions on the board. [O1]
- Ask the class to vote for the top five questions to which they would like to find the answers. [O1]

Challenge and develop

- Show students the slide of eukaryotic cells and a prokaryotic cell. Ask students to **identify** all the similarities and differences between the images, on the worksheet. [O1, O2]
- **Discuss** the features of prokaryotic cells using the PowerPoint slide. [O1, O2]
- Lower demand students could **use the Student Book** to help them.
- Ask students to **draw and label** a diagram of a prokaryotic cell on the worksheet. [O1]
- Ask students to **carry out** the mystery exercise from the card sort. They should **sort the cards** into groups of similar information. They should **use these cards** to produce a timeline of cell development from the information given. [O1, O2, O3]
- Students should **read** the section on Archaean bacteria and **answer** the questions from the Student Book. [O3]

Explain

- Students should **explain** whether prokaryotic or eukaryotic cells are more successful, with evidence to back up their ideas. [O1, O2, O3]

  Higher demand:

- Students should **explain** the advantages and disadvantages prokaryotic and eukaryotic cells have over the other. [O2, O3]

Consolidate and apply

- Students should read the section from the Student Book on how the classification system for early life developed, and **answer** the relevant questions. [O3]
- Students could **prepare** a two-minute talk on prokaryotic cells and explain how they are different to eukaryotic cells. Students should **explain why** they are thought to be more primitive. [O1, O2, O3]

Extend

Ask students who are able to progress further to:

research mitochondria and chloroplasts in greater depth and write a scientific paper about them from their research. They should **explain** their structure, their relevance and provide evidence which suggests they were once independent organisms before they became symbiotic with larger cells. Students could **explore** the advantages of this relationship. [O2, O3]

The following link could help (search terms: 'amazing cells' and 'the evolution of the cell')
http://learn.genetics.utah.edu

Plenary suggestions

- Provide students with the images from the PowerPoint slide. They should **say five things** about each image. They should be able to **classify** the objects shown on the cards as eukaryotic, prokaryotic or archaea and subdivide again into, animal, plant or bacterial, or structures in the cytoplasm. [O1, O2]
- Go back to the questions from the starter activity. See if any of these can now be answered. [O1, O2, O3]

Answers to Worksheet 1.5

1.

Cell 1	Cell 2	Cell 3
Plant cell Contains a nucleus and vacuole Contains mitochondria Contains many structures in the cytoplasm DNA is in the nucleus Has a cell wall No tail	Animal cell Contains a nucleus but no vacuole Contains mitochondria Contains many structures in the cytoplasm DNA is in the nucleus No cell wall No tail	Bacterial cell No nucleus or visible vacuole No mitochondria Contains few structures in the cytoplasm DNA is free in the cytoplasm Has a cell wall Has a tail

2. Drawing should show a flagellum, no nucleus, no mitochondria or internal structures, DNA floating freely, cell wall and capsule may be present, circular ring of DNA labelled as plasmid.

3. Mystery cards – many possibilities, but generally, the following should be grouped together:
1 on its own; 2 and 4; 3 and 5; 6 and 9; 7, 10 and 12; 8, 10 and 11;
Order of timeline: 1, 3, (2, 4), (8, 6 and 9), (7 and 10), 12

Lesson 6: Cell division

Lesson overview

AQA Specification reference

AQA 4.1.2.1; AQA 4.1.2.2

Learning objectives

- Describe the process of mitosis in growth, and mitosis as part of the cell cycle.
- Describe how the process of mitosis produces cells that are identical genetically to the parent cell.

Learning outcomes

- Recognise why cells need to be replaced, to grow and multiply. [O1]
- Describe the steps in mitosis. [O2]
- Explain which cells undergo mitosis and explain why not all cells do. [O3]

Skills development

- WS 1.2 Use a variety of models such as representational, spatial, descriptive, computational and mathematical to solve problems, make predictions and to develop scientific explanations and understanding of familiar and unfamiliar facts.
- WS 3.1 Present observations and other data using appropriate methods.

Resources needed Worksheets 1.6.1 and 1.6.2, Technician's notes 1.6

Digital PowerPoint

Key vocabulary cell division, chromosomes, DNA, growth, mitosis, nucleus

Teaching and learning

Engage

- **Show** students a model using bubbles of cell growth. See the Technician's notes for details. **Question** students about how they think cells grow. [O1]
- **Show** students a short video of bacterial cells growing. Emphasise this has been vastly speeded up. It can be found by searching 'dividing bacteria' at the following link www.cellsalive.com. **Question** students about what they think is happening and explain that they will find out during the lesson. [O1]

Challenge and develop

- Students should **look at** the images of chromosomes, nucleus and DNA from the PowerPoint slide, and, working in pairs, they should **explain** the relationship between them in the worksheet. Show students the PowerPoint slide and ask them to **correct** their diagrams. [O1, O2]
- Direct students to the images in the worksheet showing cells dividing, or the at video http://www.dailymotion.com, which can be found by entering the search terms 'stages of mitosis'. Students should **identify** five things they can see in the images. [O2]
- Students should **make models** of chromosomes using pipe cleaners. Students should use their model and other materials **to demonstrate** what is happening during the cell cycle. They could **use the information** in the Student Book to help them or the PowerPoint slide. [O2]

Explain

- Ask students to **write an explanation** about what is happening in mitosis, which cells undergo this and how often. They should **consider** which cells do not undergo mitosis and why this is so. Ask students to **identify** where the energy and raw materials for mitosis might come from. [O1–O3]

Consolidate and apply

- Ask students to **read** the section about chromosomes from the Student Book and **answer** the questions. [O1, O2]
- Students could **use the data** from the worksheet and **draw a graph** of the life span of each of the cells identified.

  Lower demand (Worksheet 1.6.1) – **draw data** from just one table and **answer** respective questions. [O1]

  Standard and higher demand (Worksheet 1.6.2) – **draw data** from both tables and **answer** respective questions. [O1, O3]

Extend

Ask students who are able to progress further to:

- **find out** the differences between binary fission and mitosis. They could **compare** the advantages and disadvantages of each process. [O1–O3]

Plenary suggestions

- Ask students to **decide** if the statements on the PowerPoint slide are true or false. [O1–O3]
- Show students images from the PowerPoint slides showing the different stages in mitosis. Ask students to **arrange** the images in order of each sequence. [O2]

Answers to Worksheets 1.6.1 and 1.6.2

1. The nucleus of the cell is made up of chromosomes.
   In every human cell, except the sex cells and red blood cells, there are 23 pairs of chromosomes inside the nucleus. Each chromosome is one strand of DNA. DNA is our genetic material. It gives instructions to the cells about the chemical reactions which keep us alive and make us human.
   Five things could be: 1) two cells are made from every one; 2) the new cells are the same size as the original; 3) inside the nucleus the chromosomes are pulling apart; 4) the dividing cells are twice as big as the parent cell; 5) the cytopasm is also replicating.

2. Explaining mitosis

   Each chromosome makes a copy of itself. This is to ensure that each new cell has identical genetic material. Each copy of a chromosome migrates to the opposite end of the cell. The cytoplasm also replicates. This ensures each new cell has all the materials it needs in order to carry out its reactions. The cell divides into two. Each cell has the same number of chromosomes as the original cell and is usually an exact copy. These are daughter cells. They are exact copies so they can carry out the same work as the original cell.

3. Low demand:

   1) Egg cell

   2) Liver cells

   3) No pattern

   Standard and high demand:

   1) Their life span is short because they have to be made by other cells. They are very active cells and, therefore, do not last very long.

   2) If cells are much more complicated, like nerve cells, it is inefficient to keep on making new ones. It is better to have cells that last much longer.

   3) They are constantly being worn away and need to be replaced.

Lesson 7: Cell differentiation

Lesson overview

AQA Specification reference

AQA 4.1.1.3; AQA 4.1.1.4

Learning objectives

- Explain the importance of cell differentiation.
- Understand size and scale in relation to cells, tissues, organs and body systems.
- Describe how cells, tissues, organs and organ systems are organised to make up an organism.

Learning outcomes

- Name some specialised cells and explain how their structure is adapted to their function. [O1]
- Explain the hierarchy of organisation in a multicellular organ system. [O2]
- Describe the scale of size in relation to cells, tissues, organs and body systems. [O3]

Resources needed Worksheet 1.7

Digital resources PowerPoint

Key vocabulary differentiation, organ, specialised, system, tissue

Teaching and learning

Engage

- Pair talk: students should **discuss** these questions: Are all cells the same? How do they know? Why might cells be different? How different can they be? [O1]
- Ask students to work in pairs and **write down** the names of as many organs, tissues and specialised cells as they can. Ask pairs to **join** another pair and **add** to their ideas, and groups of fours to join another group of four and again, **add** to their lists. They should **complete** the table in the worksheet. [O1, O2]

Challenge and develop

- If possible, show students some real slides of different types of animal and plant cells, including root hair cells, red blood cells, palisade cells and muscle cells. Alternatively, show students images of these cells on PowerPoint slides. Ask students to **identify** the cells as animal or plant cells and **justify** their reasons. [O1]
- Ask students to **read** page 26 of the Student Book to explain how cells become differentiated. Provide students with the worksheet and ask them to **explain** how each of the cells are adapted to carry out their jobs. [O1]
- Provide students with images to show how a multicellular organism is organised and of all the organ systems present. Students should **sort** the cards into their respective organ systems and order of level of organisation.

Explain

- Show students the image in the Student Book of a foetus. Students should **write** an explanation of how the different cells were formed and why different cells are needed. They should **explain** the levels of organisation in the foetus. [O1, O2]

Consolidate and apply

- Students could **build scale 3D models** of specialised cells, their tissues, organs and systems, using the data from the worksheet. [O1–O3]

- Ask students to work with another group. They should **use** their three-dimensional model or role play **to explain** the organisation in the body for two different systems. They should **explain** the features of any specialised cells in the system and the function of these cells within the system. [O1–O3]

Extend

Ask students who are able to progress further to:

- **compare and contrast** the organisation in a multicellular organism with that of a single-celled organism, like a bacterium. They could **explain** the advantages and disadvantages of each. [O2]

Plenary suggestions

Play 'lucky dip'. Have a bag with different word cards from the card sort. Ask one student to pick one card from the bag. They should **identify** the type of cell in the organ system, **explain** how it is adapted to perform its job and **describe** the tissues and organs to which it is related. [O1, O2]

Ask students to return to the list they produced in the table at the start of the lesson. They should **correct** any misconceptions and add to their list. They should **complete** any gaps and **add** to their list, so that all aspects of the hierarchy are filled for all the suggestions they made. [O1, O2]

Answers to Worksheet 1.7

1.

Skeletal system	Bone cell	Bone	Bone marrow
Reproductive system	Egg cell	Ovary	Epithelial tissue
Cardiovascular system	Red blood cell	Artery	Epithelial tissue
Respiratory system	Alveoli	Lung	Epithelial tissue
Nervous system	Motor neurone	Eye	Retina
Endocrine system		Adrenal gland	Glandular tissue
Digestive System		Stomach	Epithelial tissue
Excretory system		Kidney	
Integumentary system	Sweat glands	Skin	Connective tissue

2.

a) Sperm cell: tail for swimmining; large head packed with chromosomes; lots of mitochondria for energy.

b) Red blood cell: no nucleus so maximum area for haemoglobin; shaped to increase surface area; thin cell membrane for shortest diffusion path, so oxygen can enter quickly and be released quickly.

c) Muscle cell: long fibrous elastic strand which can expand and contract; large blood capillary network surrounding it for oxygen and glucose delivery; lots of mitochondria.

d) Nerve cell: long axon for transmitting messages; axon is protected by insulative layer, so electrical conductivity remains in the axon; cell body contains many mitochondria; synapses release chemical messages.

Lesson 8: Cancer

Lesson overview

AQA Specification reference

AQA 4.2.2.7

Learning objectives

- Describe cancer as a condition resulting from changes in cells that lead to their uncontrolled growth, division and spread.
- Understand some of the risk factors that trigger cells to become cancerous.
- Use data to analyse and evaluate the impact of cancer.

Learning outcomes

- Describe what cancer is and factors that cause it to occur. [O1]
- Explain why cancer causes damage and death. [O2]
- Use and analyse data to evaluate different types of cancer. [O3]

Skills development

- WS 1.3 Appreciate the power and limitations of science and consider any ethical issues which may arise.
- WS 3.5 Interpret observations and other data (presented in verbal, diagrammatic, graphical, symbolic or numerical form), including identifying patterns and trends, making inferences and drawing conclusions.
- WS 3.6 Present reasoned explanations including relating data to hypotheses.

Resources needed Worksheets 1.8.1, 1.8.2 and 1.8.3

Digital resources PowerPoint

Key vocabulary benign tumour, carcinogen, malignant tumour, mutation, secondary tumour

Teaching and learning

Engage

- Being aware of the sensitivities around this topic, ask students to work in pairs to **discuss** what they already know about cancer and if anyone close to them has had it. [O1]
- Show students the PowerPoint slide of tumorous tissues. **Discuss** these slides with students. Ask students what they can deduce about cancer from these images. [O1]

Challenge and develop

- Students should **look at** all the stages of cancer in the worksheet. Students should **answer** the differentiated questions in the worksheet about this flow diagram.

  Lower demand, answer questions 1–3 on Worksheet 1.8.1; standard demand answer questions 1–3 on Worksheet 1.8.2 and higher demand answer questions 1–3 on Worksheet 1.8.3. [O1, O2]
- **Discuss** with students the newspaper article on Angelina Jolie and breast cancer. Ask students why it was she might have taken this decision, even though she did not have breast cancer. [O1, O2]
- Go to www.dailymail.co.uk and enter the search terms 'Angelina Jolie' and 'double mastectomy'.
- Students should **analyse** and **evaluate** the data in the worksheet and **answer** the questions relating to this. [O3]
- **Discuss** with students the PowerPoint slides and/or some short videos relating to different types of cancers and causes of cancer.

Explain

- Students should **write a leaflet** for a doctor's surgery, explaining what cancer is, how it comes about and which people are most likely to be at risk. They should **use data** from the Student Book and/or other sources to back up their arguments. [O1, O2]

Consolidate and apply

- Students should work in groups of four. They should **role play** a consultation in the doctor's surgery using the role play information cards in the worksheet. Students should **read** the information in the cards and **use further information** from the lesson to **prepare** what they will say. The doctor **needs to decide** who needs to have an X-ray to see if cancer is present and who just needs some antibiotics. [O1, O2, O3]

Extend

Ask students who are able to progress further to:

- **research further data** to establish which type of cancer causes the most deaths in the world and why. They should **try to explain** why certain types of cancer are more common and why others are very rare. [O2, O3]

Plenary suggestions

- Ask students to **write a list** of 10 questions they could ask someone if they suspect they might have cancer. [O1, O2]
- Ask students to work in small groups to **produce** a one-minute talk to explain why breast and lung cancer are the most common forms of cancer in the UK.

Answers to Worksheets 1.8.1, 1.8.2 and 1.8.3

1.

Worksheet 1.8.1

a) Uncontrollable growth of cells

b) Mitosis

c) In a different part of the body

Worksheet 1.8.2

a) Uncontrollable growths can cause blockages in blood supplies, pressure on organs and stop organs from functioning properly.

b) More parts of the body are affected and cancer growths use up the body's oxygen and glucose in their growth. People are more tired and lose a lot of weight quickly. More blockages occur in the body, affecting more organs.

c) Through the blood

Worksheet 1.8.3

a) They need oxygen, glucose and nutrients to reproduce and grow, so they need their own blood supply to deliver these and to remove their waste.

b) The capillary would burst and cells that rely on the capillary would die.

c) Because it is caused by your own body cells. Your immune system does not recognise these cells as enemy cells.

2.

a) As the amount of animal fat increases, the death rate by breast cancer increases.

b) More developed and richer countries

c) More developed countries are likely to have a greater rate of breast cancer.

d) No; genetic factors and other health factors have not been taken into account.

e) The genetic factors related to breast cancer need to be accounted for; the age of the women in the study; and other factors like exposure to radiation, UV radiation, whether women smoke, and if they are exposed to other chemicals that cause cancer

Lesson 9: Stem cells

Lesson overview

AQA Specification reference

AQA 4.1.2.3

Learning objectives

- Identify cells as being differentiated, or as stem cells.
- Describe the function of stem cells in embryonic and adult animals.
- Explain how stem cells can be useful.

Learning outcomes

- Describe and recognise stem cells. [O1]
- Explain how stem cells can be used to cure diseases and repair damaged tissues and organs. [O2]
- Evaluate the advantages and disadvantages of using stem cells in ethical and moral terms. [O3]

Skills development

- WS 1.3 Appreciate the power and limitations of science and consider any ethical issues which may arise.
- WS 1.4 Explain everyday and technological applications of science; evaluate associated personal, social, economic and environmental implications; and make decisions based on the evaluation of evidence and arguments.
- WS 4.1 Use scientific vocabulary, terminology and definitions.

Resources needed Plants that have been grown from cuttings; or a plant from which cuttings may be taken, rooting hormone; Worksheets 1.9.1 and 1.9.2

Digital resources PowerPoint

Key vocabulary adult stem cell, culture, embryonic stem cell, *in vitro* fertilisation, therapeutic cloning

Teaching and learning

Engage

- Ask students to **name and draw** some differentiated cells. Show them the PowerPoint which looks at the cells of an embryo. Ask students to **describe** the different types of cells they see. Ask students to **discuss** in pairs how ordinary cells turn into specialised cells. [O1]
- Show students cuttings of plants that have been grown into new plants, or demonstrate taking a cutting from a plant. Ask students to **discuss** in pairs how different plant structures might develop from simple cuttings. [O1]
- Introduce the term 'stem cell' as a cell that is undifferentiated, using the PowerPoint slide. Explain that, under the right hormonal and genetic signals, stem cells can be turned into cells with particular functions. [O1]

Challenge and develop

- Ask students to **watch** the following video on the use of stem cells. [O1, O2]
- www.cells4life.co.uk (search terms: 'video' and 'how do stem cells work')
- Show students the PowerPoint slides of different types of stem cells and applications relating to them. [O1, O2]
- Ask students to **read** and **discuss** the information in Worksheet 1.9.1 (low and standard demands) or Worksheet 1.9.2 (high demand) and answer the differentiated tasks.

Low and standard demand:

- Write a list of the advantages and disadvantages of using stem cells.
- Why are some religions against the use of stem cells?

High demand:

- **Write an argument:** Do the benefits of using human embryonic stem cells outweigh the moral arguments against it? Explain your answer. [O1–O3]

Explain

- Students should **read** the section on stem cells from page 30 of the Student Book. They should **write an explanation** for another class of where different stem cells come from and how they can be used by doctors and scientists. [O1, O2]

Consolidate and apply

- Ask students to **write and present a news bulletin** describing stem cells and discussing the applications they could have for the future. They should **use information** from the lesson to **explain** the advantages and disadvantages of using stem cells. [O2, O3]

Extend

Ask students who are able to progress further to:

- **watch** the video on tissue engineering using stem cells. www.cells4life.co.uk (search terms: 'video' and 'tissue engineering')

  They should **use** this to **prepare** their own PowerPoint presentation to **present** to the class about different applications of stems cells and some of the problems scientists have had to overcome to grow tissues and organs. [O2, O3]

Plenary suggestions

Ask students to work in small groups and **make up** 10 questions about stem cells. Each group should present their questions to another group to **answer**. [O1, O2]

Students should **present their own case** for or against the use of stem cells, **using reasoned arguments**. [O1–O3]

Answers to Worksheet 1.9.1

1. **Advantages of using stem cells:**

   Replacement of blood and other tissues (for example, bone, skin and eye) repair of damage to organs and nerve cells. Millions of people may benefit.

   Disadvantages:

   Human stem cells are painful to remove from adults and may introduce disease. Embryos have the right to live; interfering with stem cells is unnatural, against some religious beliefs.

   Some religions believe that life starts at the moment of conception and it is wrong to interfere with human development.

Lesson 10: Stem cell banks

Lesson overview

AQA Specification reference

AQA 4.1.2.3

Learning objectives

- Explore the use of stem cells in medicine.
- Identify the risks in using stem cells.
- Evaluate the benefits and disadvantages of using stem cells.

Learning outcomes

- Explain the uses of stem cells in medicine. [O1]
- Describe what is meant by the term therapeutic cloning. [O2]
- Explain the advantages and disadvantages in using different techniques for producing stem cells. [O3]

Skills development

- WS 1.3 Appreciate the power and limitations of science and consider any ethical issues which may arise.
- WS 1.4 Explain everyday and technological applications of science; evaluate associated personal, social, economic and environmental implications; and make decisions based on the evaluation of evidence and arguments.
- WS 4.1 Use scientific vocabulary, terminology and definitions.

Resources needed Worksheet 1.10

Digital resources PowerPoint

Key vocabulary donor, gene, mutation, therapeutic cloning, umbilical cord

Teaching and learning

Engage

- Ask students to **write** their own true and false statements about stem cells from the previous lesson. [O1]
- Ask students to **complete** the first activity on the worksheet. They should **identify** the similarities and differences between stem cells from embryos, from plants and from adult humans. [O1]

Challenge and develop

- Show students the video on using stem cells to make new heart muscle cells to recap the benefits of using stem cells www.eusem.com (search terms: 'ComScience', 'events' and 'stem cells') [O1]
- Show students the PowerPoint slides on the uses of umbilical cord and therapeutic cloning as alternative sources for the production of stem cells, rather than using human embryos. [O1, O2]
- Provide students with a card sort on stem cells and their uses. Ask them to **group** the cards into common areas. [O1, O2, O3]
- In small groups, students should use the card sort to **produce a list** of benefits and risks for the use of different types of stem cells. [O3]

Explain

- Students should **write a letter** to their local MP either to support or to argue against the use of stem cells. They should support their arguments with evidence from the last two lessons. They should **explain** how stem cells are available from different sources. [O2, O3]

Consolidate and apply

- Ask students to work in small groups. They should **imagine** they will provide a significant sum of money to sponsor scientific developments. Ask them to **justify** which cause it will go to: adult or embryonic stem cell development? They should **explain** why they have not chosen to sponsor development in stem cell technology from the other area. [O1, O2, O3]

Extend

Ask students who are able to progress further to:

- **research the legislation** surrounding the use of stem cells and explain why this has changed over the years. Students could **evaluate** the impact this has had on scientific developments and ethical considerations. [O2, O3]

Plenary suggestions

Ask students to **summarise** the arguments for and against the use of embryonic stem cells, adult stem cells, umbilical cord stem cells and therapeutic cloning on the worksheet.

Answers to Worksheet 1.10

Card sort

Embryonic stem cells	**Therapeutic stem cells**	**Umbilical cord stem cells**	**Adult stem cells**
Many stem cells are produced which can be made into any other cell in the body.	Many stem cells can be produced by this articifical technique.	This is a rich source of stem cells that are formed from the placenta after a child is born and the structure is no longer needed.	Small numbers of stem cells are found mainly in the bone marrow, blood vessels, skin and other organs.
Egg cells need to be extracted, which can be painful and infections are possible.	The stem cells produced will closely match the patient.	No painful or intrusive techniques are applied.	Often painful techniques are required to extract them.
Storage of embryos requires a lot of energy and space. There is a time limit for how long they can be stored	Human egg cells need to be extracted, which can be painful and infections can arise.	These stem cells have been used to treat a variety of blood disorders.	Rejection of stem cell implants are a problem, if used in humans who were not the source.
Immunosuppressant drugs are needed to avoid rejection of transplanted tissues and organs.	Minimal use of immunosuppressant drugs will be needed, as the stem cells will closely match the patient.	Immunosuppressant drugs will be needed if the cells are used to develop bone and other tissues.	Immunosuppressant drugs are needed.
Embryos are produced specifically for making stem cells and then discarded. This raises ethical issues.	An embryo is produced by this technique, which is discarded once stem cells have been extracted.	There are no ethical issues relating to the use of these stem cells, but consent from the mother is needed.	No ethical issues relating to the use of these stem cells.

Summary of arguments for and against

Embryonic stem cells	**Therapeutic stem cells**	**Umbilical cord stem cells**	**Adult stem cells**
For Many stem cells available Can be turned into any type of cell All tissues and organs could potentially be grown	**For** Many stem cells available Can be turned into any type of cell All tissues and organs could potentially be grown Fewer immunosuppressant drugs and rejection issues as stem cells from the patient are used	**For** Many stem cells are available No embryos are involved Already successfully used in treating blood disorders in children	**For** Own cells can be used, so rejection issues are minimal and few immunosuppressant drugs needed No embryos are involved
Against Ethical issues with the use of human embryos Immunosuppressant drugs will be needed Technology is still developing Other stem cells could be used without using human embryos	**Against** Ethical issues with the use of human embryos Technology is still developing Other stem cells could be used without using human embryos	**Against** Immunosuppressant drugs will be needed Only successful so far in treatment of blood disorders, other types of tissues are at a developmental stage	**Against** Only a few stem cells are available Not all cell types can be grown Painful techniques needed to extract stem cells

Lesson 11: Key concept: Cell development

Lesson overview

AQA Specification reference

AQA 4.1.2.3

Learning objectives

- Revise ideas about cell structure, cell division and stem cells.
- Apply ideas about cells to unfamiliar contexts.
- Describe how plant meristems can be used in cloning.

Learning outcomes

- Make links between cell structure, cell division and stem cells. [O1]
- Compare the structure and function of fungi and viruses to normal cells. [O2]
- Compare the use of meristems in cloning with animal cell cloning. [O3]

Skills development

- WS 1.2 Use a variety of models such as representational, spatial, descriptive, computational and mathematical to solve problems, make predictions and to develop scientific explanations and understanding of familiar and unfamiliar facts.
- WS 1.4 Explain everyday and technological applications of science; evaluate associated personal, social, economic and environmental implications; and make decisions based on the evaluation of evidence and arguments.
- WS 4.1 Use scientific vocabulary, terminology and definitions.

Resources needed Worksheets 1.11.1 and 1.11.2

Digital resources PowerPoint

Key vocabulary meristems, virus, yeast

Teaching and learning

Engage

- Ask students to work in pairs to **make a word wall** on poster paper. They should **split** the paper into four sections. On each of the sections they should **discuss and write down** all the words relating to animal cells, plant cells, cell division and cloning, respectively. [O1]
- Students should **join** another pair and compare the words on their posters. Discuss and check students' ideas as a class and ask them to **add in** any omissions. [O1]

Challenge and develop

Low and standard demand

- Ask students to **read** the first section from the Student Book or use the PowerPoint to **discuss** the structure of a fungus. Ask students to **complete** task 1 from Worksheet 1.11.1 in which they **compare** this structure to that of a typical animal and plant cell. [O2]
- Check students' ideas using the PowerPoint slide. [O2]

High demand

- Students should **read** the section on viruses in Worksheet 1.11.2. They should **answer** the questions in the worksheet and **justify** the view that a virus is not living. [O2]
- Show students the PowerPoint and discuss the importance of meristems in plant cloning. They should **discuss** animal and plant cloning in pairs and **compare** the use of meristems in plant cloning with animal cloning, on the worksheet. [O3]

Explain

- Ask students to **explain** how different cells, or organisms, divide and what form they become in the end. They should **answer** this on the worksheet. [O1–O3]

Consolidate and apply

- Students could draw an annotated diagram to show what happens in the cloning of plants. They should **explain** on the diagram where mitosis occurs and **use as many key words** from their word wall as they can.

Extend

Ask students who are able to progress further to:

- **find out** more about other types of cell division, for example, meiosis. They could **compare and contrast** mitosis, meiosis, budding and forms of asexual reproduction. [O1]

Plenary suggestions

Students could be asked to write the definitions of all the words on their word wall.

They should **make up** at least five short paragraphs **connecting** words from each section of the word wall together.

Answers to Worksheets 1.11.1 and 1.11.2

1. **Worksheet 1.11.1:**

   Fungi have cytoplasm, nucleus, mitochondria and ribosomes. There are no divisions between the cells, so the cytoplasm and many nuclei run throughout the fungus. There are no chloroplasts. It does not have a cell membrane between cells. There is a cell wall.

   Worksheet 1.11.2:

   a) (Viruses posses) genetic material
   b) The virus is completely dependent on other cells for reproduction; it cannot do this by itself. It does not excrete, respire or feed.

2. How animal cells are cloned:

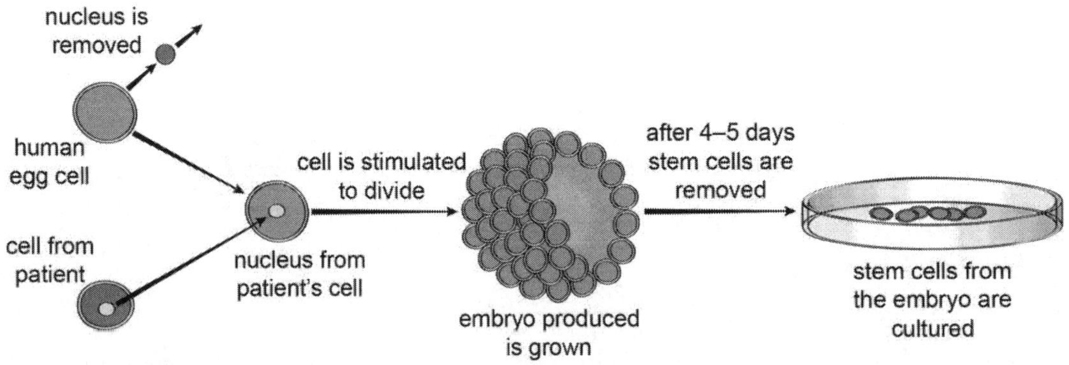

Similarities	Differences
Identical daughter cells produced Daughter cells have same number of chromosomes as parent Stem cells are used	Nuclei of meristem cells do not have to be removed No electrical shock is needed No ethical considerations with the use of meristems Only cells from one organism are involved

3.

   a) Mitosis – two new animal cells – identical daughter cells each other and to parent cell
   b) Mitosis –two now plant cells – identical daughter cells to each other and to parent cell
   c) Mitosis – differentiated cells – identical daughter cells – different to parent cell
   d) Mitosis – differentiated cells – identical daughter cells – different to parent cell
   e) Budding – two new yeast cells – identical daughter cells to each other and to parent cell

Lesson 12: Cells at work

Lesson overview

AQA Specification reference

AQA 4.4.2.1

Learning objectives

- Recognise that all organisms respire.
- Explain respiration as the process of making energy.
- Describe aerobic respiration as an exothermic reaction.

Learning outcomes

- Identify a range of different reasons why different organisms need energy. [O1]
- Explain aerobic respiration as a word equation, with glucose as the source of energy. [O2]
- Relate the activity of a cell to the amount of mitochondria present. [O3]

Skills development

- WS 2.4 Carry out experiments appropriately having due regard for the correct manipulation of apparatus, the accuracy of measurements and health and safety considerations.
- WS 2.6 Make and record observations and measurements using a range of apparatus and methods.
- WS 3.5 Interpret observations and other data (presented in verbal, diagrammatic, graphical, symbolic or numerical form), including identifying patterns and trends, making inferences and drawing conclusions.

Resources needed Worksheet 1.12; Practical sheet 1.12; Technician's notes 1.12

Digital resources PowerPoint

Key vocabulary active transport, aerobic respiration

Teaching and learning

Engage

- Ask students to work in pairs to **write a list** of why animals and plant cells need energy. They should use the pro forma in the worksheet. Ask pairs to **team up** with another pair, **comparing and adding** to their lists. Take feedback from the students and show them the PowerPoint, to see if they identified all the different reasons organisms need energy. [O1]

Challenge and develop

- Demonstrate the 'screaming jelly baby' experiment which shows energy release from glucose through a chemical reaction. See Practical worksheet 1.12 for details. See the video of this reaction on the BBC website (www.bbc.co.uk) by searching for 'aerobic and anaerobic respiration'. [O2]
- Explain that bread is made of starch, which comes from glucose. Ask students to **carry out an investigation** whereby they burn a known mass of bread over a boiling tube of water and **determine** the increase in temperature of the water. Ask students to make some inferences about glucose from the reactions. They should **deduce** that, when glucose is burnt in oxygen, an exothermic reaction takes place and lots of energy is released. Use the PowerPoint slide to explain what is happening. [O2]
- Ask students to **relate** these demonstrations to what happens inside cells. Ask them to **identify** where glucose and oxygen come from and how they get to the cells. Remind students of the role of mitochondria in cells using the PowerPoint slide. [O2, O3]

Explain

- Ask students to **annotate** the diagram of the human body on their worksheet to explain where the most sugar and oxygen are likely to be found. High demand students should write a full explanation of how these materials reach the cells, including the role of the heart and lungs. [O2]
- Ask low and standard demand students to **summarise** aerobic respiration as a word equation.
- Ask high demand students to **summarise** it as a balanced symbol equation. [O2]

Consolidate and apply

- Ask students to look at the different cell types on the worksheet. Ask them to **rank** these in order of which carries out the most aerobic respiration and ask them to say where the greatest concentration of mitochondria is likely to be found. They should **analyse the data** and use this to help them and **answer** the questions. [O2, O3]

Extend

Ask students who are able to progress further to:

- **carry out** some research and find out whether the statement 'animals respire more than plants because they move about more' is true or false. They should **use evidence** they uncover to back up their arguments.

Plenary suggestions

Present students with the key words from the PowerPoint slide. Ask them to **make up five sentences** about what they have learnt using as many key words as possible.

Answers to Worksheet 1.12

1. Energy is needed:
   - to drive the chemical reactions that are needed to keep us alive
   - to build large molecules like proteins from smaller ones, such as amino acids
   - for movement
   - to make our muscles expand and contract
   - to keep our bodies warm and at a constant temperature
   - to transport substances around the body
   - for cell division
   - for active transport, whereby molecules are moved from a low concentration to a high concentration, for example, in mineral uptake in plants
   - to transmit nerve impulses.

2. Glucose comes from food – molecules are broken down in the digestive system and absorbed by the blood. They are transported by the blood to all cells.

   Oxygen comes from the air and is absorbed by the lungs into the blood. It is transported to all cells by red blood cells in our blood. The heart pumps the blood around the body to reach all cells.

   glucose + oxygen → carbon dioxide + water

   $C_6H_{12}O_6 + 6O_2 \rightarrow 6CO_2 + 6H_2O$

3. Muscle cell; sperm cell; nerve cell; red blood cell.

   Muscle cells will have the most mitochondria.

   1) Retinal eye cells have the most mitochondria and may be the most actively respiring cells.

   2) No; bacteria cells respire but do not use mitochondria; they have alternative mechanisms.

Lesson 13: Living without oxygen

Lesson overview

AQA Specification reference

AQA 4.4.2.1

Learning objectives

- Describe the process of anaerobic respiration
- Explain when anaerobic processes occur
- Compare the processes of aerobic and anaerobic respiration

Learning outcomes

- Describe the situations when aerobic and anaerobic respiration take place [O1]
- Write a word equation for anaerobic respiration [O2]
- Compare and contrast aerobic and anaerobic respiration [O3]

Skills development

- WS 2.2 Plan experiments or devise procedures to make observations, produce or characterise a substance, test hypotheses, check data or explore phenomena.
- WS 3.7 Be objective, evaluate data in terms of accuracy, precision, repeatability and reproducibility and identify potential sources of random and systematic error.
- WS 4.1 Use scientific vocabulary, terminology and definitions.

Resources needed Worksheet 1.13

Digital resources PowerPoint

Key vocabulary anaerobic respiration, fermentation

Teaching and learning

Engage

- Show students the PowerPoint slide showing alcohol, bread and yeast. Ask students to **make a connection** between the images. Take feedback and **discuss** their ideas. [O1]

Challenge and develop

- Show students some readymade dough that has risen. Discuss what has happened to the dough and what has been added to the flour in the process. Use the PowerPoint slide in your discussion. [O1]
- Also show students a video from www.science.howstuffworks.com (search terms: 'episode 7: bread making'). [O1]
- Now show students the fermentation process of making alcohol using yeast. [O1] www.science.howstuffworks.com (search terms: 'episode 5: fermentation').
- Show the students the video 'Use of microbes in the food and drink industry': www.bbc.co.uk (search terms: 'microbes' and 'food and drink').
- Ask students to **plan** their own investigation into anaerobic fermentation by investigating conditions for making their own dough. See the worksheet for details. [O1]

Explain

- Ask students to **draw a cartoon strip** to explain the process of anaerobic respiration in fermentation. They should explain the process using a word equation. [O1, O2]

  High demand

  Students should **write a balanced symbol equation** for the process. [O2]

Consolidate and apply

Students should **compare and contrast** the key differences between aerobic and anaerobic respiration, in the worksheet. [O1–O3]

Extend

Ask students able to progress further to:

- **Draw a graph** of the data from the worksheet to **compare** the amount of carbon dioxide produced from the fermentation of yeast at different temperatures. They should **explain why** there might be differences. [O3]

Plenary suggestions

Ask students to work in pairs and **prepare their own card sort** to compare and contrast the processes of aerobic and anaerobic respiration. They should **present** their cards to another pair to **sort out**. [O1, O2, O3]

Answers to Worksheet 1.13

1. *What are the variables in the investigation?* Amount of glucose, amount of yeast, temperature

   What will you choose as the independent variable? Amount of yeast (1 g, 2 g, 3 g, 4 g, 5 g and 10 g)

   What will the dependent variable be and how will you measure this? How high the bread rises

   What are the control variables and how will you control them? Same amount of glucose (e.g. 5 g); same temperature
   (35 °C), same amount of flour and water

   How will you produce reliable results? Repeat the investigation until results are consistent

   How can you check that your procedure is reproducible? Give the procedure to someone else to carry out

   How can you check that your procedure is repeatable? Carry it out again and see if the results are the same

   Sources of random error – not all the flour is mixed in to the same extent; yeast is unevenly distributed

   Sources of systematic error – not all the dough is warmed to the same 35 °C, there will be some heat loss.The same yeast cells are not being used; there may be genetic differences which affect the rate of respiration.

   Write your procedure here:

   Mix 100 g of flour to required amount of yeast. Add about 100 ml of warm water at 35 °C to ensure a smooth dough is produced. Mix thoroughly to make a smooth dough. Leave to rise at 35 °C for half an hour. Measure the height of the dough before and after rising. Repeat with another amount of yeast.

2. 2 Glucose → ethanol + carbon dioxide (+energy released)

   $C_6H_{12}O_6 \rightarrow 2C_2H_5OH + 2CO_2$

3.

Aerobic respiration	Anaerobic respiration
Glucose used Uses oxygen Lots of energy produced CO_2 produced Water produced	Glucose used No oxygen Some energy produced Use in fermentation Alcohol produced CO_2 produced Water not produced

Extension

1. The amount of carbon dioxide increases linearly with temperature until 400 °C, after which there is a rapid decline.This is beacause enzymes in the yeast have denatured and therefore the rate of anaerobic respiration decreases. At 500 °C the yeast cells have nearly all died.
2. Place yeast at different water baths at different temperatures in the same conditions and measure the volume of carbon dioxide with a gas syringe. Use the same amount and species of yeast, the same type of sugar and the same amount of water.
3. There would be higher levels of carbon dioxide, but the overall shape of the graph would be the same.

Lesson 14: Growing microorganisms

Lesson overview

AQA Specification reference

AQA 4.1.1.6

Learning objectives

- Describe the techniques used to produce uncontaminated cultures of microorganisms.
- Describe how bacteria reproduce by binary fission.
- Calculate the number of bacteria in a population.

Learning outcomes

- Describe how to prepare an agar plate culture of microorganisms. [O1]
- Explain how binary fission differs from mitosis. [O2]
- Analyse data from agar plates and make inferences. [O3]

Skills development

- WS 1.5 Evaluate risks both in practical science and the wider societal context, including perception of risk in relation to data and consequences.
- WS 2.2 Plan experiments or devise procedures to make observations, produce or characterise a substance, test hypotheses, check data or explore phenomena.
- WS 2.4 Carry out experiments appropriately having due regard for the correct manipulation of apparatus, the accuracy of measurements and health and safety considerations.

Resources needed Worksheet 1.14; Practical sheet 1.14; Technician's notes 1.14

Digital resources PowerPoint

Key vocabulary agar plate, autoclave, bacteria, colony, incubation, inoculating loop, nutrient broth, Petri dish, sterilisation

Teaching and learning

Engage

- Show students the clip about bacteria (search terms: 'microorganisms and us') at www.bbc.co.uk. [O1]
- Ask students to work in pairs to **discuss** where they think the most number of bacteria may be found in our bodies and why. Ask students to **discuss** how they might investigate this idea. Take feedback from students' ideas. [O1]

Challenge and develop

- Show students the short clip of bacteria dividing (search terms: 'Escherichia coli') at www.cellsalive.com. [O1, O2]
- Ask students to **look at** the population growth chart of bacteria in the worksheet and on the PowerPoint slide. In pairs, ask them to **discuss** the results and why they might have this pattern. [O3]
- **Discuss** with the class how this data might have been obtained, using the PowerPoint slides to support the discussion.
- Show students the video of growing bacteria (search terms: 'microorganisms and bacteria') at www.bbc.co.uk. Ask students to **discuss** how binary fission compares to and contrasts with mitosis. Use the PowerPoint slides to help. [O1–O3]
- Demonstrate how to produce an agar plate of bacteria. Pay strict attention to the need for following safety guidelines and ensuring the dishes are not completely sealed (so some oxygen can get in to prevent the growth of pathogenic bacteria). Again the PowerPoint slides can be used to help. [O1, O2]
- Ask students to **follow the instructions** on the Practical sheet and produce their own cultures of bacteria from swabbing everyday items or their hands. [O1]

Explain

- Show students a set of results from the worksheet. Ask them to **explain the observations** and **draw conclusions** from the data. They should **answer** the differentiated questions from the worksheet. [O1, O3]

Consolidate and apply

- Ask students to **write** their own practical instructions to find out if animals carry more or less bacteria than humans. [O1]
- Ask students to **read** the section about binary fission from the Student Book and **explain how it works** in a flow diagram. [O1, O2, O3]
- Students should **compare and contrast** binary fission with mitosis. [O3]

Extend

Ask students who are able to progress further to:

- **find out** about how the first bacteria were grown and how aseptic techniques were developed so bacteria could be studied further. [O1, O2, O3]

Plenary suggestions

- **Show** students the short video about the discovery of penicillin at www.bbc.co.uk (search terms: 'Fleming and the discovery of penicillin').
- **Ask** students to explain what they see and to write some instructions for Mr Fleming about how to make up an agar plate safely. Ask them to predict the outcomes if he had been extremely safe and followed aseptic procedures. [O1, O3]

Answers to Worksheet 1.14

1. Bacteria first grow at a steady rate, which then increases directly proportionally, and then remains constant before decreasing rapidly.

   The first part is steady as there is no replication of bacteria. They are getting used to their environment. In the second part they are dividing at a constant rate as each bacteria becomes two. Then the population growth becomes steady as some begin to die off due to lack of nutrients, or lack of space, or build up of toxic waste products. Here growth rate = death rate. Finally, more bacteria begin to die due to lack of nutrients.

2.

   a) **Low demand** (*Describe in full what you see on the agar plates*):

   First plate shows uneven distribution of red bacterial growth with more growing on the edges than in the centre. Second plate shows relatively even distribution of two different types of bacteria, yellow and white. They are growing round the centre and hardly any colonies grow round the edges. The final slide shows lots of different species of bacteria growing in specific places, where they have been placed.

   Standard and high demand:

   In the first plate, the way the plate was innoculated suggests that the edges were swiped first, before the centre. Or there is something in the centre that is killing off the bacteria. In the second plate, there are even amounts of white and yellow, but not as much growth has taken place as in the first, so the bactreria are in clumps. Maybe as they are growing, each type is producing something toxic to the other type. In the third, the sample contains lots of different types of bacteria; some are growing faster than others. This is why some spots are larger than others.

   b) Prepare three agar plates using the techniques shown in the lesson. Take a sterilised swab from a hand and dab it onto the agar plater; do the same with an animal paw and leave the third as a control. Seal the plates, leaving an air gap and incubate for 2–3 days at 25 °C.Take care to disinfect all equipment after use.

3.

Binary fission	Mitosis
Cells divide into two Circular strand of DNA replicates Circular strands migrate to opposite ends of the cell Cytoplasm replicates and divides into two Two identical daughter cells are formed, also identical to the parent	Cells divide into two All chromosomes replicate Chromosomes line up at the centre before parting to opposite ends Cytoplasm replicates and divides into two Two identical daughter cells are formed, also identical to the parent

4. Instructions as per practical sheet, with special attention to using sterilised equipment. If Fleming had used these instructions, antibiotics would not have been discovered.

Lesson 15: Testing new antibiotics

Lesson overview

AQA Specification reference

AQA 4.3.1.8; AQA 4.3.1.9

Learning objectives

- Use appropriate apparatus to investigate the effect of antibiotics on bacterial growth.
- Use microorganisms safely.
- Apply sampling techniques to ensure that samples are representative.

Learning outcomes

- Identify reasons why new antibiotics need to be produced and tested. [O1]
- Describe the method for testing new antibiotics and interpret results. [O2]
- Explain how safety is ensured and representative samples produced. [O3]

Skills development

- WS 1.2 Use a variety of models such as representational, spatial, descriptive, computational and mathematical to solve problems, make predictions and to develop scientific explanations and understanding of familiar and unfamiliar facts.
- WS 1.6 Recognise the importance of peer review of results and of communicating results to a range of audiences.
- WS 3.5 Interpret observations and other data (presented in verbal, diagrammatic, graphical, symbolic or numerical form), including identifying patterns and trends, making inferences and drawing conclusions.

Maths focus

- 4a Translate information between graphical and numeric form

Resources needed Video 'Antibiotics – the beginning or the end?' from the Business Reporter section of www.telegraph.co.uk. Worksheet 1.15.1 (low demand); Worksheet 1.15.2 (standard demand); Worksheet 1.15.3 (high demand)

Digital resources

Key vocabulary antibiotic, pathogen, sampling techniques

Teaching and learning

Engage

- Show students the data on antibiotic resistance from section 1 of the worksheet. Ask students to work in pairs and **discuss** what the data is showing. Ask students to **identify** potential solutions to the problems shown and **answer** the questions on the worksheet. [O1]
- **Discuss** the need to produce new antibiotics and show students the first 5 minutes of the video clip on antibiotic resistance ('Antibiotics – the beginning or the end?'). [O1]

Challenge and develop

- Show students the results of an experiment to test a new antibiotic from section 2 of the worksheet. Ask students to work in small groups to **discuss** how this data might have been collected. [O2]
- Students should **complete** section 2 of Worksheet 1.15.1 (low demand), Worksheet 1.15.2 (standard demand) or Worksheet 1.15.3 (high demand). [O2]

Explain

- Students should **read** the section on how scientists test antibiotics in the Student Book (pages 42-43). They should **explain** how scientists ensure that representative samples are given. They should **explain** how the experiments are peer-reviewed and why this is so important. [O2, O3]

Consolidate and apply

- Students should **compare** and **contrast** how a scientist carries out an investigation to test a new antibiotic as opposed to how this might be carried out in school. They should **record** their ideas in section 3 of their worksheet. [O2, O3]

Extend

Ask students able to progress further to:

- **Find out** about some of the new antibiotics being developed today and the time scale needed. They should **explain why** it take so long for a new antibiotic to be made. [O1,O2, O3]

Plenary suggestions

- Students could **imagine** they are scientific researchers who are planning to test a new antibiotic. They should **write a bid** to the government in order to fund their work. The bid should include why the work is needed, how they go about their research, and what they will achieve in the end. They should **explain** the impact a successful result may have. [O1, O2, O3]

Answers to Worksheet 1.15

1 Antibiotic resistance

1. The graph shows that resistance of bacteria A to antibiotics has risen sharply from 1980 to 1990, from 5% to 40%. This remained at 40% for about 5 years, after which it increased steeply again to 57% in 2000. Resistance of bacterias B and C show a similar trend. They have both risen steeply from 1987 from 2% to about 30% in 2000. There are no indications from either graph of the percentage incidence diminishing.
2. The possible solutions are to find new antibiotics or find some way of treating the bacteria so they are no longer resistant to the current antibiotics.

2 Testing a new antibiotic

Low demand: Antibiotics A, C and E affected the bacteria. C (greatest effect), E, A (least effect)

Standard demand: The results around F were anomalous. This could be because the bacterial colony had not grown uniformly in this area, or the antibiotic had not spread evenly through the agar medium.

High demand: Area around A = 19.64 cm^2; Area around C = 78.55 cm^2; area around E = 38.49cm^2

3 Comparing methodologies

Scientific research	School experiment
The medium used is called Mueller-Hinton agar and is a mixture of blood, beef and milk protein. The temperature of incubation is 37 °C More pathogenic bacteria are likely to be tested	Normal agar jelly is used 25 °C is used as the incubation temperature as this prevents pathogens from forming. Safe bacteria will be used

Similarities:

Identical discs containing different concentrations of the antibiotic will be placed on the culture of bacteria.

The clear area around the discs will indicate how sensitive the bacteria are to the antibiotic. The greater the area, the more sensitive the bacteria will be.

Lesson 16: Required practical: Investigating disinfectants

Lesson overview

AQA Specification reference

AQA 4.1.1.6

Learning objectives

- Carry out experiments with due regard to health and safety.
- Present and process data, identifying anomalous results.
- Evaluate methods and suggest further investigations.

Learning outcomes

- Describe how to investigate the effect of disinfectants on bacteria. [O1]
- Present and analyse data. [O2]
- Evaluate methods and suggest improvements and methods to reduce the growth of bacteria. [O3]

Resources needed Worksheet 1.16; Practical sheet 1.16; Technician's notes 1.16

Key vocabulary antiseptic, bacteria, diffusion, incubation, sterilisation

Teaching and learning

Engage

- Look at the agar plates students produced during the last lesson, or ones prepared for this lesson. **Identify** which plates produced the greatest number of bacterial colonies and which produced the greatest variety. [O1]
- Tell students that now we know bacteria are everywhere, ask questions such as 'How can we keep ourselves safe from bacteria?' **Discuss** their ideas. [O1]
- Show students a range of different disinfectants. Ask students to work in small groups and **plan** an investigation on the worksheet that would enable them to find out which disinfectant was the best value for money. [O1]
- Take feedback from the students and **discuss** their ideas. [O1]

Challenge and develop

- Ask students to **swap** their plans with another group. Ask them to **look at** the procedure from the practical worksheet and **peer assess** the practical plans. They should **highlight** the strengths and weaknesses from the plans. [O1]
- Students should **follow** the given procedure with utmost care and precision, paying great attention to safety, and produce their own agar plates to test different disinfectants. [O1]
- Ask students to **read** the labels of the disinfectants and **predict** which are the most likely to kill off the most bacteria. Ask students to **draw a diagram** to **predict** what the plate would look like after a few days and then after a few weeks. [O2]

Explain

- Show students a set of results from the worksheet after a few hours and then after a few days. Ask students to **explain** these results using ideas about binary fission. [O2]
- Students should answer the differentiated questions on the worksheet (low demand: describe the pattern in the data; standard demand: explain the differences in the results; high demand: predict what the results would look like in a few weeks and justify your reasons). [O2]

- Students should **explain** any sources of error in their investigation and **make some suggestions** for improvement. This should be done on the practical sheet. [O3]

Consolidate and apply

- Students should **use** their results to **promote** the brand of disinfectant they believe to be the best value for money. They should **produce** a short commercial for the brand, **explaining** what a disinfectant does and how we know it works. [O1–O3]

Extend

Ask students who are able to progress further to:

- **find out** about how aseptic techniques have developed over time and **produce** a short presentation to the class about the history of disinfectants. [O1–O3]

Plenary suggestions

Tell students it is important that hospitals keep bacterial levels to a minimum. Ask students to **imagine** they are in charge of hygiene in a hospital. They should **write a leaflet** explaining what tests should be done to ensure the hospital is safe from harmful bacteria. They should **write an information leaflet** for patients to **explain why** it is important to wash hands regularly, particularly after using the toilet and before preparing or eating food. They should explain why it is important to keep washing hands throughout the day and why doing this once a day is not enough. [O3]

Answers to Worksheet 1.16

1. See practical instructions.

   Predictions – after a few days there will be clear patches around the circles of disinfectant, with bacterial growth elsewhere.

2. Circle 4 has the largest area of clear patch around it, whereas circle 12 has the smallest area. Large areas are seen for spots 2, 3, 5, 7, 9 and 10, whereas only small areas are seen for circles 6, 8 and 11.

   Standard demand:

   Circle 4 contains the strongest disinfectant, whereas circle 12 contains the weakest.

   Spots 2, 3, 5, 7, 9 and 10 contain similar strengths of disinfectant.

   Spots 6, 8 and 11 contain weaker strengths, or the bacteria are most resistant to these disinfectants.

   High demand:

   Predictions after a few weeks: the area around the clear patches may be smaller, as some resistant bacteria develop and grow. Also the strength of the disinfectant will reduce after a few weeks.

3. Areas that are used regularly by patients should be sterilised every couple of hours; swabs should be taken and samples grown on an agar plate. The types of bacterial colonies should be identified, particularly any pathogenic bacteria. If harmful bacteria are observed, areas should be sterilised again using stronger disinfectants and the procedure repeated until no pathogenic bacteria are detected.

Lesson 17: Maths skills: Size and number

Lesson overview

Learning objectives

- To make estimates of the results of simple calculations, without using a calculator.
- To use ratio and proportion to calibrate a microscope.
- To recognise and use numbers in decimal and standard form.

Learning outcomes

- Make estimates and explain why they may be important. [O1]
- Describe how to use ratio and proportion to calibrate a microscope. [O2]
- Convert numbers from decimal to standard form, and vice versa. [O3]

Skills development

- WS 4.2 Recognise the importance of scientific quantities and understand how they are determined.
- WS 4.4 Use prefixes and powers of 10 for orders of magnitude (e.g. tera, giga, mega, kilo, centi, milli, micro and nano).

Maths focus

Making estimates, ratio and proportion, standard and decimal form

Resources needed Worksheets 1.17.1 and 1.17.2

Key vocabulary cell size, estimate, order of magnitude, stage micrometer

Teaching and learning

Engage

- Introduce the learning outcomes for the lesson. Tell students they are going to make some estimates of sizes. Show students a metre rule and ask them to find as many objects in the classroom that are approximately 1 m long. Repeat the activity for 30 cm and 1 mm. Ask students to **identify why** making estimates is sometimes useful, before making accurate measurements. [O1]
- Ask students to work in pairs and **write a list** of things it is useful to estimate first, before accurately working out. [O1]
- Recap units of measurement and model how to convert millimetres to metres, and vice versa. Remind students of how to use standard form and ask them to **complete the task** in Worksheets 1.17.1 (low demand) and 1.17.2 (standard and high demand). They may need reminding of how to interconvert various units. [O3]

Challenge and develop

- Use the Student Book to show students microscope image of cells. Ask students to **identify why** it is important to know the size of a cell. **Explain how** to make an estimate of a cell's size using the example given in the Student Book. [O1]
- Ask students **to estimate** the size of the cells given in the worksheet. [O1]
- Explain how to use a stage micrometer to make accurate microscope measurements using the Student Book. Use the example given to show how the size of a cell may be accurately determined. Ask students to **answer the questions** in the worksheet. [O2, O3]

Explain

- Ask students to **write a help sheet** for another student to explain how to convert numbers into standard form, and how to make accurate measurements using a stage micrometer. [O2, O3]

Consolidate and apply

- Students could **answer** all the questions in the Student Book .They should complete the task in the worksheet. [O1–O3]

Extend

Ask students who are able to progress further to:

- **find out** how blood samples are analysed and **identify** any estimates that are given. They **could prepare** a talk for the class to **explain how** to determine the number of red blood cells in the blood, which measurements are made accurately and which might be estimated, and how this is done. [O1–O3]

Plenary suggestions

Ask students to review the skills from the learning outcomes using the 'traffic light' system.

Answers to Worksheets 1.17.1 and 1.17.2

1.

Worksheet 1.16.1 (low demand)

a) 1×10^1, 1×10^2, 1×10^3, 1×10^{-6}, 1×10^4, 1×10^{-3}

b) 10 000, 1/100 000, 100 000 000, 1/100

Worksheet 1.16.2 (standard and higher demand)

a) 3×10^3, 5.2×10^2, 7.3×10^{-3}, 4.2×10^1, 5.3×10^{-4}

b) 200, 700 000, 1/4000, 1/7 500 000

μm	mm	cm	m
1 000 000	1000	100	1
1	1×10^{-3}	1×10^{-4}	1×10^{-6}
1×10^4	1×10^1	1	1×10^{-2}
1×10^3	1	1×10^{-1}	1×10^{-3}

2.

a) 12 (cells across the field of view)

b) 500/12 = 41.7 μm

(High demand)

c) 120

d) 24

3.

a) 2.8 μm

b) 20

c) 56 μm

d) 40 μm

When and how to use these pages: Check your progress, Worked example and End of chapter test

Check your progress

Check your progress is a summary of what students should know and be able to do when they have completed the chapter. Check your progress is organised in three columns to show how ideas and skills progress in sophistication. Students aiming for top grades need to have mastered all the skills and ideas articulated in the final column (shaded pink in the Student book).

Check your progress can be used for individual or class revision using any combination of the suggestions below:

- Ask students to construct a mind map linking the points in Check your progress
- Work through Check your progress as a class and note the points that need further discussion
- Ask the students to tick the boxes on the Check your progress worksheet (Teacher Pack CD). Any points they have not been confident to tick they should revisit in the Student Book.
- Ask students to do further research on the different points listed in Check your progress
- Students work in pairs and ask each other what points they think they can do and why they think they can do those, and not others

Worked example

The worked example talks students through a series of exam-style questions. Sample student answers are provided, which are annotated to show how they could be improved.

- Give students the Worked example worksheet (Teacher Pack CD). The annotation boxes on this are blank. Ask students to discuss and write their own improvements before reviewing the annotated Worked example in the Student Book. This can be done as an individual, group or class activity.

End of chapter test

The End of chapter test gives students the opportunity to practice answering the different types of questions that they will encounter in their final exams. You can use the Marking grid provided in this Teacher Pack or on the CD Rom to analyse results. This shows the Assessment Objective for each question, so you can review trends and see individual student and class performance in answering questions for the different Assessment Objectives and to highlight areas for improvement.

- Questions could be used as a test once you have completed the chapter
- Questions could be worked through as part of a revision lesson
- Ask Students to mark each other's work and then talk through the mark scheme provided
- As a class, make a list of questions that most students did not get right. Work through these as a class.

Marking Grid for End of Chapter 1 Test

	Getting Started [Foundation Tier]					Going Further [Foundation and Higher Tiers]				More Challenging [Higher Tier]					Most demanding [Higher Tier]			Total marks	Percentage
Student Name	Q. 1 (A01) 2 marks	Q. 2 (A01) 2 marks	Q. 3 (A02) 1 mark	Q. 4 (A02) 1 mark	Q. 5 (A02, A03) 4 marks	Q. 6 (A01) 1 mark	Q. 7 (A01) 1 mark	Q. 8 (A02) 2 marks	Q. 9 (A02. A03) 6 marks	Q. 10 (A01) 1 mark	Q. 11 (A01) 1 mark	Q. 12 (A01) 2 marks	Q. 13 (A01) 2 marks	Q. 14 (A02) 4 marks	Q. 15 (A01) 2 marks	Q. 16 (A02) 4 marks	Q. 17 (A02) 4 marks		

Check your progress

You should be able to:

describe the functions of the sub-cellular structures found in eukaryotic cells →	understand the size and scale of cells and be able to use and convert units →	carry out order of magnitude calculations when comparing cell size; calculate with numbers in standard form
calculate magnification used by a light microscope using eyepiece and objective lens magnifications →	calculate the magnification of a light or electron micrograph →	explain limitations of light microscopy and advantages of electron microscopy
describe the structure of a prokaryotic cell →	describe the differences between eukaryotic and prokaryotic cells →	explain why scientists have now separated organisms into three domains using evidence from chemical analysis
recall that cells must divide for growth and replacement of cells →	describe how chromosomes double their DNA and are pulled to opposite ends of the cell, before the cytoplasm divides, during mitosis →	describe the events of the cell cycle and explain the synthesis of new sub-cellular components and DNA
recall that organism development is based on cell division and cell specialisation →	explain the importance of differentiation and explain how cells are specialised for their functions →	understand size and scale in the components of organ systems
recall where stem cells are found →	understand the potential of stem cell therapies →	evaluate scientific and ethical issues involved with stem cell therapies
recall that organisms can respire with oxygen (aerobic respiration) or without oxygen (anaerobic respiration) →	use word equations to describe the processes of aerobic and anaerobic respiration →	use symbol equations for aerobic and anaerobic respiration and be able to compare the two processes
describe equipment, materials and procedures required to work with microorganisms →	describe the process of binary fission →	be able to calculate numbers of microorganisms produced given the mean generation time

Worked example

Some students see a newspaper article on a European stem cell clinic. The clinic uses stem cell therapy to treat diabetes and other conditions.

1 **What is a stem cell?**

An unspecialised cell that can differentiate, or can be made to differentiate, into many different cell types.

The part about differentiation is important.

2 **The article contains some information on the clinic's treatment of diabetes.**

It includes data on the 55 patients the clinic has treated so far, which it claims is a success.

Type of diabetes	one month after treatment, number of patients who...		
	showed an improvement	showed no change	became worse
Type 1	8	13	2
Type 2	20	9	3
Total	28	22	5

a Calculate the overall percentage of patients who:

- showed an improvement 50.9%
- showed no change 40%
- became worse 9.1%

The calculations are correct.

The answers have not been recorded consistently, however. Think about significant figures. All the numbers in the table have a maximum of two figures, so you cannot have more in the answer.

b Does the data support the newspaper article's claims? Explain your answer.

My percentages show there is no difference between the overall percentage of patients who showed an improvement and those who did not.

The article is certainly untrue for Type 1 diabetes, where only 34.8% improved.

It might be true for Type 2, because 62.5% improved.

The answer draws correct conclusions from the table and uses data to support these.

With the small number of patients (23 for Type 1; 32 for Type 2), we cannot draw any firm conclusions from the small differences in numbers. Sample size is important in any scientific study.

The heading of the table says 'one month after treatment…'. This may be too early to draw any conclusions. Also, the patients may believe they feel better because they've had treatment. Importantly, we do not know how the 'improvement' was judged, another point to make in your answer.

You will notice that this question draws on your knowledge of scientific methods as well as stem cells.

You may also have to use relevant knowledge from another topic.

When you've learnt about diabetes, you will be aware that Type 2 diabetes can usually be controlled by diet and exercise, so the 'improvements' suggested by the data may be the result of this.

Photosynthesis: Introduction

When and how to use these pages

This chapter builds on ideas of photosynthesis from KS3 – that plants use energy from the Sun to form glucose, which is then used as an energy source in respiration. It links to lessons 1.3 Looking at cells in more detail and 1.11 Cells at work. There are also links to lesson 8.9 Adapting for survival in plants.

Overview of the chapter

In this chapter, students will learn about the process of photosynthesis, including the many uses of glucose in the plant. They will also test a leaf for the presence of starch and investigate how to change the rate of photosynthesis. Students will investigate how leaves are adapted for photosynthesis. They will then find out how the process of photosynthesis is linked to increasing food production. Students will also learn about the basic processes of diffusion and how this is related to transport in plants. Finally, transport of water and sugars around the plant are investigated.

This chapter offers a number of opportunities for the students to use mathematics to carry out surface area to volume ratio calculations, plan and carry out investigations into factors affecting photosynthesis and evaluate investigations on transpiration.

Obstacles to learning

Students may need extra guidance with the following terms and concepts:

- Students need to appreciate that the process of photosynthesis only produces sugars. All the other substances that a plant needs are made from this sugar and plants often need additional minerals to supply other elements.
- Many students, even at A level, think that plants respire at night and then switch to photosynthesis during the day. The continual nature of respiration needs to be emphasised.
- The topic of leaf structure is not one that is found particularly interesting by students. By moving fairly swiftly through the topic and introducing practical work, they should stay focused.
- Students have always thought of chlorophyll as being a single chemical substance. For higher tier they must appreciate that it is a combination of pigments.
- Many students think that plants just take up water for photosynthesis. It must be emphasised that this only makes up a small part of the water moving through a plant.
- Many students think that transpiration rate increases if light intensity or temperature increases because it is directly linked to photosynthesis rather than increased diffusion and evaporation.

Practicals in this chapter

In this chapter students will do the following practical work.

- Test a leaf for the presence of starch
- Investigate the movement of molecules by diffusion
- Investigate the distribution of stomata on a leaf
- Investigate root hairs and xylem
- Investigate rate of transpiration
- Required practical: Investigate the effect of light intensity on the rate of photosynthesis using an aquatic organism such as pond weed

	Lesson title	Overarching objectives
1	Explaining photosynthesis	Describe the process of photosynthesis.
2	Looking at photosynthesis	Explain how plants use the products of photosynthesis.
3	Investigating leaves	Explain how the structure of a leaf is adapted for photosynthesis.
4	Required practical	Investigate the effect of light intensity on the rate of photosynthesis using an aquatic organism such as pond weed.
5	Increasing photosynthesis	Identify factors that affect the rate of photosynthesis and explain the interaction of factors in limiting the rate of photosynthesis.
6	Increasing food production	Explain how to control factors linked to photosynthesis that can increase food production.
7	Diffusion in living systems	Explain the importance of diffusion in the movement of molecules around the plant.
8	Looking at stomata	Describe the process of transpiration and the relationship between stoma and transpiration.
9	Moving water	Describe how xylem and roots are adapted to take in water and move it around the plant.
10	Investigating transpiration	Explain how transpiration is affected by different factors.
11	Moving sugar	Explain how the structure of phloem is adapted to its function of moving sugars around the plant.
12	Maths skills: Surface area to volume ratio	To know how to calculate surface area to volume ratio and how to apply ideas about surface area and volume.

Lesson 1: Explaining photosynthesis

Lesson overview

AQA Specification reference

AQA 4.4.1.1

Learning objectives

- Identify the raw materials and products of photosynthesis.
- Describe photosynthesis by an equation.
- Explain gas exchange in leaves.

Learning outcomes

- State that carbon dioxide and water are the raw materials used in photosynthesis and that glucose and oxygen are the products. [O1]
- State the word and symbol equations for photosynthesis. [O2]
- Interpret photosynthesis investigations to explain the direction of gas exchange in leaves. [O3]

Skills development

- WS 1.2 Use a variety of models such as representational, spatial, descriptive, computational and mathematical to solve problems, make predictions and to develop scientific explanations and understanding of familiar and unfamiliar facts.
- WS 2.2 Plan experiments or devise procedures to make observations, produce or characterise a substance, test hypotheses, check data or explore phenomena.
- WS 2.3 Apply a knowledge of a range of techniques, instruments, apparatus and materials to select those appropriate to the experiment.
- WS 2.4 Carry out experiments appropriately having due regard for the correct manipulation of apparatus, the accuracy of measurements, and health and safety considerations.

Resources needed Images of unusual plants. These can be found by searching 'amazing plants'. Sort cards for photosynthesis equations. Worksheet 2.1; Practical sheet 2.1; Technician's notes 2.1

Digital resources

Key vocabulary chlorophyll, carbohydrate

Teaching and learning

Engage

- Show students images of some very different plants that they may not have seen before. These could include *Lithops*, corpse flower, Venus flytrap or pitcher plant.
- Ask students to **discuss** why all the plants seen need carbon dioxide. [O1]
- Ask students to **predict** what would happen if all the plants in the world died. Students should provide reasons for their prediction, which can then lead into the next part of the lesson. [O1, O2]

Challenge and develop

- Provide the students with the words and symbols that make up the equation for photosynthesis. This can be in the form of sort cards or written on the whiteboard. Students then **rearrange** them to produce the correct word equation (standard demand) or symbol equation (high demand). [O2]
- Show students a variegated leaf. Explain that the white part of the leaf does not have any chlorophyll so cannot photosynthesise. Ask the students to **predict** if starch would be found in the white part of the leaf. [O1]
- Standard and high demand students should **plan** how they could test their prediction using the method in the Student Book. They could use Worksheet 2.1 investigation plan to help them plan their investigation. [O1]

- Low demand students should be provided with Practical sheet 2.1; this gives the method for them.
- Students then **test** the variegated leaves for starch. [O1]

Explain

- Standard and high demand students should **complete** the questions on interpreting results on Worksheet 2.1. [O1, O2]
- Low demand students should **complete** the questions on Practical sheet 2.1.

Consolidate and apply

- Ask students to look at the gas exchange investigation on spread 2.1 of the Student Book. In groups ask students to **interpret** the results of the investigation. Then as individuals they **complete** the questions. [O3]

Extend

Ask students able to progress further to:

- (high demand) **predict** what would happen if a snail had been placed in tube 2 with the leaf and **explain** their prediction. (The indicator is likely to stay orange or turn yellow depending on how fast the plant uses carbon dioxide released by the snail.) [O3]

Plenary suggestions

Ask students to **suggest** reasons why our understanding of the reactions of photosynthesis has changed over time.

Answers to Worksheet 2.1

Investigation plan

1. Students should predict that there will be no starch present as there are no chloroplasts.
2. This should include the following: beakers, boiling tube, kettle or heating apparatus, ethanol, white tile, iodine solution, leaves, tongs, forceps, eye protection.
3. Ideas may include:
   - Ethanol is flammable, so do not use near a Bunsen flame.
   - Iodine solution stains, so do not get onto hands.
   - Using harmful chemicals, so wear eye protection.
   - Boiling tube may get hot, so use tongs to remove from hot water.
4. Methods may vary but the basic idea is the one seen in the Student Book. For example:
   - Put the leaf in a beaker of hot water and leave for 1 minute.
   - Take out the leaf, put it into a boiling tube and cover it with ethanol.
   - Put the tube into a beaker of hot water.
   - Pour the ethanol into a waste beaker and dip the leaf back into the hot water to remove the ethanol.
   - Spread the leaf out on the white tile.
   - Add about five drops of iodine to the leaf and leave for 2 minutes.

Interpreting results

1. This will depend on their prediction and results but the iodine solution on the white area should remain orange
2. Photosynthesis is when the plant uses light energy to make carbohydrates. Starch will be found in the cells that photosynthesise. The cells in the white area have no chloroplasts. The chlorophyll in the chloroplasts is needed to absorb the light. No starch is made as there is no photosynthesis taking place inside these cells.

Answers to Practical sheet 2.1

1. green
2. chlorophyll
3. Photosynthesis only takes place in the green part where chlorophyll is present.

Lesson 2: Looking at photosynthesis

Lesson overview

AQA Specification reference

AQA 4.2.3.2; AQA 4.4.1.1

Learning objectives

- Explain the importance of photosynthesis.
- Explain how plants use the glucose they produce.

Learning outcomes

- Understand that photosynthesis is an endothermic reaction that converts light energy to chemical energy. [O1]
- Explain that glucose can be converted to other substances in plants. [O2]
- Describe the functions of fats, oils, cellulose, starch and proteins in the plant. [O3]

Skills development

- WS 1.1 Understand how scientific methods and theories develop over time.
- WS 2.4 Carry out experiments appropriately having due regard for the correct manipulation of apparatus, the accuracy of measurements, and health and safety considerations.
- WS 2.6 Make and record observations and measurements using a range of apparatus and methods.

Maths focus

- 3a Understand and use the symbols: $=, <, <<, >>, >, \propto, \sim$.
- 3d Solve simple algebraic equations.
- 4a Translate information between graphical and numeric form.

Resources needed Time-lapse photography of growing plants (search Internet for 'plant growth time lapse'). Sort cards for photosynthesis equations. Image of a woodland in autumn. Worksheet 2.2; Technician's notes 2.2

Digital resources

Key vocabulary endothermic, veins

Teaching and learning

Engage

- Show students a time-lapse video of a plant growing.
- Ask students to **discuss** what the plant needs for growth. [O1]
- Ask students to **predict** what would happen if the plant could not get any light. [O1]

  Students should provide reasons for their prediction, which can then lead into the next part of the lesson.

Challenge and develop

- Show students a leaf that has been partially covered in black card and another from a plant that has had no carbon dioxide. Ask the students to **predict** if starch would be found in any parts of either leaf. [O1]
- Ask students to **test** the leaves for starch. They should use the method from the previous lesson. Each group could be given a leaf that has been given a different condition (partially covered by black card / left without carbon dioxide / left in full light with carbon dioxide). They should not be told the conditions the leaves were left in. [O1]

Explain

- Ask each group to inform the rest of the class which of the conditions the leaves were left in and to **explain** how they know this. [O1]

- Discuss the idea of light energy being needed and that the energy is taken in and converted to chemical energy. Ask them to **decide** if that makes the process an endothermic or exothermic reaction. [O1]
- Ask students to **complete** 'Testing a leaf for starch' on Worksheet 2.2. [O1]
- Ask students to **suggest** what happens to the glucose produced in photosynthesis other than being stored as starch. [O2]
- Discuss the different uses of glucose in the plant with the students. Explain to high demand and standard demand students why the relative solubility of glucose and starch requires glucose to be converted to starch for storage. [O2, O3]

Consolidate and apply

- Low demand: List the following words on the whiteboard: cellulose, oxygen, starch, protein. Read out a use for one of these substances and ask the students to choose the substance that matches the use. [O2]
- Standard demand: Students should complete the table 'Using glucose' on Worksheet 2.2. [O3]
- High demand: List the following words on the whiteboard: glucose, nitrates, amino acids, proteins. Ask the students to write a paragraph that links all the words together and **explains** how plants make proteins (detail of chemical reactions and molecular structure is not required at this level). [O2, O3]

Extend

Ask students able to progress further to:

- (high demand) **write** a summary that **explains** why deforestation may be linked to global warming. [O1]

Plenary suggestions

Explain to students that to maximise growth a gardener may add carbon dioxide to their greenhouse. Ask them to suggest why. Then ask them to explain why lighting a greenhouse with green light would not be much use.

Answers to Worksheet 2.2

1 Testing a leaf for starch

1.

	Observations		
Questions	left in full light with carbon dioxide	partially covered by black card	left without carbon dioxide
What colour did the leaf turn when iodine was added?	blue-black	area covered is orange; uncovered is blue-black	orange
Did any part of the leaf contain starch?	all of it	uncovered areas	no

2. Needs light and carbon dioxide

2 Using glucose

1.

Substances	How it is made	Uses
starch	by building glucose into large starch molecules	storage of energy
proteins	by combining sugars with nitrate ions and other minerals to make amino acids for protein synthesis	growth repair
lipids	by turning sugar molecules into fats and oils for storage	energy store
cellulose	by building sugars into large cellulose molecules	make cell walls

2. Respiration

Lesson 3: Investigating leaves

Lesson overview

AQA Specification reference

AQA 4.2.3.1

Learning objectives

- Identify the internal structures of a leaf.
- Explain how the structure of a leaf is adapted for photosynthesis.
- Recall that chlorophyll pigments in chloroplasts absorb light energy for photosynthesis.

Learning outcomes

- Name and locate the parts of a leaf. [O1]
- State that chlorophyll pigments in chloroplasts absorb light energy for photosynthesis. [O2]
- Explain how the cellular structure of a leaf is adapted for efficient photosynthesis. [O3]

Skills development

- WS 2.6 Make and record observations and measurements using a range of apparatus and methods.
- WS 4.1 Use scientific vocabulary, terminology and definitions.

Resources needed Videos showing leaf structure using an electron microscope. Worksheet 2.3.1 (low demand); Worksheet 2.3.2 (standard and high demand); Technician's notes 2.3

Digital resources

Key vocabulary epidermal tissues, palisade mesophyll, spongy mesophyll, vascular bundle

Teaching and learning

Engage

- Show the students images or videos showing leaf structure using an electron microscope (search videos for 'electron microscope images leaf').
- Ask students to **discuss** what they are looking at and why it is linked to photosynthesis.

Challenge and develop

- Provide the students with prepared microscope slides of a transverse section of a leaf. [O1]
- Students **observe** the slides using a light microscope. [O1]
- Ask students to **draw** a plan of what they see using the low power objective on the microscope. They should not draw individual cells but just identify the different areas of cells. [O1]
- Students then use the Student Book to **label** the different areas of cells in their plan. [O1]

Explain

- **Discuss** the different types of cells and their functions with the students.
- (Standard and high demand) Students **complete** the questions on Worksheet 2.3.2. [O1, O2, O3]
- Low demand students should be given Worksheet 2.3.1. This has the same questions as Worksheet 2.3.2 as well as a 'hints' section included (on second page), which they can use to help them **complete** Question 2 (by matching the part of the plant to the correct adaptation).

Consolidate and apply

- Inform students that on a tree the leaves growing in the shade are larger and thinner, and can also be a darker green. Ask them to **suggest** why this would be the case. They should be able to conclude that there will be more surface area and chloroplasts to absorb more light. [O2, O3] Extend

Ask students able to progress further to:

- (High demand) Use the structure of the chloroplast to **explain** why starch is only present in the green areas of variegated leaves. [O3]

Plenary suggestions

Show students an image of the leaf with no labels. Write a part of the leaf on the board and invite students up to label the part on the diagram. Repeat this with other parts.

Answers to Worksheet 2.3.2

Leaf structure

1. Clockwise from top right: upper epidermis, palisade cell, spongy mesophyll cell, lower epidermis, guard cell, vascular bundle
2. Upper epidermis: thin and transparent to allow light to pass through to the mesophyll layer

   Palisade cells: regular-shaped cells, arranged end on near upper surface; contain many chloroplasts, mostly at the top of the cells.They can absorb the maximum amount of light possible

   Spongy cells: irregular-shaped cells with many air spaces to allow gases to diffuse through

   Lower epidermis: single layer with stomata to allow gases to diffuse through

   Vasuclar bundle: contains xylem and phloem for transport

   Guard cells: open and close stomata

Lesson 4: Required practical: Investigate the effect of light intensity on the rate of photosynthesis using an aquatic organism such as pondweed

Lesson overview

AQA Specification reference

AQA 4.4.1.2

Learning objectives

- Identify and manage variables.
- Process data and identify outliers.
- Evaluate an experimental process.

Learning outcomes

- Identify the variables when investigating the effect of light on photosynthesis. [O1]
- Construct a graph to show the effect of light on photosynthesis. [O2]
- Evaluate the investigation and suggest improvements. [O3]

Skills development

- WS 1.5 Evaluate risks both in practical science and the wider societal context, including perception of risk in relation to data and consequences.
- WS 2.4 Carry out experiments appropriately having due regard for the correct manipulation of apparatus, the accuracy of measurements, and health and safety considerations.
- WS 2.6 Make and record observations and measurements using a range of apparatus and methods.
- WS 2.7 Evaluate methods and suggest possible improvements and further investigations.
- WS 3.1 Presenting observations and other data using appropriate methods.
- WS 3.2 Translating data from one form to another.
- WS 3.5 Interpreting observations and other data (presented in verbal, diagrammatic, graphical, symbolic or numerical form), including identifying patterns and trends, making inferences and drawing conclusions.
- WS 3.7 Being objective, evaluating data in terms of accuracy, precision, repeatability and reproducibility, and identifying potential sources of random and systematic error.
- WS 3.8 Communicating the scientific rationale for investigations, methods used, findings and reasoned conclusions through paper-based and electronic reports and presentations using verbal, diagrammatic, graphical, numerical and symbolic forms.
- WS 4.3 Use SI units (e.g. kg, g, mg; km, m, mm; kJ, J) and IUPAC chemical nomenclature unless inappropriate.

Maths focus

- 1a Recognise and use expressions in decimal form.
- 2b Find arithmetic means.
- 4a Translate information between graphical and numeric form.
- 4c Plot two variables from experimental or other data.

Resources needed Worksheet 2.4; Practical sheet 2.4.1 (low demand); Practical sheet 2.4.2 (standard demand); Practical sheet 2.4.3 (high demand); Technician's notes 2.4

Digital resources

Key vocabulary light intensity, range, value, variable

Teaching and learning

Engage

- Show students a video of pond weed producing bubbles of oxygen (search videos for 'pond weed oxygen production'). The video should not show how to investigate the rate of photosynthesis just oxygen being produced. Alternatively show them the actual apparatus they will be using in their investigation.

- Explain to students that they can use the apparatus to measure the rate of photosynthesis. [O1]
- Ask students in groups to list on mini-whiteboards the variables that might affect the rate of photosynthesis. As a prompt tell them to think about the equation for photosynthesis. [O1]
- Groups report back on their lists to compile a complete list on the main whiteboard.

Challenge and develop

- Explain to students that they are going to use the apparatus to measure the rate of photosynthesis and how it may be affected by light intensity. [O1]
- Ask students to **list** on mini-whiteboards the variables that they would change, measure and control. [O1]
- Ask students to **state** a question they could test about light intensity and photosynthesis. This could be: 'Will increasing the light intensity increase the rate of photosynthesis?' [O1, O2]
- Provide students with Practical sheet 2.4.1 (low demand), Practical sheet 2.4.2 (standard demand) or Practical sheet 2.4.3 (high demand). Ask them in groups to measure the rate of photosynthesis. As this can take a long time, it is suggested that each group starts with a different distance. [O1]

Explain

- Low demand students should complete questions on Practical sheet 2.4.1.
- Standard demand students should complete questions on Practical sheet 2.4.2.
- High demand students should complete the questions on Practical sheet 2.4.3. [O1, O2, O3]

Consolidate and apply

- Ask students to **sugges**t how they could change their investigation to find out about the effect of temperature or carbon dioxide levels on the rate of photosynthesis. [O3]

Extend

Ask students able to progress further to:

- (high demand) **complete** the questions on Worksheet 2.4. [O2, O3]

Plenary suggestions

Collate the class results. Ask students to decide if the class results were reproducible. Ask them to explain the difference between reproducible and repeatable.

Answers to Worksheet 2.4

1. 0.2 = 25, 0.4 = 6.25, 0.6 = 2.78, 0.8 = 1.56
2. The idea that there are other factors involved in photosynthesis so it does not follow a simple rule; e.g. when get close to the lamp it is already at a maximum and cannot increase anymore

Answers to Practical sheet 2.4.3

NB. Low demand: questions 1–4 and 6; standard demand: questions 1–4, 6 and 7; high demand: all questions

1. Number of bubbles produced
2. Distance or lamp / light intensity
3. Size of pond weed, lamp, amount of carbon dioxide, temperature (to some extent if large beaker is used)
4. Water spillage could result in somebody slipping so clear up any spillages; lamp could get hot, let it cool down before moving it
5. The x-axis should be 'distance of light' and the y-axis 'average number of bubbles per minute'; line of best fit should be a curve through most points – the shape should show more bubbles when the lamp is closer.
6. For example, the closer to the pond weed you place the lamp, the more bubbles you get.
7. More light absorbed so more energy for photosynthesis and more oxygen released; levels off as you get closer as other factors have an effect
8. Not very accurate as bubbles will be different sizes so they may suggest measuring the volume of gas produced
9. How repeatable the results were will depend on their results; it is unlikly that the repeated results are close together for each distance.

Lesson 5: Increasing photosynthesis

Lesson overview

AQA Specification reference

AQA 4.4.1.2

Learning objectives

- Identify factors that affect the rate of photosynthesis.
- Interpret data about the rate of photosynthesis.
- Explain the interaction of factors in limiting the rate of photosynthesis.

Learning outcomes

- State that the rate of photosynthesis is affected by temperature, light intensity, carbon dioxide levels and the amount of chlorophyll. [O1]
- Interpret graphs of photosynthesis rate involving one limiting factor. [O2]
- Explain graphs of photosynthesis rate involving two or three factors and decide which one is the limiting factor. [O3]

Skills development

- WS 1.2 Use a variety of models such as representational, spatial, descriptive, computational and mathematical to solve problems, make predictions and to develop scientific explanations and understanding of familiar and unfamiliar facts.
- WS 3.2 Translating data from one form to another.
- WS 4.1 Use scientific vocabulary, terminology and definitions.

Maths focus

- 4a Translate information between graphical and numeric form

Resources needed Worksheet 2.5; Technician's notes 2.5

Digital resources

Key vocabulary limiting factor, tundra

Teaching and learning

Engage

- Show students the images on spread 2.5 of the Student Book – Plants in different habitats.
- Ask students to **discuss** Questions 1 and 2 in pairs. [O1]
- Then, as a class, **produce** a combined list of factors affecting photosynthesis in each habitat.

Challenge and develop

- (Low demand) **Discuss** the idea that growth of a plant will depend on the rate of photosynthesis.
- (Low demand) Ask students if they have noticed when grass grows the most – summer or winter? Students then **complete** 'Photosynthesis and growth' on Worksheet 2.5. [O1, O2]
- (Standard and high demand) Show students a graph of light intensity against rate of photosynthesis. There is one on spread 2.5 of the Student Book. Ask the students to **compare** the graph with their results from the previous lesson. Why is it a different shape? (They plotted number of bubbles not rate so the curve was the other way round.) [O1, O2]
- (Standard and high demand) Ask students to **suggest** why the graph levels off. Introduce the term 'limiting factor' and explain that at the start light is the limiting factor; at the end it is something else, such as carbon dioxide levels. [O1, O2]

Explain

- Ask students to **complete** the 'Limiting factors 1' questions on Worksheet 2.5. [O1, O2]

Consolidate and apply

- Ask students to **produce** a summary of what they know about photosynthesis. They should include ideas from all five lessons. The summary can be in the form of bullet points or a mind map. (High demand candidates should miss out this activity and complete the Extend instead.)

Extend

Ask students able to progress further to:

- (high demand) **complete** 'Limiting factors 2' and 'Limiting factors 3' on Worksheet 2.5. [O3]

Plenary suggestions

Try the 'true or false' activity given in the Technician's notes.

Answers to Worksheet 2.5

1 Photosynthesis and growing

1. 65
2. June. Temperatures are getting warmer, there is maximum light and also water is available so the grass photosynthesises more and grows more.
3. Temperatures are too low so the grass does not grow.

2 Limiting factors 1

1. carbon dioxide concentration
2. Carbon dioxide is no longer the limiting factor; it might be light or temperature or the amount of chlorophyll.

3 Limiting factors 2

1. light intensity
2. For line C there is a higher temperature so rate can increase faster than for D as light intensity increases; line C levels off at a higher rate as D will be limited by the lower temperature.
3. For line A there is a higher level of carbon dioxide so rate can increase faster than for C as light intensity increases; line A levels off at a higher rate as C will be limited by the lower levels of carbon dioxide.

4 Limiting factors 3

1. 130 units into the plant
2. The light intensity is not the limiting factor; some other factor is limiting the reaction.
3. about 200 arbitrary units
4. They are equal.

Lesson 6: Increasing food production

Lesson overview

AQA Specification reference

AQA 4.4.1.2

Learning objectives

- Identify the factors that increase food production.
- Explain how these factors can be controlled.
- Evaluate the benefits of manipulating the environment to increase food production.

Learning outcomes

- State that heating and lighting a greenhouse can increase food production. [O1]
- Describe ways to increase food production by controlling limiting factors. [O2]
- Use data to relate limiting factors to the cost effectiveness of adding heat, light or carbon dioxide to greenhouses. [O3]

Skills development

- WS 1.4 Explain everyday and technological applications of science. Evaluate associated personal, social, economic and environmental implications, and make decisions based on the evaluation of evidence and arguments.
- WS 3.8 Communicating the scientific rationale for investigations, methods used, findings and reasoned conclusions through paper-based and electronic reports and presentations using verbal, diagrammatic, graphical, numerical and symbolic forms.
- WS 4.1 Use scientific vocabulary, terminology and definitions.

Resources needed Images showing how cold it is in the area by having snow on the ground and if possible evidence of geothermal activity (search images for 'Icelandic geothermal greenhouses'). Worksheet 2.6; Technician's notes 2.6

Digital resources

Key vocabulary hydroponics, yield

Teaching and learning

Engage

- Show students the images of greenhouses in Iceland.
- Ask students to discuss why the plants have to be grown inside and why the geothermal energy is important. [O1]
- Ask students to **suggest** other factors that could help the plants in the greenhouse grow. [O1]

  Students should provide reasons for their suggestions, which can then lead into the next part of the lesson.

Challenge and develop

- Continue the discussion on factors that can increase growth. **Ask** students which suggestions that they made are directly linked to the rate of photosynthesis. They may have suggested water, nutrients or fertilisers and pesticides, which are not limiting factors of photosynthesis. [O1]
- Once the class has a list that includes light, carbon dioxide, temperature and chlorophyll levels, ask them to **suggest** which ones can be controlled by the farmer and how. [O1, O2]

Explain

- In groups students **produce** a poster about growing crops in greenhouses. Use Worksheet 2.6 to help them. [O1, O2]

Consolidate and apply

- Ask standard and high demand students to swap their posters with another group and **evaluate** them using the questions on Worksheet 2.6. [O1, O2]

Extend

Ask students able to progress further to:

- (high demand) **complete** the 'Greenhouse costs' section on Worksheet 2.6. [O3]

Plenary suggestions

Ask students to comment on the evaluation of their poster. Do they think the comments from the evaluating group were fair? How could they improve the poster?

Answers to Worksheet 2.6

1 Greenhouses

2. Students would be expected to add lighting, heating and a source of carbon dioxide to their greenhouse.
   Lighting: lamps could be used to provide light during the night so plants could photosynthesise 24 hours a day.
   Heating: electrical or paraffin heaters to maintain optimum temperature for photosynthesis.
   Carbon dioxide: a paraffin heater will release carbon dioxide; as long as the greenhouse is ventilated this will increase the rate of photosynthesis.
   Water: may be mentioned as without it plants would die; however, it is not classed as a limiting factor.

2 Evaluation

Answers will depend on students' work.

3 Greenhouse costs

1. Solar energy = £1350 profit; propane = loss of £100 so solar energy would be better even though yield is not as high.
2. Propane heater may increase the concentration of carbon dioxide, which is limiting the rate of photosynthesis in the greenhouse, so by increasing the concentration of carbon dioxide you can increase the rate of photosynthesis even if temperatures and light intensities are the same.
3. There will be no cost for the construction of the growing beds so the profit could increase to £2850.

Lesson 7: Key concept: Diffusion in living systems

Lesson overview

AQA Specification reference

AQA 4.1.3.1

Learning objectives

- Use concentration gradients to explain the direction of diffusion.
- Apply the principles of diffusion to movement of different substances in plants.

Learning outcomes

- Describe and explain how molecules move by diffusion. [O1]
- Explain how the rate of diffusion through cells is increased by various factors. [O2]
- Explain how leaves are adapted to increase the rate of diffusion of gases. [O3]

Skills development

- WS 1.2 Use a variety of models such as representational, spatial, descriptive, computational and mathematical to solve problems, make predictions and to develop scientific explanations and understanding of familiar and unfamiliar facts.
- WS 2.6 Recognise when to apply a knowledge of sampling techniques to ensure any samples collected are representative.
- WS 3.5 Interpreting observations and other data (presented in verbal, diagrammatic, graphical, symbolic or numerical form), including identifying patterns and trends, making inferences and drawing conclusions.
- WS 4.1 Use scientific vocabulary, terminology and definitions.

Resources needed Worksheet 2.7.1 (low demand); Worksheet 2.7.2 (standard demand); Worksheet 2.7.3 (high demand); Practical sheet 2.7; Technician's notes 2.7

Digital resources

Key vocabulary concentration gradient, diffusion

Teaching and learning

Engage

- Using forceps, drop a crystal of potassium permanganate into a beaker of cold water.
- Ask students to **describe** what they see. [O1]
- Ask students to **predict** how the result would be different if hot water was used. [O2]
- Test their prediction by dropping a crystal of potassium permanganate into a beaker of hot water.

Challenge and develop

- Low demand: Develop the discussion about the potassium permanganate by asking students to **explain** why the purple colour spreads throughout the water. [O1]
- Introduce the idea of a concentration gradient. Provide the students with a definition of diffusion and explain that it can be responsible for the movement of substances across the cell membrane if the molecules are small enough. [O1]
- Class demonstration – moving molecules. Set up the demonstration shown on Practical sheet 2.7.

  Discuss with the class the relative sizes of the iodine and starch molecules and ask them to **predict** the direction of movement linked to diffusion and concentration of molecules. [O1]

  Ask the class to **answer** the 'Prediction' questions on Practical sheet 2.7. Leave the apparatus for 10 minutes or use apparatus set up before the lesson. [O1]

Explain

- Discuss the results with the class. Link the results to the movement of molecules though membranes of cells.
- Students then **complete** the 'Conclusion' questions on Practical sheet 2.7. [O1, O2]

Consolidate and apply

- Ask students to **suggest** which gases diffuse in and out of a leaf. Then discuss with them the direction of movement of these gases and how the leaf is adapted to increase the rate of diffusion. [O3]
- Ask low demand students to complete the questions on Worksheet 2.7.1. [O3]
- Ask standard demand students to complete the questions on Worksheet 2.7.2. [O3]
- Ask high demand students to complete the questions on Worksheet 2.7.3. [O3]

Extend

Ask students able to progress further to:

- (high demand) **explain** why molecules still move across membranes even when the concentration on both sides is equal. [O3]

Plenary suggestions

Ask students to use ideas about diffusion to explain why plants store carbohydrates as starch but transport them as sugars.

Answers to Worksheet 2.7.3

NB. Low demand: questions 1–2; standard demand: questions 1–3; high demand: questions 1–4

1. Arrow should show water moving from cells out through stomata
2.
   a. Arrow (red) should show oxygen moving through the stomata from the outside then into the cells
   b. Arrow (green) should show carbon dioxide moving from cells out through stomata
3. Thin membranes, large surface area and air spaces between allow the gasses to easily diffuse and reach all the cells
4. Higher concentration of water molecules in the air spaces compared to the air, so water molecules diffuse out of the leaf through the stomata

Answers to Practical sheet 2.7

Prediction

1. in beaker
2. diffusion
3. inside Visking tubing

Results

Inside Visking tubing will be blue-black; no colour change in beaker.

Conclusion

1. into the Visking tubing
2. higher concentration outside than in, so they diffused into the Visking tubing where the concentration is lower; the movement is down the concentration gradient
3. start: in beaker; end: concentration will be even both in and out of the Visking tubing
4. The Visking tubing represents the membrane of a cell; molecules will diffuse through membranes down concentration gradients in the same way as the idodine molecules.

Lesson 8: Looking at stomata

Lesson overview

AQA Specification reference

AQA 4.2.3.1; AQA 4.2.3.2

Learning objectives

- Describe transpiration in plants.
- Explain the structure and function of stomata.
- Explain the relationship between transpiration and leaf structure.

Learning outcomes

- Describe transpiration as the movement of water through a plant. [O1]
- Explain that transpiration involves evaporation and diffusion. [O2]
- Explain how stomata can be opened and closed to control the rate of transpiration. [O3]

Skills development

- WS 2.4 Carry out experiments appropriately having due regard for the correct manipulation of apparatus, the accuracy of measurements, and health and safety considerations.
- WS 2.6 Make and record observations and measurements using a range of apparatus and methods.
- WS 3.5 Interpreting observations and other data (presented in verbal, diagrammatic, graphical, symbolic or numerical form), including identifying patterns and trends, making inferences and drawing conclusions.
- WS 4.1 Use scientific vocabulary, terminology and definitions.

Resources needed Celery that has been sitting in a beaker of red dye for 2 days; image of a water lily; Worksheet 2.8; Practical sheet 2.8; Technician's notes 2.8

Digital resources

Key vocabulary flaccid, guard cell, transpiration, turgid, xylem

Teaching and learning

Engage

- Show students a photograph of a giant redwood from California. Tell them that they are the tallest trees in the world. The tallest is about 90 m tall.
- Ask students to **write down** (perhaps using mini-whiteboards) any problems that being so tall might cause for the plant.

Challenge and develop

- Use the Engage activity to lead into a discussion about how the plant gets the water it needs for photosynthesis.
- Show the students a stick of celery that has been left in a beaker of red dye for 2 days.
- Ask the students to **suggest** how the dye has managed to colour the leaves. [O1]
- Use the demonstration to explain to students that most of the water that reaches the leaves is evaporated from the leaves, this draws the water (and the dye) up the stem. This movement of water through a plant is called transpiration. [O1, O2]
- High demand: Explain transpiration in terms of evaporation from the cells into the air spaces and diffusion out of the stomata. [O1, O2]
- Low demand: Students **complete** 'Movement of water' on Worksheet 2.8. [O1]

- Ask all students to **think** back to the lesson on leaf structure and **suggest** how the leaves could control the amount of water that evaporates from their surface. [O2, O3]
- Students then **investigate** stomata using Practical sheet 2.8

Explain

- Discuss the observations from the practical.
- Students then **complete** the questions on Practical sheet 2.8. [O1, O2, O3]

Consolidate and apply

- Show the students an image of a water lily.
- Ask students to **suggest** why water lilies have their stomata on the upper surface of the leaf.

Extend

Ask students able to progress further to:

- (high demand) **complete** 'Stomatal distribution' on Worksheet 2.8. [O2, O3]

Plenary suggestions

Ask students to suggest why pine trees only have small narrow leaves and plants in rainforests have very large leaves. They should include ideas about transpiration and photosynthesis in their answer.

Answers to Worksheet 2.8

1 Movement of water

1. pathway: soil to roots to stem (inside xylem) to leaves then out of the stomata
2. The upper surface has a waxy cuticle to prevent water loss; no stomata on upper surface, plant loses more water from under surface.

2 Stomatal distribution

A – normal leaf as stomata on lower surface, no other adaptation

B – grass leaves because they tend to grow vertically so there are even numbers of stomata on each surface

C –water lily leaf because it has stomata on the upper surface

D – cactus leaf because it is shaped as a spine and has few stomata to reduce water loss

E – heather leaf because it has sunken stomata to reduce water loss

F – marram grass leaf because it has special adaptations to reduce water loss

Answers to Practical sheet 2.8

1. There should be more stomata on the under surface of the leaf.
2. This will depend on the leaf.
3. Guard cells take in water and become **turgid** (swollen). The inner walls are thickened and rigid so they pull apart and the pore opens. To close the stomata the guard cells lose water and become flaccid – the rigid walls move together.
4. to prevent water loss when they are not photosynthesising
5. when transpiration / evaporation is high, such as on very hot, windy days

Lesson 9: Moving water

Lesson overview

AQA Specification reference

AQA 4.2.3.1; AQA 4.2.3.2

Learning objectives

- Describe the structure and function of xylem and roots.
- Describe how xylem and roots are adapted to absorb water.
- Explain why plants in flooded or waterlogged soil die.
- Explain how wilting occurs.

Learning outcomes

- Recall the structure and function of xylem and roots. [O1]
- Describe how root hair cells and xylem tissue are adapted for the efficient uptake and transport of water. [O2]
- Explain that wilting can protect a plant from further water loss. [O3]

Skills development

- WS 2.4 Carry out experiments appropriately having due regard for the correct manipulation of apparatus, the accuracy of measurements, and health and safety considerations.
- WS 2.6 Make and record observations and measurements using a range of apparatus and methods.
- WS 3.2 Translating data from one form to another.
- WS 3.5 Interpreting observations and other data (presented in verbal, diagrammatic, graphical, symbolic or numerical form), including identifying patterns and trends, making inferences and drawing conclusions.
- WS 4.3 Use SI units (e.g. kg, g, mg; km, m, mm; kJ, J) and IUPAC chemical nomenclature unless inappropriate.
- WS 4.5 Interconvert units.

Resources needed Worksheet 2.9.1 (low demand); Worksheet 2.9.2 (standard and high demand); Practical sheet 2.9; Technician's notes 2.9

Digital resources

Key vocabulary lignin, lumen, phloem, translocation, xylem

Teaching and learning

Engage

- Show students a photograph of a plant that has wilted. Or show them an actual wilted plant – basil plants from the local supermarket would work.
- Ask students them to **write down** (perhaps using mini-whiteboards) what has happened to the plant and why.
- Discuss with the students what they have written down and ask them to **suggest** how the plant can be saved.
- If the students have been shown a live plant, water the plant and leave it. Basil left to wilt can recover within the lesson if given water.

Challenge and develop

- Ask students to **observe** root hair cells using Practical sheet 2.9. [O1]
- Students **complete** the questions on Practical sheet 2.9. [O1, O2]
- Ask students to **recall** from the previous lesson how the water moves up the stem of the plant. Students then **investigate** xylem using Practical sheet 2.9. [O1, O2]

Explain

- Students then **complete** the questions on Worksheets 2.9.1 (low demand) and 2.9.2 (standard and high demand). [O1, O2]

Consolidate and apply

- Remind students of the plant from the beginning of the lesson.
- Discuss with the students why plants wilt and why overwatering can kill plants. [O3]
- Ask students to write a summary explaining why it is important to get watering of house plants correct. [O3]

Extend

Ask students able to progress further to:

- (high demand) **compare** and **contrast** the distribution of xylem and phloem in roots and stems. [O1, O2]

Plenary suggestions

Key words bingo. Put a list of about 12 key words from Chapter 2 on the whiteboard.

Ask students to draw a 2 × 3 grid on paper. They **choose** six key words from the list and place one in each space of the grid.

Read out a definition of one the key words. If a student thinks the word is in their grid they **cross** it out.

Continue this until a student has won by correctly crossing out all the words in their grid.

Answers to Worksheet 2.9.2

NB. Low demand: up to Question 2 of 'Moving water'; standard and high demand: all questions

1 Xylem structure

1. Xylem tubes are hollow tubes with thick, reinforced walls containing lignin. The lignin helps make the walls strong. The cell walls are also waterproof so they can hold the water inside the lumen.

2 Moving water

1. 11 leaves = 6.7; 7 leaves = 3.5; 5 leaves = 2.1
2. The more leaves, the more mass is lost; this is because there is a larger surface area for evaporation of water so the plant loses more water.
3.
   a. Draws water up from the soil to where it is needed in the leaves for photosynthesis, evaporation of the water cools the plant down in hot conditions.
   b. If too much water is lost, the plant can wilt, which could result in death of the plant.

Answers to Practical sheet 2.9

Investigating root hairs (Conclusion)

2. Root hairs have a large surface area to absorb more water.

Investigating xylem (Conclusion)

1. Water only flows though part of the stems and root. In the root it should be the centre and in the stems it should be near the outside. The areas of dye should match the positions of the xylem.
2. This will depend on how far the dye has moved. Students must measure the distance moved in metres – this may cause confusion for low demand students.
3. The force needed to pull the water to such a height and the fact that it takes such a long time means the tree may not get enough water to the leaves.

Lesson 10: Investigating transpiration

Lesson overview

Learning objectives

- Describe how transpiration is affected by different factors.
- Explain the movement of water in the xylem.

Learning outcomes

- Recall that transpiration rate can be affected by light intensity, temperature, air flow and humidity. [O1]
- Describe the effect on transpiration rate of an increase in light intensity, temperature and air flow, and decrease in humidity. [O2]
- Explain that transpiration produces a tension from above, creating a water potential gradient that moves water up the plant. [O3]

Skills development

- WS 2.2 Plan experiments or devise procedures to make observations, produce or characterise a substance, test hypotheses, check data or explore phenomena.
- WS 2.4 Carry out experiments appropriately having due regard for the correct manipulation of apparatus, the accuracy of measurements, and health and safety considerations.
- WS 2.6 Make and record observations and measurements using a range of apparatus and methods.
- WS 3.2 Translating data from one form to another.
- WS 3.3 Carrying out and representing mathematical and statistical analysis.
- WS 4.3 Use SI units (e.g. kg, g, mg; km, m, mm; kJ, J) and IUPAC chemical nomenclature unless inappropriate.

Maths focus

- 2a Use an appropriate number of significant figures
- 2c Construct and interpret frequency tables and diagrams, bar charts and histograms
- 2f Understand the terms mean, mode and median
- 4c Plot two variables from experimental or other data

Resources needed Images of plants that are living in hostile environments. Worksheet 2.10; Practical sheet 2.10; Technician's notes 2.10

Digital resources

Key vocabulary potometer, water potential gradient

Teaching and learning

Engage

- Show students images of plants that are living in hostile environments such as deserts or waterlogged soil. Ask students to **write down** (perhaps using mini-whiteboards) how the environment might affect the rate of transpiration within the plants – do they expect it to be fast or slow? [O1]
- Discuss with the students what they have written down and ask them to **suggest** what problems this could cause the plants. [O1]

Challenge and develop

- Show students a potometer that has been set up and explain how it can be used to measure the rate of water uptake.
- Ask standard and high demand students to **suggest** how they could use the apparatus to look at the effects of different factors on transpiration rate. [O1, O2]
- As a class, **list** the different factors and how the apparatus would be altered to measure the affect. Students can use a potometer to investigate the effects of different conditions. [O1, O2]

- Discuss with all students the main precautions that need to be taken when setting up the potometer. Also discuss the fact that it measures water uptake not water loss. Remind students that these figures are very similar. Part of the difference being the water used for photosynthesis. [O1, O2]
- Students then **investigate** water loss using Practical sheet 2.10. There are many different types of potometer they can use. Students find this experiment difficult to set up. Sometimes the plants have been wet, there are air bubbles in the xylem cells or the apparatus leaks. A spare set of results might be useful. [O1, O2]

Explain

- Discuss the results with the students. Ask them to **explain** the results in terms of evaporation.
- (High demand) Explain the idea of water potential gradient with the students. This should be kept as simple as possible but explain that cells with higher water potentials will be cells that have high concentrations of solutes (dilute solutions). When water evaporates from the cells of the spongy mesophyll it decreases the water potential of the cell. This means the cells have a lower water potential than the xylem vessels close by. This pulls the water from the xylem which is then replaced by water in the roots. Movement of water relies on there always being a lower water potential in the mesophyll than the xylem. The bigger the difference the faster the water moves, increasing rate of evaporation increases the water potential gradient.
- Students then **complete** the questions on Practical sheet 2.10. [O1, O2]

Consolidate and apply

- Ask students to complete Worksheet 2.10. [O1, O2]

Extend

Ask students able to progress further to:

- (High demand) **Write** a summary to explain why transpiration rate changes in terms of water potential gradient. [O3]

Plenary suggestions

Ask students a series of questions (e.g. 'What is the name of the pores through which plant leaves lose water?'). Students could indicate their 'certainty of response' by holding up 1 (very uncertain) to 5 (certain) fingers.

Answers to Worksheet 2.10

1 Interpreting results

1. 3.5, 4.1, 4.3, 1.3, 3.7
2. Check they have drawn a bar chart, labelled axis and bar heights correspond to means
3. The lamp increases the rate the most: increases temperature which will increase transpiration and increased light intensity could result in more stomata opening as the rate of photosynthesis will increase. The black bag has the lowest rate as stomata will close in the dark reducing transpiration.
4. Group 4 as all their results are lower, suggesting less surface area

2 Adapting to reduce water loss

1. Trapping moisture increases the humidity and the air flow so reduces evaporation, so less water lost through transpiration through stomata; waxy cuticle reduces water loss through the cuticle by evaporation. More leaves so more mass lost, as there is a larger surface area for evaporation so the plant loses more water
2. Stomata are open at night when it is cooler so less water is lost by transpiration.

Answers to Practical sheet 2.10

1. Depends on results
2. Depends on results
3. Light intensity and humidity: in the dark the stomata will close reducing evaporation

   Humidity: less water evaporates when humidity is higher so less water diffuses through the stomata (high demand may understand that the diffusion gradient is reduced)

   Air flow: lowers the humidity around the stomata so increases diffusion and evaporation

   Temperature and air flow: increase in temperature, increased evaporation rate
4. Some of the water is used for other processes such as photosynthesis.

Lesson 11: Moving sugar

Lesson overview

AQA Specification reference

AQA 4.2.3.2

Learning objectives

- Describe the movement of sugar in a plant as translocation.
- Explain how the structure of phloem is adapted to its function in the plant.
- Explain the movement of sugars around the plant.

Learning outcomes

- Recall that translocation is the movement of sugars through phloem. [O1]
- Explain how the structure of phloem is adapted for translocation. [O2]
- Explain that the movement of sugars requires energy and the direction can be up or down the plant. [O3]

Skills development

- WS 1.2 Use a variety of models such as representational, spatial, descriptive, computational and mathematical to solve problems, make predictions and to develop scientific explanations and understanding of familiar and unfamiliar facts.
- WS 3.8 Communicating the scientific rationale for investigations, methods used, findings and reasoned conclusions through paper-based and electronic reports and presentations using verbal, diagrammatic, graphical, numerical and symbolic forms.
- WS 4.1 Use scientific vocabulary, terminology and definitions.

Resources needed Scanning electron microscope image of a greenfly feeding. Worksheet 2.11; Technician's notes 2.11

Digital resources

Key vocabulary companion cells, sieve plates, translocation

Teaching and learning

Engage

- Show students a scanning electron microscope image of a greenfly feeding (search images for 'greenfly feeding scanning electron microscope').
- Ask students to **suggest** what the greenfly is feeding on and which part of the plant it is feeding from.
- Discuss with the students that the greenfly is feeding by pushing its mouth parts into the phloem where it can get at the sugars produced by the plants. [O1]

Challenge and develop

- Develop the discussion about the aphids by asking the students to **suggest** why there is sugar in the phloem. This should lead into a discussion about the sugar being transported from the leaves (source) where it is produced to the other parts of the plant for use of storage (sink). [O1]
- Introduce the term translocation and **explain** that this is the movement of sugars around the plant and that unlike transpiration it can occur in both directions. [O1]
- Ask students to think back to the work they did on the uses of glucose from photosynthesis to **identify** the areas of storage in the plant. [O1]
- Ask standard and high demand students to **explain** why carbohydrates are stored as starch but transported as sugar. [O1]

- High demand: Explain the idea of energy being used to move sugars into the phloem at the source (leaf) forcing water into the phloem. Explain that it is important to convert the sugars to starch in the sink to maintain the differences in concentrations between the source and sink that drives the movement. [O3]
- Students then observe images of phloem vessels either from prepared microscope slides or images from the Internet. Discuss with the students how they are similar to and different from xylem. [O1, O2]

Explain

- Students then **complete** the questions on Worksheet 2.11. [O1, O2]

Consolidate and apply

- Low demand students **complete** a summary poster of xylem and phloem. They should include ideas about: position of xylem and phloem in roots and shoots, structure and function of xylem and phloem, what is transported and the direction in which it is transported.
- Standard demand students **complete** a summary poster of transport in plants. They should include ideas about: position of xylem and phloem in roots and shoots, structure and function of xylem and phloem, what is transported and the direction in which it is transported, how environmental factors affect transpiration.
- High demand students should **complete** the Extend activity.

Extend

Ask students able to progress further to:

- (high demand) **write** a summary to explain how mass flow is used to move sugars around the plant. [O3]

Plenary suggestions

Read out statements about either xylem or phloem: for example, ‘They are a hollow tube’.

Ask students to identify which one the statement is about.

Answers to Worksheet 2.11

1 Phloem

1. move sugars around the plant
2. so the cell sap can move from one cell to the next
3. source is where the sugar is made, sink is where it is used or stored
4. idea that this is where the sugar is

2 Phloem and xylem

Feature	**Xylem**	**Phloem**
Made of living or dead cells?	dead	living
Are the cells completely hollow?	yes	no
Are the cells continuous tubes or are there end walls?	continuous tubes	end walls
Direction of flow of contents	upwards	both ways
Main substances transported	water & minerals	sugars
Name of process that moves substances	transpiration	translocation

Lesson 12: Maths skills: Surface area to volume ratio

Lesson overview

Learning objectives

- Be able to calculate surface area and volume.
- Be able to calculate surface area to volume ratio.
- Know how to apply ideas about surface area and volume.

Learning outcomes

- Understand how to calculate the surface area and volume of different objects. [O1]
- Understand how to calculate the surface area to volume ratio of different objects. [O2]
- Explain how surface area and volume ratios may affect biological factors, such as heat loss. [O3]

Skills development

- WS 2.3 Apply knowledge of a range of techniques, instruments, apparatus, and materials to select those appropriate to the experiment.
- WS 2.6 Make and record observations and measurements using a range of apparatus and methods.
- WS 3.3 Carrying out and representing mathematical and statistical analysis.
- WS 3.5 Interpreting observations and other data (presented in verbal, diagrammatic, graphical, symbolic or numerical form), including identifying patterns and trends, making inferences and drawing conclusions.

Maths focus

- 1a Recognise and use expressions in decimal form
- 1c Use ratios, fractions and percentages
- 2c Construct and interpret frequency tables and diagrams, bar charts and histograms
- 4c Plot two variables from experimental or other data

Resources needed Pieces of paper shaped like leaves or rectangles, material blocks (any wood or metal block that can be used to measure volume), supply of 30 cm rulers and squared paper. Worksheet 2.12

Digital resources

Key vocabulary sphere, surface area to volume ratio

Teaching and learning

Engage

- Divide the class into small groups and provide each group with a piece of paper cut into different shapes.
- Low demand students could be given a square or rectangle; other students should be given an irregular shape such as a leaf shape.
- Ask students to **discuss** how they can measure the surface area of the piece of paper they have been given. If they need help in suggesting ways to measure area of an irregular object, provide them with some squared paper as a hint. [O1]
- Ask students to then **calculate** the surface area of their piece of paper. [O1]

Challenge and develop

- Use the Engage activity to determine how well students understand how to calculate surface area. Ask the students to **suggest** why surface area is important in plants. They should refer back to the work on leaves and the idea that larger surface area means they can absorb more light. [O1]

- Ask students if they know how to **calculate** the volume of a cube. Most students at this level should be able to do this; however, low demand students may need some practice so provide the students with material blocks and ask them to calculate the volume and total surface area of the block. [O1]
- Introduce the idea of surface area to volume ratio and **explain** how this can be calculated. Students then **complete** Question 1 of Section 1 on Worksheet 2.12. [O1, O2]

Explain

- **Discuss** with the students the applications of surface area to volume ratio. From their graph they should see that as the size of an object increases the surface area to volume ratio decreases. This means that large organisms have a small surface area compared to their volume. [O3]
- **Explain** that this can cause problems with function in the body. For example, animals in hot countries like the elephant in the picture need to lose heat from the body – they have a very small surface area compared to their volume so heat loss is a problem. [O3]
- Ask students to **suggest** why larger organisms need a transport system but single celled organism can rely on diffusion of gases through their surface. Explain that they will come back to this idea when they look at transport systems. [O3]

Consolidate and apply

- Low demand students should complete the rest of Section 1 on Worksheet 2.12. [O3]
- Standard and high demand students should complete the rest of Worksheet 2.12. [O3]

Extend

Ask students able to progress further to:

- (high demand) **write** a summary that **explains** why calculating surface area to volume ratio is relevant to biology. [O3]

Plenary suggestions

Ask students to suggest why babies lose heat faster than adults.

Answers to Worksheet 2.12

1 Calculating surface area to volume ratio

1.

Side length in cm	Total surface area in cm^2	Volume in cm^3	Surface area to volume ratio
1	1 × 1 × 6 = 6	1 × 1 × 1 = 1	6/1 = 6
2	2 × 2 ×6 = 24	2 × 2 × 2 = 8	24/8 = 3
3	3 × 3 × 6 = 54	3 × 3 × 3 =27	54/27 = 2
4	4 × 4 × 6 = 96	4 × 4 × 4 = 64	96/64 = 1.5
5	5 × 5 × 6 = 150	5 × 5 × 5 = 125	150/125 = 1.2
6	6 × 6 × 6 = 216	6 × 6 × 6 = 216	216/216 = 1

2. As the organism increases in length, the surface area to volume ratio decreases.

3. idea that volume is too great compared with surface area for substances to diffuse from across the surface and through the whole volume

2 Heat loss

1. A = 33.25; B = 42; C = 3.76
2. Animal C as they have a small surface area to volume ratio, which means their volume is large compared to their surface area; they will lose less body heat.

When and how to use these pages: Check your progress, Worked example and End of chapter test

Check your progress

Check your progress is a summary of what students should know and be able to do when they have completed the chapter. Check your progress is organised in three columns to show how ideas and skills progress in sophistication. Students aiming for top grades need to have mastered all the skills and ideas articulated in the final column (shaded pink in the Student book).

Check your progress can be used for individual or class revision using any combination of the suggestions below:

- Ask students to construct a mind map linking the points in Check your progress
- Work through Check your progress as a class and note the points that need further discussion
- Ask the students to tick the boxes on the Check your progress worksheet (Teacher Pack CD). Any points they have not been confident to tick they should revisit in the Student Book.
- Ask students to do further research on the different points listed in Check your progress
- Students work in pairs and ask each other what points they think they can do and why they think they can do those, and not others

Worked example

The worked example talks students through a series of exam-style questions. Sample student answers are provided, which are annotated to show how they could be improved.

- Give students the Worked example worksheet (Teacher Pack CD). The annotation boxes on this are blank. Ask students to discuss and write their own improvements before reviewing the annotated Worked example in the Student Book. This can be done as an individual, group or class activity.

End of chapter test

The End of chapter test gives students the opportunity to practice answering the different types of questions that they will encounter in their final exams. You can use the Marking grid provided in this Teacher Pack or on the CD Rom to analyse results. This shows the Assessment Objective for each question, so you can review trends and see individual student and class performance in answering questions for the different Assessment Objectives and to highlight areas for improvement.

- Questions could be used as a test once you have completed the chapter
- Questions could be worked through as part of a revision lesson
- Ask Students to mark each other's work and then talk through the mark scheme provided
- As a class, make a list of questions that most students did not get right. Work through these as a class.

Marking Grid for End of Chapter 2 Test

	Getting Started [Foundation Tier]	Going Further [Foundation and Higher Tiers]	More Challenging [Higher Tier]	Most demanding [Higher Tier]	Total marks	Percentage
Student Name	Q. 1 (A01, AO2, AO3) 10 marks	Q. 2 (AO1, A02) 10 marks	Q. 3 (AO1, AO2, A03) 10 marks	Q. 4 (A01, AO2, AO3) 13 marks		

Check your progress

You should be able to:

use the word equation to describe photosynthesis	→	recall and use the symbol equation for photosynthesis	→	recall and use the balanced symbol equation for photosynthesis
know that chloroplasts absorb light and convert it to chemical energy	→	describe the use of light and chloroplasts in photosynthesis	→	explain that chloroplasts absorb energy to drive chemical reactions
understand that photosynthesis is an endothermic reaction	→	explain why photosynthesis is an endothermic reaction		
understand that plants respire and photosynthesise	→	explain why plants carry out respiration	→	describe the difference in gas exchange in plants during day and night
name the factors that affect photosynthesis	→	describe how the rate of photosynthesis can be increased	→	explain the effects of limiting factors on photosynthesis
identify the parts of a leaf and their function	→	describe how leaves are adapted for efficient photosynthesis	→	explain how the leaf's structure is adapted for photosynthesis
know the definition of diffusion	→	explain diffusion using the idea of particles	→	explain how substances pass in and out of cells
describe how water travels in plants	→	describe adaptations in xylem and phloem	→	explain adaptations of xylem and phloem
describe experiments on the rate of transpiration	→	describe how different factors affect transpiration	→	explain how different factors affect transpiration
recall that the movement of sugars is called translocation	→	describe how proteins and carbohydrates are transported in plants	→	explain how concentration gradients affect processes

Worked example

Laura is growing tomatoes in her greenhouse.

Laura knows that to increase her crop of tomatoes, she must increase photosynthesis.

1 Complete the equation for photosynthesis

$6CO_2 + H_2O \rightarrow C_6H_{12}O_6 + H_2O$

Water (H_2O) has been correctly identified as a reactant, but has also been given as the product. The equation has not been balanced. The correct answer is:

$6CO_2 + H_2O \rightarrow C_6H_{12}O_6 + 6O_2$

2 Laura heats her greenhouse. She wants to know the best temperature to grow the tomatoes in. Look at the graph.

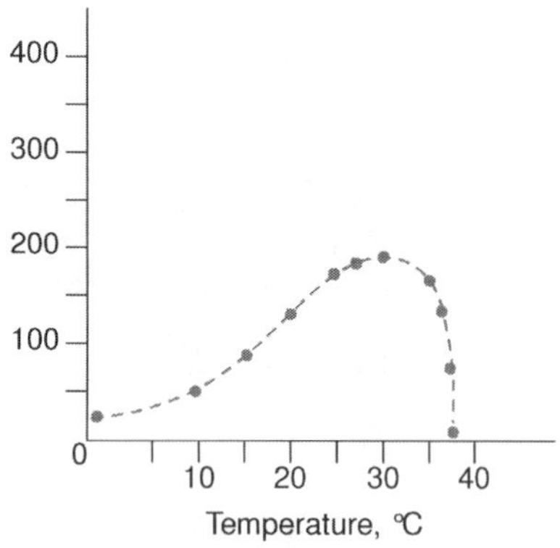

Explain what the graph shows

The rate increases from 5-30°C and then decreases to 0 at 37°C because the enzymes denature.

The pattern has been given (increases and then decreases) and the explanation for the decrease in rate is correct.

An alternative answer would be to explain that the rate increases from 5–30°C as the enzymes gain more energy or collide more often.

3 What is the best temperature for Laura to grow the tomatoes?

30°C

Correct answer given.

How does water for photosynthesis get to the leaves?

Through the xylem by osmosis.

One correct answer is given (xylem).

Another answer would be '...by transpiration'.

4 Laura notices that when she turns the heaters up, she needs to water the plants more often. Explain why.

More water evaporates.

One correct answer is given

A better answer would be: Transpiration increases because more water evaporates through the stomata.

Moving and changing materials: Introduction

When and how to use these pages

This chapter builds on ideas about cells from Chapter 1. It looks at ideas about the movement of substances in and out of these cells and the transport of substances around the body. This chapter links to Chapter 2 Photosynthesis, especially lessons 2.8 Looking at stomata and 2.9 Moving water. There are also links to lesson 4.3 Exploring non-communicable diseases and lesson 2.12 Maths skills: Surface area to volume ratio.

Overview of the chapter

In this chapter, students will learn about the movement of substances in and out of cells. They will investigate osmosis in animal and plant cells as well as active transport. They will then discover why enzymes are so important and investigate the effect of pH on enzyme activity. Students will learn about the role of enzymes in digestion and how the digestive system is adapted to its function. They will then find out more about gas exchange surfaces in plants and animals. Finally, students will learn more about transport in animals and research facts about coronary heart disease.

This chapter offers a number of opportunities for the students to use mathematics to carry out surface area to volume ratio calculations and practice their graph skills. They also carry out investigations into factors affecting enzyme activity and analyse investigations on osmosis.

Obstacles to learning

Students may need extra guidance with the following terms and concepts:

- Students struggle to describe the movement of water in terms of concentration. Try to avoid using the term 'concentration' when referring to the water molecules. Instead use the terms 'dilute' and 'concentrated' as they appear in the specification.
- Many students find it difficult to visualise the concept of surface area to volume ratio. They need to understand that it is not the total surface area that is important but how this surface area compares to the volume.
- Most students will understand that enzymes speed up reactions but few will understand why. Opportunities to interpret enzyme-controlled reactions will help with this understanding.
- Students often struggle in examinations when they are shown unfamiliar respiratory surfaces and asked to describe adaptations. Explaining that most exchange surfaces have similar adaptations may help.
- The spreads on blood and the heart contain many terms and concepts that students find difficult to recall. They should be given plenty of opportunity to use the key words. It is also important to show them different diagrams of the heart; if they only see one they may be confused by a different one in the exam.

Practicals in this chapter

In this chapter students will do the following practical work.

- Observe osmosis in plant cells
- Investigate the effect of changing the surface area to volume ratio on the rate of diffusion
- Investigate the effect of amylase on starch
- Observe blood vessels using a microscope
- Dissect a heart
- Compare inhaled and exhaled air
- Required practical: Investigate the effect of a range of concentrations of salt or sugar solutions on the mass of plant tissue
- Required practical: Investigate the effect of pH on the rate of reaction of amylase enzyme

	Lesson title	Overarching objectives
1	Explaining water movement	Describe and explain how water moves by osmosis in living tissues.
2	Required practical: investigating osmosis	Investigate the effect of a range of concentrations of salt or sugar solutions on the mass of plant tissue.
3	Learning about active transport	Explain how active transport is different from diffusion and osmosis.
4	Investigating the need for transport systems	Relating the need for transport systems to surface area to volume ratio.
5	Explaining enzymes	Describe what enzymes are and how they work.
6	Required practical: investigating enzymes	Investigate the effect of pH on the rate of reaction of amylase enzyme.
7	Learning about the digestive system	Explain how the digestive system is adapted to its function.
8	Explaining digestion	Investigating the digestive enzymes.
9	Required practical: Testing for carbohydrates, lipids and proteins	Use qualitative reagents to test for a range of carbohydrates, lipids and proteins
10	Looking at more exchange surfaces	Describe the adaptations of different gas exchange surfaces.
11	Learning about plants and minerals	Explain how root hair cells are adapted for efficient osmosis, and the function of minerals in plants.
12	Investigate how plants use minerals	Explain the effects of mineral deficiencies on plant growth, and the need for fertilisers to increase food production.
13	Learning about the circulatory system	Identify and explain how the structure of each part of the circulatory system relates to its function.
14	Exploring the heart	Describe the structure and function of the heart.
15	Studying blood	Identify the parts of the blood and explain how red blood cells and haemoglobin transport oxygen efficiently.
16	Investigating gas exchange	Explain how the human gas exchange surface is adapted to its function.
17	Learning about coronary heart disease	Identify the causes and symptoms of coronary heart disease. Describe and evaluate possible treatments of coronary heart disease.
18	Maths skills: Extracting and interpreting information	Extract and interpret information from tables, charts and graphs.

Lesson 1: Explaining water movement

Lesson overview

AQA Specification reference

AQA 4.1.3.2

Learning objectives

- Describe how water moves by osmosis in living tissues.
- Identify factors that affect the rate of osmosis.
- Explain what the term 'partially permeable membrane' means.

Learning outcomes

- Recognise that water moves in and out of plant cells by osmosis through the cell membrane. [O1]
- Explain what a partially permeable membrane is. [O2]
- Explain the effects of osmosis on plant and animal cells. [O3]

Skills development

- WS 1.2 Use a variety of models such as representational, spatial, descriptive, computational and mathematical to solve problems, make predictions and to develop scientific explanations and understanding of familiar and unfamiliar facts.
- WS 2.4 Carry out experiments appropriately having due regard for the correct manipulation of apparatus, the accuracy of measurements, and health and safety considerations.
- WS 2.6 Make and record observations and measurements using a range of apparatus and methods.
- WS 4.1 Use scientific vocabulary, terminology and definitions.

Resources needed Worksheet 3.1; Practical sheet 3.1; Technician's notes 3.1

Digital resources

Key vocabulary flaccid, osmosis, partially permeable membrane, plasmolysis, turgid

Teaching and learning

Engage

- Show students images of fresh cod steaks and salt cod (search 'cod steaks' and 'salt cod').
- Ask students to **discuss** the differences they can see and to **suggest** any difference in the feel of the fish there may be. [O1]
- Ask students to **explain** why the salt cod looks dryer and would feel harder. [O1]

Challenge and develop

- Explain that fish is salted to dry it out as a way of preserving it. Explain that the salt draws the water out of the cells and that movement of water is by osmosis. [O1]
- Show the students the demonstration of osmosis (see Technician's notes 3.1). Explain why the level of sugar solution has changed in terms of osmosis and partially permeable membranes. [O1, O2]
- Ask students to **predict** what would happen if the beaker contained sucrose solution and the Visking tubing contained distilled water. [O1]
- Students then **investigate** the effect of osmosis on onion cells using Practical sheet 3.1. [O1, O2, O3]

Explain

- Discuss the observations the students made during the practical.

- Ask students to **explain** why the cells in distilled water were different from those in sucrose solution. They should do this by answering the conclusion question on the practical sheet. [O1, O2, O3]

Consolidate and apply

- Show students some images of blood cells in distilled water and salt solution. These can be found in the Student Book (page 89). [O3]
- Ask low demand students to **describe** the change and identify in which direction the water has moved.
- Ask standard and high demand students to **explain** the change in terms of osmosis.
- Students then complete the questions on Worksheet 5.1. [O1, O2, O3]

Extend

Ask students able to progress further to:

- (high demand) **complete** the extend questions on Worksheet 3.1. [O1, O2, O3]

Plenary suggestions

Show students a video of contractile vacuoles in single-celled organisms. Explain that the vacuole is used to get rid of excess water in the organisms and that the organisms live in fresh water.

Ask students to explain why water would enter the cells and what would happen if their contractile vacuole stopped working.

Answers to Worksheet 3.1

1 Demonstrating osmosis

1. Water has entered the Visking sac because the contents are more dilute than the surrounding liquid.
2. The sugar molecules will stay on the right-hand side; some water molecules will have moved from the left to the right, so they are about equal.
3. Membrane that will only let some substances through
4. Root hair cells

2 Osmosis and animal cells

1. Distilled water: C; concentrated salt solution: B; solution with the same salt concentration as the cytoplasm of the cells: A
2. A: the movement of water in will equal the movement of water out so cells keep their shape

   B: water moves out by osmosis as the solution is more concentrated, cells shrink

   C: water in by osmosis as the solution is less concentrated, cells burst when they become full

3 Extend Osmosis

1. Root hair, as the water moves out of these cells into the other cells
2. Concentration difference from soil to xylem. Soil is the most dilute and xylem the least so the water moves by osmosis from cell to cell. Some students may refer to transpiration pulling the water up the xylem to ensure that the xylem next to the root cells is always less dilute than the root cells – this will depend on whether they have covered Chapter 2.

Answers to Practical sheet 3.1

Results

1. Onion cell in water should be turgid
2. Onion cell in sugar solution should be flaccid or plasmolysed

Conclusion

1. Cells become turgid because there is a more concentrated solution inside the cell, water will move in by osmosis. Cells become flaccid because the water moves out of the cell into the more concentrated sugar solution by osmosis. If enough water moves out, the membrane pulls away from the cell wall and the cell becomes plasmolysed.

Lesson 2: Required practical: Investigate the effect of a range of concentrations of salt or sugar solutions on the mass of plant tissue

Lesson overview

AQA Specification reference

AQA 4.1.3.2

Learning objectives

- Use scientific ideas to develop a hypothesis.
- Plan experiments to test a hypothesis.
- Draw conclusions from data and compare these with hypotheses made.

Learning outcomes

- Use appropriate apparatus to measure changes in plant tissue due to osmosis. [O1]
- Interpret data to describe the effect of different concentrations on plant tissue. [O2]
- Explain why the rate of osmosis changes depending on concentration gradients. [O3]

Skills development

- WS 2.4 Carry out experiments appropriately having due regard for the correct manipulation of apparatus, the accuracy of measurements, and health and safety considerations.
- WS 2.6 Make and record observations and measurements using a range of apparatus and methods.
- WS 3.1 Presenting observations and other data using appropriate methods.
- WS 3.2 Translating data from one form to another.
- WS 3.5 Interpreting observations and other data (presented in verbal, diagrammatic, graphical, symbolic or numerical form), including identifying patterns and trends, making inferences and drawing conclusions.

Maths focus

- 1c Use ratios, fractions and percentages
- 1d Make estimates of the results of simple calculations
- 4c Plot two variables from experimental or other data
- 4d Determine the slope and intercept of a linear graph

Resources needed Worksheet 3.2; Practical sheet 3.2.1 (low and standard demand); Practical sheet 3.2.2 (high demand); Technician's notes 3.2

Digital resources

Key vocabulary osmosis, partially permeable membrane, plasmolysed

Teaching and learning

Engage

- Show students some potato chips that have been soaking in concentrated salt water.
- Ask students to **feel** the potato chips and discuss the texture and firmness. [O1]
- Ask students to **predict** what would happen if you placed the potato chips into pure water. [O1]

Challenge and develop

- Students then **investigate** the effect of osmosis on potato tissue using Practical sheet 3.2.1 (low and standard demand) or Practical sheet 3.2.2 (high demand). [O1, O2] (NB. It is important that this investigation is set up as soon as possible so that results can be measured in the same lesson.)
- While the potato is left, students should **complete** part 1 of Worksheet 3.2.

- Low and standard demand students **calculate** the change in mass of the potato chips. [O2]
- High demand students **calculate** the percentage change in mass of the potato chips. [O2]
- Ask students to **plot** the mean changes on a graph. They may need to be shown how to plot both positive and negative changes.

Explain

- Ask students to **explain** their results by completing the conclusion on their practical sheet. [O1, O2, O3]

Consolidate and apply

- Ask students to **evaluate** their investigation by completing part 2 of Worksheet 3.2. [O1, O2, O3]

Extend

Ask students able to progress further to:

- Complete questions 5 to 10 on page 91 of the student book.

Plenary suggestions

Place the following words on the board: risk, variable, accuracy, precision, repeatability, random error, anomalous. Select individual students to **choose** a word and **construct** a sentence about their investigation using that word.

Answers to Worksheet 3.2

1 Investigating osmosis

1. concentration of solution
2. type of tissue, length or size of tissue, time left in solution
3. It is difficult to measure the exact mass as liquid on the surface of the potato chip can affect the mass, especially if they are not dried after removing from solution.
4. Larger chips will have more surface area for uptake of water so would not be a fair test.
5. Cutting finger when using knife – take care when handling the knife; slipping on a wet floor – try not to spill any of the solutions or drop the potato chips on the floor

2 Evaluating the investigation

1. No, measurements are precise; if they cluster closely with only two results it is difficult to judge how accurate they are.
2. The main problem will be not measuring accurately or not leaving the chips long enough.
3. doing more repeats, extending the range in concentration, leaving the solutions longer

Answers to Practical sheet 3.2

NB High demand: all questions below; low and standard demand: omit Question 6 under 'Conclusion'.

Conclusion

1. higher concentrations
2. lower concentrations
3. lower concentrations
4. higher concentrations
5. Increases are due to water moving in by osmosis and decreases due to water moving out by osmosis; the bigger the change the bigger the concentration gradient between the cells and the solutions.
6. Point where concentrations are equal
   a. Concentrations inside and outside the cells are equal
   b. Yes, there is still random movement in and out but no net movement

Lesson 3: Learning about active transport

Lesson overview

AQA Specification reference

AQA 4.1.3.3

Learning objectives

- Describe active transport.
- Explain how active transport is different from diffusion and osmosis.
- Explain why active transport is important.

Learning outcomes

- Describe active transport as transport that requires energy. [O1]
- Explain that active transport moves substances against the concentration gradient. [O2]
- Explain that active transport is needed for the movement of mineral ions in plants and sugar in animals. [O3]

Skills development

- WS 1.2 Use a variety of models such as representational, spatial, descriptive, computational and mathematical to solve problems, make predictions and to develop scientific explanations and understanding of familiar and unfamiliar facts.
- WS 3.8 Communicating the scientific rationale for investigations, methods used, findings and reasoned conclusions through paper-based and electronic reports and presentations using verbal, diagrammatic, graphical, numerical and symbolic forms.
- WS 4.1 Use scientific vocabulary, terminology and definitions.

Resources needed Ball and a slope – the slope could just be a textbook leaning against other books; Worksheet 3.3; Technician's notes 3.3

Digital resources

Key vocabulary active transport

Teaching and learning

Engage

- Show students a ball rolling down a slope. Explain that the top of the slope represents the level of oxygen in the lungs and the bottom the level of oxygen in the blood. The slope is the difference in concentrations or the concentration gradient, and the ball is the process of transporting the oxygen from the lungs to the blood.
- (Low demand) Ask students to **suggest** what process the rolling ball represents – osmosis or diffusion. [O1]
- (Standard demand) Ask students to **suggest** what process the rolling ball represents linked to lungs, oxygen and blood. [O1]
- Ask students to **describe** how the ball could be moved back up the slope. [O1]

Challenge and develop

- Develop the discussion to **explain** that pushing the ball back up the slope or placing it back to the top would require energy. **Explain** that this is also true of the movement of some substances in living organisms: if you want to move something against the concentration gradient it will require energy. [O1, O2]
- Introduce the term active transport and **explain** that energy from respiration is used to move the substances from a more dilute solution to a more concentrated solution. For example, movement of ions across the root hair cells and sugar molecules from the gut to the blood. If the supply of energy is stopped so is active transport. [O1, O2, O3]

Explain

- (Standard and high demand) Ask students to **produce** a poster about active transport and some of its uses in living organisms. They should use Worksheet 3.3 to help them. They can use the Student Book (pages 92–93) and find more information on the Internet.
- (Low demand) Students should complete the task just using the ideas in the Student Book (pages 92–93). They can then add illustrations by hand or from the Internet. [O1, O2, O3]

Consolidate and apply

- Ask all students to **complete** the 'Uptake of minerals' section on Worksheet 3.3 – Questions 1 and 2. [O3]

Extend

Ask students able to progress further to:

- (high demand) **complete** the 'Uptake of minerals' section on Worksheet 3.3 – Questions 3 and 4. [O3]

Plenary suggestions

Read out the sentences on Technician's notes 3.3. For each sentence ask students to indicate if it is about diffusion, osmosis, active transport or more than one of these.

Answers to Worksheet 3.3

2 Uptake of minerals

1. More sulfate ions are absorbed when oxygen is present (almost doubles)
2. respiration
3. more oxygen so more respiration, which provides the energy for the active transport of sulfate ions
4. Diagram should be silmilar to Figure 3.11 in the Student Book (page 93) but the molecule should be labelled as 'sulfate ion'.

Lesson 4: Key concept: Investigating the need for transport systems

Lesson overview

AQA Specification reference

AQA 4.1.3.1

Learning objectives

- Describe the need for transport systems.
- Describe how the effectiveness of an exchange surface can be increased.
- Explain in terms of surface area to volume ratios the need for transport systems.

Learning outcomes

- Understand that larger organism need transport systems to allow sufficient molecules to be transported to and from all cells. [O1]
- Describe efficient exchange surfaces as having large surface areas and thin membranes. [O2]
- Calculate and compare surface area to volume ratios to explain the need for transport systems. [O3]

Skills development

- WS 1.2 Use a variety of models such as representational, spatial, descriptive, computational and mathematical to solve problems, make predictions and to develop scientific explanations and understanding of familiar and unfamiliar facts.
- WS 2.4 Carry out experiments appropriately having due regard for the correct manipulation of apparatus, the accuracy of measurements, and health and safety considerations.
- WS 2.6 Make and record observations and measurements using a range of apparatus and methods.
- WS 3.3 Carrying out and represent mathematical and statistical analysis.
- WS 4.1 Use scientific vocabulary, terminology and definitions.

Maths focus

- 1a Recognise and use expressions in decimal form
- 1c Use ratios, fractions and percentages

Resources needed Worksheet 3.4; Practical sheet 3.4; Technician's notes 3.4

Digital resources

Key vocabulary surface area to volume ratio, transport systems

Teaching and learning

Engage

- Show students images of large and small potatoes.
- Ask students to **discuss** which they think would be quicker to peel – 1 kg of small or 1 kg of large potatoes.
- Ask students to **explain** their choice in terms of surface area.

Challenge and develop

- Develop the discussion on surface area to **explain** the idea of surface area to volume ratio. Link this to the need for transport systems in larger animals by explaining that a single-celled organism has a relatively large surface area to volume ratio. This allows sufficient transport of molecules into and out of the cell by simple diffusion across the membrane. In multicellular organisms, the smaller surface area to volume ratio means specialised surfaces and organ systems are needed for transporting molecules to all the cells. [O1]

- Ask students to **name** organ systems in the body that are involved in transporting molecules or exchanging molecules with the external environment. They should be able to mention the circulatory system and the lungs. [O1]
- Using the lungs as an example of an exchange system, discuss the idea of how efficiency of gas exchange is increased because the lungs have a large surface area and thin membranes. [O2]
- Show the students two cubes and ask them to **suggest** which has the largest surface area to volume ratio. [O3].
- (Standard demand) Explain how to **calculate** surface area to volume ratio by working through the examples in the Student Book.
- (High demand) Ask students to **complete** the calculations without help. [O3]
- Students then **investigate** the effect of surface area to volume ratio on diffusion rate, Practical sheet 3.4.

Explain

- Ask students to **complete** the questions on Practical sheet 3.4. [O2, O3]

Consolidate and apply

- Ask students to **complete** section 1 of Worksheet 3.4. Low demand students may struggle with rate calculation, so they can answer the questions based on time only. [O1, O2, O3]

Extend

Ask students able to progress further to:

- (high demand) **complete** section 2 of Worksheet 3.4. [O1, O2, O3]

Plenary suggestions

Inform students that one effect of physical fitness training is an increase in the surface area for gas exchange in the lungs.

Ask students to explain why this would be an advantage to an athlete when it came to a race.

Answers to Worksheet 3.4

1 Specialised gas exchange surfaces

1. B, C, A, D
2. Large surface area
3. Their surface area to volume ratio is not large enough for efficient gas exchange so without these surfaces they would not get enough oxygen into their bodies.

2 Diffusion equation

1. surface area and difference in concentration gradient
2. thickness of membrane
3. large surface area; ventilation of lungs maintains the difference in concentration gradient; surface has a very thin membrane
4. increases surface area for gas exchange across the skin

Answers to Practical sheet 3.4

Rate of diffusion should increase with the surface area to volume ratio.

Lesson 5: Explaining enzymes

Lesson overview

AQA Specification reference

AQA 4.2.2.1

Learning objectives

- Describe what enzymes are and how they work.
- Explain the lock-and-key theory.
- Use the collision theory to explain enzyme action.

Learning outcomes

- Understand that enzymes are biological catalysts that speed up chemical reactions in living organisms. [O1]
- Use the lock-and-key theory to explain why enzymes are specific. [O2]
- Explain the effects of temperature and pH on enzyme activity. [O3]

Skills development

- WS 2.4 Carry out experiments appropriately having due regard for the correct manipulation of apparatus, the accuracy of measurements, and health and safety considerations.
- WS 2.6 Make and record observations and measurements using a range of apparatus and methods.
- WS 3.5 Interpreting observations and other data (presented in verbal, diagrammatic, graphical, symbolic or numerical form), including identifying patterns and trends, making inferences and drawing conclusions.
- WS 4.1 Use scientific vocabulary, terminology and definitions.

Maths focus

- 4a Translate information between graphical and numeric form

Resources needed Worksheet 3.5; Practical sheet 3.5; Technician's notes 3.5

Digital resources

Key vocabulary biological catalyst, collision theory, denatured, lock and key, metabolism, optimum

Teaching and learning

Engage

- Show the students a beaker containing 20 cm^3 of hydrogen peroxide. Inform them that if left, the hydrogen peroxide will eventually turn into oxygen and water. [O1]
- Add a small piece of liver or yeast to the hydrogen peroxide.
- Ask the students to **suggest** what has caused the reaction to speed up. [O1]

Challenge and develop

- Discuss the reaction with students and introduce the terms 'enzyme', 'substrate', 'active site' and 'biological catalyst'. [O1]
- (Low demand) Ask students to **complete** section 1 of Worksheet 3.5. [O1]
- (Standard and high demand) Ask students to **write** a summary explaining what they have just seen using the terms 'enzyme', 'substrate', 'active site' and 'biological catalyst'. [O1]
- Students then **investigate** the effect of amylase on starch. They should set up the experiment, which then needs to be left for at least 10 minutes. [O1]

Explain

- While the test tubes are left in the water bath for 10 minutes, discuss with the students the idea of the lock-and-key theory. Ask the students to **suggest** why the enzyme in the liver is different from the one they are using with the starch. Link this to the terms 'substrate' and 'active site' mentioned earlier. [O1, O2]
- Ask students to **complete** their investigation and answer the questions on Practical sheet 3.5. [O1, O2]

Consolidate and apply

- Ask students to **suggest** why the investigation was carried out at 37 °C. If the investigation had been done at room temperature or at 100 °C, **predict** how the results might have been different. [O3]
- Ask students to **look** at the graphs on page 97 of the Student Book that show the effect of temperature and pH. Discuss the graphs with the students and explain the effects in terms of denaturing and collisions. [O3]
- Students then **complete** section 2, questions 1 and 2 of Worksheet 3.5. [O2, O3]

Extend

Ask students able to progress further to:

- (standard demand) **suggest** how they could change the investigation from the lesson to investigate the effect of temperature on amylase. [O3]
- (high demand) **complete** Question 3 on Worksheet 3.5. [O2, O3]

Plenary suggestions

Sketch a graph on the board that shows the optimum pH for three different enzymes; peaks should be at approximately pH 2, pH 7 and pH 8. Ask the class to describe and explain the patterns they see in the graph.

Answers to Worksheet 3.5

1 Enzymes and chemical reactions

1. enzyme, oxygen, protein
2. enzyme, active (site)

2 Optimum conditions

1. 35 °C
2. become denatured, active site changes shape, substrate can no longer fit into active site
3. temperatures below 35 °C: molecules have less energy so fewer collisions, as temperature increases then more collisions occur so rate increases; high temperatures above 35°C: enzymes become denatured, active site changes shape, substrate can no longer fit into active site

Answers to Practical sheet 3.5

1. a reducing sugar
2. no
3. Active site on the catalase is the wrong shape for amylase to fit into; enzymes are specific.

Lesson 6: Required practical: Investigate the effect of pH on the rate of reaction of amylase enzyme

Lesson overview

AQA Specification reference

AQA 4.2.2.1

Learning objectives

- Describe how safety is managed, apparatus is used and accurate measurements are made.
- Explain how representative samples are taken.
- Make and record accurate observations.
- Draw and interpret a graph from secondary data using knowledge and observations.

Learning outcomes

- Use a water bath safely to control the temperature of an enzyme-controlled reaction. [O1]
- Measure the rate of the reaction by the colour change of an iodine indicator. [O2]
- Explain the effects of pH on the activity of amylase. [O3]

Skills development

- WS 2.6 Make and record observations and measurements using a range of apparatus and methods.
- WS 3.5 Interpreting observations and other data (presented in verbal, diagrammatic, graphical, symbolic or numerical form), including identifying patterns and trends, making inferences and drawing conclusions.
- WS 3.7 Being objective, evaluating data in terms of accuracy, precision, repeatability and reproducibility and identifying potential sources of random and systematic error.

Maths focus

- 1c Use ratios, fractions and percentages

Resources needed Worksheet 3.6.1 (low demand); Worksheet 3.6.2 (standard demand); Worksheet 3.6.3 (high demand); Practical sheet 3.6; Technician's notes 3.6

Digital resources

Key vocabulary amylase, iodine, rate of reaction, starch

Teaching and learning

Engage

- Show the following pieces of apparatus: boiling tube containing starch solution, test tube of amylase solution, both tubes should be sitting in a water bath (electronic or beaker of water), dropping pipette, spotting tile containing drops of iodine solution.
- Mix the amylase and starch solutions and put them back into the water bath. Take a drop of the mixture and add it to one of the wells on the spotting tile; at this point the iodine solution should turn blue-black.
- Ask the students to **explain** what would happen if you took another sample from the mixture every 30 seconds for about 10 minutes.

Challenge and develop

- Show the students a bottle containing buffer solution. Explain that the buffer solution will maintain a constant pH. [O1, O2]
- Ask students to **suggest** how they could use the buffer solution and the demonstration they have just seen to investigate the effect of pH on amylase activity. [O1, O2]

- Students then **investigate** the effect of pH on amylase enzyme activity using Practical sheet 3.6. Each group should be given a pH to start at. If they complete that pH, they just move onto the next one. In this way a range of pH values should be covered by the class with some repeats. The pH values used will depend on the those available in the school. However, the range should include strong acid, strong alkali and neutral. [O1, O2]

Explain

- Collect together the class results. Students then **complete** the questions about their investigation on Worksheet 3.6.1 (low demand), Worksheet 3.6.2 (standard demand) or Worksheet 3.6.3 (high demand). [O3]

Consolidate and apply

- Ask students to **evaluate** their investigation by answering the evaluation on their worksheet. [O3].

Extend

Ask students able to progress further to:

- (high demand) **complete** the questions on Q_{10} values on Worksheet 3.6.3. [O3]

Plenary suggestions

Select individual students to **identify** one problem with their investigation and **explain** how they could improve the investigation to help solve the problem.

Answers to Worksheet 3.6

NB. Low demand: questions 1 and 2 of section 1 and all of section 2; standard demand: up to the end of section 2; high demand: all questions

1 Enzyme experiment

2.
   a. should be pH 7 or close to this value
   b. Low and high pH may not have worked, as extreme pH may dentaure amylase.
   c. As the pH increases or decreases from optimum of pH 7, the time to breakdown the starch decreases
3. At high or low pH the shape of the active site changes so the starch molecule can no longer fit into it – amylase has become denatured.
4. Small intestine as lower pH in stomach would denature amyalse

2 Evaluating

1. Depends on the results
2. Main problem will be maintaining the temperature or identifying end point
3. Several repeats are needed so that anomalies can be identified

3 Calculating Q_{10}

1. 0.017/0.005 = 3.4, much higher than expected; inaccuracy in the investigation, such as maintaining temperature. Also they are measuring the change every 30 seconds this will not give them an accurate time for the investigation in which to measure rate.

Lesson 7: Learning about the digestive system

Lesson overview

AQA Specification reference

AQA 4.2.2.1

Learning objectives

- Identify and locate the organs in the digestive system, and describe their functions.
- Describe how the products of digestion are absorbed into the body.
- Explain why the small intestine is an efficient exchange surface.

Learning outcomes

- Label a diagram of the digestive system to show the different parts and their function. [O1]
- Describe the role of bile in the digestion of fats. [O2]
- Explain how the small intestine is adapted for absorption of the products of digestion. [O3]

Skills development

- WS1.2 Use a variety of models such as representational, spatial, descriptive, computational and mathematical to solve problems, make predictions and to develop scientific explanations and understanding of familiar and unfamiliar facts.
- WS 3.8 Communicating the scientific rationale for investigations, methods used, findings and reasoned conclusions through paper-based and electronic reports and presentations using verbal, diagrammatic, graphical, numerical and symbolic forms.
- WS 4.1 Use scientific vocabulary, terminology and definitions.

Resources needed Worksheet 3.7; Practical sheet 3.7 (teacher demonstration); Technician's notes 3.7

Digital resources

Key vocabulary absorption, emulsify, lymphatic system, peristalsis

Teaching and learning

Engage

- Ask students to **list** all they know about the digestive system on a piece of paper, then screw up the paper and throw it into the middle of the room. They then collect one of the 'snowballs', read, add more knowledge, then repeat.
- Ask one student to **read out** what is on the piece of paper they have and **write** this on the whiteboard. Invite other students to add to this to build up a picture of what the class remember about digestion from KS3.

Challenge and develop

- Show students a short video about what happens in the digestive system; search videos for 'the digestive system'.

Explain

- **Demonstrate** digestion using Practical sheet 3.7.
- **Discuss** the idea of the different parts of the digestive system and their function. **Explain** why bile is needed for fat digestion. [O1, O2]
- Show students an image of the small intestine and **discuss** the features that enable efficient absorption. [O3]

Consolidate and apply

- Ask low demand students to **complete** section 1 of Worksheet 3.7. [O1, O2]

- All students should then **complete** section 2 of Worksheet 3.7. [O3]
- Standard demand students should then **write** a summary of what happens to food in the digestive system. At this stage they do not need to include the names of all the enzymes. [O1, O2, O3]
- High demand students should write a summary **comparing** and **contrasting** the demonstration with digestion. [O1, O2, O3]

Extend

Ask students able to progress further to:

- **complete** section 3 of Worksheet 3.7. [O2]

Plenary suggestions

'Yes and no' questions. Students work in pairs. One student writes down a section of the digestive system on a sticky note and sticks it to the forehead of their partner so that the partner cannot see the words. The partner must then ask questions of the first student that can only be answered 'yes or no'. The object is to work out what word is on the sticky note.

Answers to Worksheet 3.7

1 Digestive systems

1. mouth, stomach, small intestine, large intestine
2. soluble, enzymes, fats, small

2 Small intestine

1. Long, folds, villi and microvilli that all increase the surface area for diffusion and a rich blood supply and lymph vessels to take the food away and maintain concentration gradients, very thin permeable membrane for easy diffusion, very long to allow time for absorption (high demand students may refer to ideas about active transport).
2. Less surface area so less absorption of food, in particular amino acids that are used to make proteins for growth.

3 Digesting fat

1. Olestra has sugar units, fats do not; it is also a larger molecule with more fatty acids.
2. Lipase or enzymes in digestive system will not digest it because it has a different structure to fats.
3. It is not digested and so cannot be absorbed into the bloodstream, which means it cannot be stored in the body as excess fat.

Lesson 8: Explaining digestion

Lesson overview

AQA Specification reference

AQA 4.2.2.1

Learning objectives

- Describe how physical digestion helps to increase the rate of chemical digestion.
- Name the sites of production and action of specific enzymes.
- Interpret data about digestive enzymes.

Learning outcomes

- Understand that physical digestion increases the surface area of the food. [O1]
- Recall the enzymes, substrates and products of digestion in the different parts of the digestive system. [O2]
- Interpret data to link optimum pH of an enzyme with conditions inside the digestive system. [O3]

Skills development

- WS 2.6 Make and record observations and measurements using a range of apparatus and methods.
- WS 3.8 Communicating the scientific rationale for investigations, methods used, findings and reasoned conclusions through paper-based and electronic reports and presentations using verbal, diagrammatic, graphical, numerical and symbolic forms.
- WS 4.1 Use scientific vocabulary, terminology and definitions.

Resources needed Internet access; materials for making posters. Worksheet 3.8

Digital resources

Key vocabulary carbohydrase, lipase, protease

Teaching and learning

Engage

- Show students an image of a beef burger and ask them to **suggest** the food types found within the burger.
- Ask them to describe what would happen to the burger in the digestive system.

Challenge and develop

- Ask students to chew slowly on a piece of bread. They should then **describe** what happened to the bread in their mouths. [O1]
- Describe the difference between physical and chemical digestion and how the mouth can carry out both of these processes. Link back to the demonstration from the last lesson and their work on enzymes. [O1, O2]
- Ask low demand students to **complete** section 1 of Worksheet 3.8.

Explain

- **Discuss** the idea of the different enzymes in the digestive system and their function; include the substrate and products of each enzyme. [O2]

Consolidate and apply

- Students should then **produce** a PowerPoint presentation using the instructions on Worksheet 3.8 (section 2). [O1, O2, O3]
- Low demand students could be asked to **produce** a poster instead of the PowerPoint using the instructions on Worksheet 3.8 (section 3). [O1, O2, O3]

Extend

Ask students able to progress further to:

- **evaluate** the presentation of another group. [O1, O2, O3]

Plenary suggestions

- Ask individual students to present their PowerPoints to the class.

Answers to Worksheet 3.8

1 Digestion

1. broken, physical, swallow, stomach
2.

Food type	Enzyme	Products of digestion	Part of digestive system where digestion takes place
starch	carbohydrase / amylase	simple sugars / glucose	mouth and small intestine
protein	protease	amino acids	stomach and small intestine
fat	lipase	fatty acids and glycerol	small intestine

Lesson 9: Required practical: Use qualitative reagents to test for a range of carbohydrates, lipids and proteins

Lesson overview

AQA Specification reference

AQA 4.2.2.1

Learning objectives

- Suggest appropriate apparatus for the procedures.
- Describe how safety is managed and apparatus is used.
- Describe how accurate measurements are made.
- Interpret observations and make conclusions.

Learning outcomes

- Able to use apparatus safely when completing food tests. [O1]
- Identify appropriate reagents used to test for carbohydrates, lipids and proteins. [O2]
- Interpret qualitative observations to identify the types of substances present in food. [O3]

Skills development

- WS 2.2 Plan experiments or devise procedures to make observations, produce or characterise a substance, test hypotheses, check data or explore phenomena.
- WS 2.4 Carry out experiments appropriately having due regard for the correct manipulation of apparatus, the accuracy of measurements, and health and safety considerations.
- WS 2.6 Make and record observations and measurements using a range of apparatus and methods.
- WS 2.7 Evaluate methods and suggest possible improvements and further investigations.
- WS 3.1 Presenting observations and other data using appropriate methods.
- WS 3.5 Interpreting observations and other data (presented in verbal, diagrammatic, graphical, symbolic or numerical form), including identifying patterns and trends, making inferences and drawing conclusions.
- WS 4.1 Use scientific vocabulary, terminology and definitions.

Resources needed Image of some food. Practical sheet 3.9.1 (low demand); Practical sheet 3.9.2 (standard and high demand); Technician's notes 3.9

Digital resources

Key vocabulary Benedict's test, carbohydrates, qualitative reagents

Teaching and learning

Engage

- **Show** students a picture of some food; this could be a beef burger to follow on from the previous lesson. [O2]
- **Discuss** with the students the chemicals within the food linking back to the ideas on digestion from the previous lesson. Can they remember the terms protein, carbohydrate and lipid and what these foods groups are needed for in the body. [O2]
- Ask the students to suggest how we can find out which food types are in the burger; they should have come across food tests in KS3 – they will have tested a leaf for starch so should know the iodine test.

Challenge and develop.

- Provide the students with Practical sheet 3.9.1 (low demand) or 3.9.2 (standard and high demand). **Discuss** the idea that we can use different chemicals to test for the different food groups. Some students can get confused by the terms food and food groups; explain that we eat food and that food is made up of the chemicals our body needs; we divide those chemicals into different groups, carbohydrates, lipids and protein. Carbohydrates are further split into starch and sugar. [O1, O2]
- Ask the students to **complete** the practical. They will be testing different food for sugar, starch, proteins and lipids. [O1, O2]
- Before they start they should **produce** a results table that can show all their observations; Practical sheet 3.9.1 has a template that low demand students could use. [O1, O2]
- Low demand students will find it difficult to cope with four methods and different foods. They should be given just the one food substance to start with and asked to do the tests on that alone. [O1, O2]

Explain

- **Discuss** the results of the practical with the class. [O2, O3]
- Ask students to **suggest** any improvements they could have made to their investigation. [O2, O3]

Consolidate and apply

- Low demand should **complete** questions 1–4 on pages 104 & 105 of the Student Book. [O2]
- Standard demand students should **complete** questions 1–6 on page 104 & 105 of the Student Book. [O2]
- High demand students should **explain** to the students that Benedict's reagent produces a series of colours depending on the amount of sugar present. Ask the students to **discuss** how they could compare the sugar content of two different foods. Ideas should include the need for controlling temperature in a water bath and time, as well as the mass of each food sample and the volume of Benedict's reagent used. [O2, O3]

Extend

Ask students able to progress further to:

- **complete** the 'Interpreting observations' section on page 105 of the Student Book. [O3]

Plenary suggestions

Put the following words on the whiteboard: starch, glucose, protein, lipid, Benedict's, biuret, ethanol, iodine, purple, blue-black, red, milky white. Ask the students to match the words into four groups.

Answers to Practical sheet 3.9

Evaluation

1. They should find all the food types present; they may not all be found in each food tested as this will depend on the plant material
2. Some observations are difficult as the food colour may affect the colour observed; biuret change can be difficult to spot
3. Improvements could include crushing the food, dissolving the food in water first, holding the tubes against a white background to see the colour change more clearly

Lesson 10: Looking at more exchange surfaces

Lesson overview

AQA Specification reference

AQA 4.1.3.1

Learning objectives

- Identify the structures responsible for gas exchange in fish, amphibians and insects.
- Describe the adaptations of different gas exchange surfaces.
- Explain the gas exchange surfaces in amphibians.

Learning outcomes

- Recall that fish have gills, insects have trachea and amphibians have lungs but also use their skin as a gas exchange surface. [O1]
- Understand that all gas exchange surfaces have similar adaptations. [O2]
- Explain how the gas exchange surfaces of tadpoles change as the tadpole becomes a frog. [O3]

Skills development

- WS 3.2 Translating data from one form to another.
- WS 3.8 Communicating the scientific rationale for investigations, methods used, findings and reasoned conclusions through paper-based and electronic reports and presentations using verbal, diagrammatic, graphical, numerical and symbolic forms.
- WS 4.1 Use scientific vocabulary, terminology and definitions.

Resources needed Internet access, materials for making posters. Worksheet 3.10.1 (low demand); Worksheet 3.10.2 (standard demand); Worksheet 3.10.3 (high demand)

Digital resources

Key vocabulary gills, spiracles, trachea

Teaching and learning

Engage

- Show students a short video of a frog and a tadpole (search 'frog and tadpole' video).
- Ask them to **describe** how the lifestyle of the frog is different to the tadpole and how that might affect their need for different respiratory surfaces.

Challenge and develop

- Ask students to **read** the section on different gas exchange surfaces in the Student Book. [O1, O2]
- Low demand students should then **complete** the exchange surfaces questions on Worksheet 3.10.1. [O1]
- Ask standard and high demand students to **complete** Questions 4 and 5 in the Student Book (page 107). [O1, O2]

Explain

- **Discuss** the idea that all gas exchange surfaces will have three main adaptations: a large surface area, a thin permeable membrane and a moist exchange surface. [O2]
- Ask students to **explain** why the gills of a tadpole can be outside but lungs of the frog have to be inside the body. [O3]

Consolidate and apply

- Students should then **produce** a poster or PowerPoint presentation about the different gas exchange surfaces. They should use Worksheet 3.10 to help them. The worksheets are tiered: low demand: Worksheet 3.10.1; standard demand: Worksheet 3.109.2; high demand: Worksheet 3.10.3. [O1, O2, O3]

Extend

Ask students able to progress further to:

- **complete** the section 'Fish gills' on Worksheet 3.10.3 (high demand). [O1, O2]

Plenary suggestions

Call out the names of different animals and ask students to name the type of respiratory exchange surface. The animals should be familiar to the students and easy to class.

Answers to Worksheet 3.10.3

NB. Low demand: section 1 only; high demand: section 3

1 Exchange surfaces

1. amoeba – cell membrane of single cell, earthworm – moist external body surface, human – lungs, shark – gills
2. Moving increases the need for oxygen, and diffusion through the skin does not supply enough.

3 Fish gills

1. The water is always next to blood with a lower oxygen concentration, as the blood that has already passed through the gill plate will meet fresh water with higher oxygen concentration than the water leaving the gill at the other end.
2. This enables comparisons to be made between fish of different sizes.
3. The more active fish have a larger surface area to their gills, which are also thinner. This will allow more efficient diffusion of oxygen and a higher rate of respiration.

Lesson 11: Learning about plants and minerals

Lesson overview

AQA Specification reference

AQA 4.1.3.3; AQA 4.2.3.2

Learning objectives

- Describe how mineral ions from the soil help plants to grow.
- Explain how root hair cells are adapted for efficient osmosis.
- Describe the function of different mineral ions in a plant.

Learning outcomes

- Understand that plants need nitrogen to make proteins for growth. [O1]
- Understand that root hairs are exchange surfaces and have similar adaptations to other exchange surfaces. [O2]
- Describe how plants use nitrates, phosphates, potassium and magnesium from the soil. [O3]

Skills development

- WS 1.4 Explain everyday and technological applications of science. Evaluate associated personal, social, economic and environmental implications, and make decisions based on the evaluation of evidence and arguments.
- WS 3.1 Presenting observations and other data using appropriate methods.
- WS 3.2 Translating data from one form to another.
- WS 3.8 Communicating the scientific rationale for investigations, methods used, findings and reasoned conclusions through paper-based and electronic reports and presentations using verbal, diagrammatic, graphical, numerical and symbolic forms.
- WS 4.1 Use scientific vocabulary, terminology and definitions.

Maths focus

- 4a Translate information between graphical and numeric form
- 4c Plot two variables from experimental or other data

Resources needed Worksheet 3.11

Digital resources

Key vocabulary fertiliser, nitrates, phosphates

Teaching and learning

Engage

- Show students a short video of time-lapse images showing root hair growth.
- Ask students to **write down** on mini-whiteboards the function of root hairs and how they are adapted. This links to Chapter 2 lesson 9 Moving water and how much they know will depend on whether or not they have completed that part of the course.
- They should then **compare** their list to that on page 109 of the Student Book.

Challenge and develop

- Ask low demand students to **complete** section 1 of Worksheet 3.11. [O2]
- Discuss with students that root hairs not only take in water but they also take in mineral ions. Discuss with students the main mineral ions needed by the plant and their use in the plant (see the table on page 109 of the Student Book). [O1, O2]

- Ask students to **suggest** how farmers could increase the amount of minerals available in the soil. Introduce the idea of fertilisers. [O1, O2]

Explain

- Ask students to **write** a summary of why plants need mineral ions from the soil.

Consolidate and apply

- Ask low demand students to **analyse** data about using natural and artificial fertilisers by **completing** section 2 of Worksheet 3.11. [O1, O2]
- Ask all students to **analyse** data about nitrogen fertiliser by **completing** section 3 of Worksheet 3.11. [O1, O2]
- Ask high demand students to **analyse** data about investigating the use of fertilisers by **completing** section 4 'The Rothamsted experiment' of Worksheet 3.11. [O1, O2]

Extend

Ask students able to progress further to:

- **explain** why plants need oxygen in order to take up minerals from the soil (their answer should link back to active transport). [O1, O2]

Plenary suggestions

Organic farmers do not use artificial fertilisers. Ask students to suggest why their foods are more expensive to buy.

Answers to Worksheet 3.11

1 Root hairs

A – mitochondria provide energy for active transport

B – large surface area for absorption of water

C – thin wall to reduce the distance for osmosis

D – large permanent vacuole to absorb as much water as possible

2 Looking at fertilisers

1. natural: cow manure / wood ash / bonemeal / pig manure / horse manure / fishmeal
2. artificial: superphosphate / ammonium sulfate / ammonium phosphate
3. wood ash
4. Fertilisers are quicker acting.
5. The phosphorus is very concentrated.

3 Nitrogen fertiliser and growth

1. 0.7 tonnes/hectare
2. This uses 'the law of diminishing returns' – by doubling the amount of fertiliser from 0 to 50, the crop increases from 2.5 to 7 tonnes/hectare, almost three times the size. If the amount of fertiliser is again doubled from 50 to 100 tonnes/hectare, the crop only increases from 7 to 8.4 tonnes/hectare. It is tempting to state the highest amount of fertiliser, as this would give the largest crop; however by spending more and more money, the farmer's increase in yield becomes less and less; the correct amount of fertiliser depends partly on its cost. Most farmers would choose 50 kg of fertiliser/hectare.
3. wood ash
4. They are used to make amino acids and so proteins.

4 The Rothamsted experiment

1. as a control / comparison
2. Potatoes need less nitrogen than wheat.
3. Field 4; a fertiliser containing nitrogen, phosphorus and potassium would be expensive – the increase in crop yield may not justify the expense.
4. nitrogen = proteins; phosphorus = DNA, potassium = enzymes

Lesson 12: Investigating how plants use minerals

Lesson overview

AQA Specification reference

AQA 4.1.3.3

Learning objectives

- Describe why plants need different mineral ions.
- Explain the effects of mineral deficiencies on plant growth.
- Explain the importance of fertilisers.

Learning outcomes

- Understand that plants need minerals for healthy growth. [O1]
- Recognise and explain mineral deficiency symptoms of plants. [O2]
- Understand that fertilisers are used to increase crop yield and grow crops on areas with limited minerals. [O3]

Skills development

- WS 2.2 Plan experiments or devise procedures to make observations, produce or characterise a substance, test hypotheses, check data or explore phenomena.
- WS 2.3 Apply a knowledge of a range of techniques, instruments, apparatus, and materials to select those appropriate to the experiment.
- WS 3.5 Interpreting observations and other data (presented in verbal, diagrammatic, graphical, symbolic or numerical form), including identifying patterns and trends, making inferences and drawing conclusions.

Resources needed Internet access. Worksheet 3.12; Practical sheet 3.12; Technician's notes 3.12

Digital resources

Key vocabulary culture solution, deficiency

Teaching and learning

Engage

- Ask students to think about the possibility of growing crops on the moon or another planet. They should then **discuss** what would happen if we only supplied the plants with water and carbon dioxide.

Challenge and develop

- Show the students a short video about slash and burn agriculture (search 'slash and burn agriculture'). Ask the students to think back to the previous lesson and recall how farmers can solve the problem of few minerals in the soil. [O1, O3]
- Discuss the need for minerals and show students the set of apparatus on Practical sheet 3.12. [O1, O2]
- Ask low demand students to **complete** section 1 on Worksheet 3.12. [O1, O2]
- Ask standard and high demand students to **plan** an investigation to show the effect the lack of individual minerals would have on plant growth. They should use Practical sheet 3.12 to help them. As this investigation takes several weeks, the students will not be setting up the investigation but analysing a provided set of results. See Worksheet 3.12. [O1, O2]

Explain

- Ask students to **analyse** the results on Worksheet 3.12. [O1, O2]

Consolidate and apply

- Ask low demand students to **suggest** the advantages and disadvantages of using fertilisers. [O3]

- Ask standard and high demand students to **research** farming methods by **completing** section 3 of Worksheet 3.12. Low demand students could answer Question 1 only. [O1, O2]

Extend

Ask students able to progress further to:

- **explain** the importance of inorganic fertilisers in developing countries. [O1, O2, O3]

Plenary suggestions

Ask students to discuss the statement 'All farmers in the UK should use organic methods to grow crops'.

Answers to Worksheet 3.12

1 Mineral deficiency predictions

1.

Tube **2** – poor growth and yellow leaves

Tube **3** – poor fruits and flowers and discoloured leaves

Tube **4** – poor root growth and discoloured leaves

2 Analysing the investigation

1. act as a control / for comparison
2. to stop algae growing
3. taken in by the root hairs using active transport
4. A = nitrates, B = potassium, C = none, D = phosphate

3 Researching organic and inorganic farming

1. Manure provides minerals without harming the environment and can also reduce soil erosion and improve water retention in the soil; crop rotation helps maintain mineral levels in the soil without using fertilisers; hedgehogs and frogs will eat pests such as slugs; weeding reduces competition for minerals and weedkillers can harm the environment.
2. Inorganic fertilisers are artificial. They come from concentrated sources of minerals, are used in small amounts and can be absorbed immediately.
3. Runoff into rivers can cause eutrophication – increases in algae growth.
4. better for the environment, possible health risks of chemicals in food

Lesson 13: Learning about the circulatory system

Lesson overview

AQA Specification reference

AQA 4.2.2.2; AQA 4.2.2.3

Learning objectives

- Identify the parts of the circulatory system.
- Describe the functions of the parts of the circulatory system.
- Explain how the structure of each part of the circulatory system relates to its function.

Learning outcomes

- Recall that blood moves through arteries, veins and capillaries. [O1]
- Understand how arteries veins and capillaries are adapted for their function. [O2]
- Understand the difference between a single and a double circulatory system. [O3]

Skills development

- WS 1.2 Use a variety of models such as representational, spatial, descriptive, computational and mathematical to solve problems, make predictions and to develop scientific explanations and understanding of familiar and unfamiliar facts.
- WS 2.4 Carry out experiments appropriately having due regard for the correct manipulation of apparatus, the accuracy of measurements, and health and safety considerations.
- WS 2.6 Make and record observations and measurements using a range of apparatus and methods.
- WS 4.1 Use scientific vocabulary, terminology and definitions.

Maths focus

- 4a Translate information between graphical and numeric form

Resources needed Worksheet 3.13.1 (low demand); Worksheet 3.13.2 (standard demand); Worksheet 3.13.3 (high demand); Practical sheet 3.13.1 (low demand); Practical sheet 3.13.2 (standard and high demand); Technician's notes 3.13

Digital resources

Key vocabulary double circulation, lumen

Teaching and learning

Engage

- Ask students to work in small groups (perhaps using mini-whiteboards) to:

  list all they know about blood and circulation

  describe how blood gets from the heart to the liver and back again.

Challenge and develop

- Remind students that some animals (such as *Amoeba*) do not have a blood circulatory system. Ask the students to **suggest** why *Amoeba* do not have a circulatory system. This should link back to the work on surface area and exchange systems. [O1]
- Explain to the students that some animals (such as insects) have an open circulatory system and some animals (such as humans) have a closed circulatory system. Ask students to **suggest** how the blood flow in an open system is different from that of a closed system. [O1]
- Explain that in a closed system, blood flow is continuous, flowing from arteries (away from the heart) to capillaries (small blood vessels) to veins, which carry blood back to the heart. [O1, O2]

- All students should then **observe** the different types of blood vessels using Practical sheet 3.13.1 (low demand) or 3.13.2 (standard and high demand) and prepared microscope slides. [O1, O2]

Explain

- The students **draw** the three types of blood vessels. They should be outline plans only, showing the differences in size and thickness of the vessel walls. The drawings should be similar to the pictures in the table on page 113 of the Student Book. A short description of the function of each vessel should be written next to it. [O1, O2]
- Low demand students may find this task difficult, so they have been given drawings to copy on page 2 of Practical sheet 3.13.1. [O1, O2]

Consolidate and apply

- Low demand students should **complete** the different circulatory systems questions on Worksheet 3.13.1. [O1]
- Standard and high demand students should use Figure 3.33 from page 113 of the Student Book to **explain** why the human circulation is called a double circulatory system. This can be compared to the single circulatory system of fish. [O3]
- Standard demand students then **complete** the circulatory systems question on Worksheet 3.13.2. [O1, O3]
- High demand students then **complete** the circulatory system questions on Worksheet 3.13.3. [O1, O3]

Extend

Ask students able to progress further to:

- **explain** the advantage of a double circulatory system compared with a single circulatory system. [O3]

Plenary suggestions

Read out the true and false statements on Technician's sheet 3.13. Students can put their thumbs up if they are true and down if they are false.

Answers to Worksheet 3.13

Different circulatory systems

Low demand (Worksheet 3.13.1)

1. *Amoeba*
2. insect
3. shark and human
4. capillaries → veins

Circulation

Standard demand: Question 1 (Worksheet 3.13.2); high demand: both questions (Worksheet 3.13.3)

1. Fish have a single circulation, mammals have a double circulation, the blood passes through the heart twice on each circuit in the mammal and only once in the fish.
2.
   a. capillaries
   b. The pressure is higher in the arteries and there is a pulse.

Answers to Practical sheet 3.13.2

1. Include ideas about size of larger lumen in veins, thicker walls in arteries, and capillary wall only one cell thick.
2. Pressure in vein is a lot lower so larger lumen allows the blood to flow more easily; arteries need thicker walls to cope with the higher pressure. Pressure in capillary is lower but high enough to push fluid through the thin wall.
3. to allow substances to be exchanged across its surface

Lesson 14: Exploring the heart

Lesson overview

AQA Specification reference

AQA 4.2.2.2

Learning objectives

- Describe the structure and functions of the heart.
- Identify the functions and adaptations of the parts of the heart.
- Explain the movement of blood around the heart.

Learning outcomes

- Recall that the heart pumps blood to the lungs and the rest of the body and that the blood flows because of pressure differences. [O1]
- Label a diagram of the heart and explain that the left ventricle has a thicker muscle wall to pump the blood at higher pressure so that it reaches all parts of the body. [O2]
- Describe the cardiac cycle and explain how the natural resting heart rate is controlled by a group of cells located in the right atrium that act as a pacemaker. [O3]

Skills development

- WS 2.4 Carry out experiments appropriately having due regard for the correct manipulation of apparatus, the accuracy of measurements, and health and safety considerations.
- WS 3.8 Communicating the scientific rationale for investigations, methods used, findings and reasoned conclusions through paper-based and electronic reports and presentations using verbal, diagrammatic, graphical, numerical and symbolic forms.
- WS 4.1 Use scientific vocabulary, terminology and definitions.

Resources needed Images or video clips of blood passing through the heart; access to the Internet. Worksheet 3.14; Practical sheet 3.14; Technician's notes 3.14

Digital resources

Key vocabulary aorta, atrium (plural: atria), coronary artery, pacemaker, vena cava, ventricle

Teaching and learning

Engage

- Show the students images or video clips of blood passing through the heart (There are several on the Internet. Search 'video of blood circulation in heart').
- Ask students to **recall** why the heart has two sides.

Challenge and develop

- Show the class a pig heart or model heart. **Discuss** with the class the function of the heart. [O1]
- Low demand students then **complete** section 1 of Worksheet 3.14. [O1]
- **Demonstrate** or ask the class to dissect a pig heart or lamb heart. [O1, O2]
- During the dissection, **discuss** with the class the function of each part and the advantage of the thicker wall of the left ventricle. Students should also be made aware of the function of the heart valves. Instructions for the dissection are on Practical sheet 3.14. [O1, O2]

Explain

- Students **complete** section 3 of Worksheet 3.14. [O1, O2]

Consolidate and apply

- **Describe** to the students the journey of the blood through the heart from vena cava to aorta. **Explain** how the pacemaker controls the contraction of the atria followed by the ventricles. [O3]
- Low demand students should **complete** section 2 of Worksheet 3.14. [O3]
- Standard and high demand students should use Figure 3.38 from page 115 of the Student Book to write a summary **describing** the cardiac cycle in their own words. [O3]
- High demand students should **include** ideas about the pacemaker in their description. [O3]

Extend

Ask students able to progress further to:

- **complete** section 4 of Worksheet 3.14. [O3]

Plenary suggestions

Ask the class to add labels onto a projected diagram of the heart.

Answers to Worksheet 3.14

1 The function of the heart

blood, lungs, body, pressure, high, low

2 The journey of a red blood cell

oxygen, away, pressure, vein, valves, vein, vena cava, atrium, ventricle, pulmonary artery, lungs, pulmonary vein, atrium, ventricle, aorta

3 The heart

1. See Figure 5.37 on page 115 of the Student Book.
2. right ventricle
3. left ventricle
4. right atrium
5. left atrium
6. stop the blood flowing back into right ventricle
7. has to pump the blood further

4 The cardiac cycle

1. The atria are contracting and pumping blood down into the ventricles.
2. The fish heart only has one atrium and one ventricle rather than two of each, this is because the blood is all deoxygenated and there are not two systems that need to be kept separate.

Lesson 15: Studying blood

Lesson overview

AQA Specification reference

AQA 4.2.2.3

Learning objectives

- Identify the parts of the blood and their functions.
- Explain the adaptations of red blood cells.
- Explain how red blood cells and haemoglobin transport oxygen efficiently.

Learning outcomes

- Recall the function of red blood cells, white blood cells and platelets. [O1]
- Explain how their size, shape and lack of nucleus help red blood cells carry oxygen. [O2]
- Describe how haemoglobin reacts with oxygen to form oxyhaemoglobin. Students should also be able to explain how arteries, veins and capillaries are adapted to their function. [O3]

Skills development

- WS 1.2 Use a variety of models such as representational, spatial, descriptive, computational and mathematical to solve problems, make predictions and to develop scientific explanations and understanding of familiar and unfamiliar facts.
- WS 3.8 Communicating the scientific rationale for investigations, methods used, findings and reasoned conclusions through paper-based and electronic reports and presentations using verbal, diagrammatic, graphical, numerical and symbolic forms.
- WS 4.1 Use scientific vocabulary, terminology and definitions.

Resources needed Animation showing blood moving through vessels, microscopic image of blood cells showing red and white blood cells, platelets and plasma; access to the Internet. Worksheet 3.15.1 (low and standard demand); Worksheet 3.15.2 (high demand)

Digital resources

Key vocabulary haemoglobin, oxyhaemoglobin, plasma, red blood cells

Teaching and learning

Engage

- Show an animation of blood (search ‘animation of blood cells in vessel’).
- Ask students if they can identify the different parts they see.

Challenge and develop

- Show students a microscopic image of blood. This can be found on the Internet (search ‘microscopic image of blood’), and project it onto the whiteboard. [O1]
- Invite students to go to the board and **label** the parts they can **identify**. Students are likely to identify red and white blood cells but may not identify the plasma or platelets. [O1]
- Discuss the function of the different parts of the blood with the students. [O1]

Explain

- Remind students that red blood cells carry oxygen around the body. Ask them to **suggest** ways the red blood cells are adapted to their function. [O1, O2]
- High demand students – **explain** how oxygen combines with haemoglobin to form oxyhaemoglobin. [O3]

Consolidate and apply

- Divide the class into groups. Ask each group to **produce** a PowerPoint presentation about blood. They should use Worksheet 3.15.1 (low and standard demand) or Worksheet 3.15.2 (high demand) to help them. If the Internet and computers are unavailable, the Student Book will provide them with most of the information which they can present as a poster. [O1, O2]
- High demand students should **include** the extension section of Worksheet 3.15.2 in their PowerPoint presentation. [O3]

Extend

Ask students able to progress further to:

- **research** why athletes train at altitude, and add the information to their presentation. [O2, O3]

Plenary suggestions

Write the word 'blood' in the centre of the whiteboard with a circle around it. Then ask individual students to add information to the board to build up a mindmap of all that the class knows about blood and circulation from the three lessons.

Lesson 16: Investigating gas exchange

Lesson overview

AQA Specification reference

AQA 4.1.3.1; AQA 4.2.2.2

Learning objectives

- Identify the parts of the human gas exchange system and know their functions.
- Explain how gas exchange occurs in humans.
- Explain the adaptations of the gas exchange surfaces.

Learning outcomes

- Describe the functions of the main parts of the human respiratory system. [O1]
- Describe the direction of exchange of gases in the lungs and in tissues. [O2]
- Explain how the alveoli are adapted for efficient gaseous exchange. [O3]

Skills development

- WS 1.2 Use a variety of models such as representational, spatial, descriptive, computational and mathematical to solve problems, make predictions and to develop scientific explanations and understanding of familiar and unfamiliar facts.
- WS 2.4 Carry out experiments appropriately having due regard for the correct manipulation of apparatus, the accuracy of measurements, and health and safety considerations.
- WS 3.5 Interpreting observations and other data (presented in verbal, diagrammatic, graphical, symbolic or numerical form), including identifying patterns and trends, making inferences and drawing conclusions.
- WS 3.8 Communicating the scientific rationale for investigations, methods used, findings and reasoned conclusions through paper-based and electronic reports and presentations using verbal, diagrammatic, graphical, numerical and symbolic forms.
- WS 4.1 Use scientific vocabulary, terminology and definitions.

Maths focus

- 1a Recognise and use expressions in decimal form
- 4a Translate information between graphical and numeric form

Resources needed Worksheet 3.16; Practical sheet 3.16; Technician's notes 3.16

Digital resources

Key vocabulary alveolus (plural: alveoli), ventilate

Teaching and learning

Engage

- Ask students to sit still and count the number of times they breathe in over a 30 second period.
- The students should then either stand up or walk round the room for a minute then **count** their breaths again.
- Discuss with the students any changes they noticed and why these changes occurred.

Challenge and develop

- Develop the discussion on breathing to **explain** the difference between breathing and respiration. Explain that breathing involves taking air in and out of the lungs and is better described as ventilation. Gas exchange then takes place at the respiratory surface while respiration is a chemical reaction that takes place in the cells. Make sure students know the difference between inspiration and expiration. [O1]
- Show the students a model of the lungs or diagrams from the Internet. Ask them to **name** as many parts as they can and **discuss** the function of these parts. [O1]
- Students then use the Student Book to **label** the diagram of the lungs on Worksheet 3.16. [O1]
- Students then **investigate** the effect of breathing in and out into limewater using Practical sheet 3.16.

Explain

- Explain to students how inhalation and exhalation use the intercostal muscles and diaphragm to change volume and pressure inside the chest. This can be **illustrated** using the bell jar and balloons demonstration. [O1, O2]
- Low and standard demand students can **complete** section 2 'Ventilation' of Worksheet 3.16. [O1, O2]
- High demand students should **evaluate** the demonstration by **writing** a summary comparing the demonstration with ventilation in the lungs; they should include a description of ventilation shown by Figure 3.42 in the Student Book. [O1, O2]
- Link back to the practical to **explain** to all students about the gaseous exchange of carbon dioxide and oxygen across the alveoli. [O3]
- Look at a diagram showing alveoli (Figure 3.44 in the Student Book). Ask students to **explain** how the alveoli are adapted for efficient gas exchange. [O3]

Consolidate and apply

- Students **complete** section 3 'Alveoli' of Worksheet 3.16. [O3]

Extend

Ask students able to progress further to:

- **complete** the spirometer questions (section 4) on Worksheet 3.16.

Plenary suggestions

Write a list of key terms on the whiteboard. Students choose six terms and enter them into a 3 × 2 'Bingo' grid. Read out the definitions and students cross off terms as they identify them from the definitions. When a student has a grid all correct, they shout 'Bingo!', then read the terms back, giving the definitions as well.

1 Answers to Worksheet 3.16

1 The human respiratory system

1. Lung diagram labels: a) trachea, b) lung, c) intercostal muscles, d) rib, e) diaphragm, f) pleural membrane, g) bronchus, h) bronchiole, i) air sac
2. a. trachea → bronchus → bronchiole → air sac
   b. air sac → bronchiole → bronchus → trachea

2 Ventilation in the lungs

1. a. balloons inflate
   b. balloons deflate
2. Similar – movement of rubber sheet is similar to diaphragm.
3. Different – lungs are made of many air sacs not just two big ballons, no blood vessels, ventilation also involves ribs and muscles.
4. a. ribs, up, out, contracts, down, increases
   b. relax, down, in, relaxes, up, decreases

3 Alveoli

Many alveoli so large surface area for rapid diffusion, rich blood supply to maintain diffusion gradient, small distance between air and blood so less distance for diffusion, moist surface for gases to dissolve.

4 Breathing rate

1. 16
2. 0.4 litres
3. 6.4 litres
4. 24 × 0.6 = 14.4 litres
5. During exercise there's increased muscle contraction and so increased respiration producing more carbon dioxide and a greater oxygen demand.

Answers to Practical sheet 3.16

1. A
2. B
3. turns cloudy
4. stays clear
5. B
6. Tube B is exhaled air, carbon dioxide diffuses into the air of the alveoli, this air is then exhaled so contains more carbon dioxide than inhaled air.

Lesson 17: Learning about coronary heart disease

Lesson overview

AQA Specification reference

AQA 4.2.2.4

Learning objectives

- Identify the causes and symptoms of coronary heart disease and heart failure.
- Describe possible treatments of coronary heart disease and heart failure.
- Evaluate the possible treatments of coronary heart disease and heart failure.

Learning outcomes

- Understand that in coronary heart disease layers of fatty material build up inside the coronary arteries and that faulty heart valves can cause heart failure. [O1]
- Understand that treatments include stents, statins, replacement valves and replacement hearts. [O2]
- Evaluate the use of stents, replacement valves and replacement hearts. [O3]

Skills development

- WS 1.4 Explain everyday and technological applications of science. Evaluate associated personal, social, economic and environmental implications, and make decisions based on the evaluation of evidence and arguments.
- WS 3.8 Communicating the scientific rationale for investigations, methods used, findings and reasoned conclusions through paper-based and electronic reports and presentations using verbal, diagrammatic, graphical, numerical and symbolic forms.
- WS 4.1 Use scientific vocabulary, terminology and definitions.

Resources needed Animation of coronary heart disease, video clip of blood moving through a heart valve. Worksheet 3.17.1 (low demand); Worksheet 3.17.2 (standard demand); Worksheet 3.17.3 (high demand)

Digital resources

Key vocabulary artificial pacemaker, coronary heart disease, statin, stent

Teaching and learning

Engage

- Ask students to work in small groups (perhaps using mini-whiteboards) to **make** a list of all they know about heart disease.
- Each group should then **report** back to the class to determine how much they already know.

Challenge and develop

- Use the starter activity to lead into a discussion about what causes coronary heart disease. Make it clear that the cause of coronary heart disease is a lack of blood supply to the heart muscle (not just the heart itself). Remind the students of the heart dissection from Lesson 14 and that the coronary artery could be seen on the outside of the heart. The students are then told that the coronary blood vessels can develop small deposits or plaques inside on their walls. Plaques are made largely of cholesterol. Any factor that makes it more likely for these plaques to grow will increase the risk of heart disease. To help illustrate this, show the students an animation of coronary heart disease (search 'animation of coronary heart disease'). [O1]
- Explain to students what a stent is and that statins reduce cholesterol. Ask them to **explain** why these two things could help somebody with coronary heart disease. [O1, O2]
- Show the students a video clip of blood moving through a heart valve (search 'blood heart valve'). Ask students to **recall** the function of the valves and **suggest** why a faulty valve would cause problems for an individual and how the problem may be solved. [O1, O2]

Explain

- Low demand students should **complete** the questions on heart disease on Worksheet 3.17.1. [O1, O2]
- Standard and high demand students should **produce** a leaflet about heart disease by following the instructions on Worksheet 3.17.2 (standard demand) or Worksheet 3.17.3 (high demand). [O1, O2, O3]
- High demand students are asked to **include** extra ideas in their leaflet. [O3]

Consolidate and apply

- Discuss the difference between donor heart transplants and artificial hearts. [O2, O3]
- Low and standard demand students should then **complete** the questions on heart transplants. [O2, O3]
- High demand students should **write a summary** that explains why artificial hearts (like the one in Figure 5.47) are so important. [O3]

Extend

Ask students able to progress further to:

- **debate** the question 'Should England introduce an opt out donor system?' Inform the students that an opt out system means that people are automatically put on the donor register unless they state otherwise. This is opposite to the system we have now where people opt to be put on the register.

Plenary suggestions

Ask students to list some of the risks that may cause heart disease and to suggest lifestyle choices that could reduce the risk.

Answers to Worksheet 3.17

Heart disease

1. artery, muscle, oxygen, respiration, pump
2. small mesh tube that is placed inside the artery to open it up again
3. Valves stop the blood going the wrong way in the heart.
4. The valve may not open fully, or the heart valve might develop a leak.
5. The blood is not pumped around the body efficiently so the cells do not get enough oxygen, they may easily get out of breath.
6. Biological are valves from humans or other mammals; mechanical valves are made from artificial materials e.g. plastics.
7.
   a. area of the atrium that regulates the heartbeat
   b. sends electrical impulses to the heart to control heartbeat

Transplant

1. Waiting for a suitable heart as somebody has to die and there are not many people on the donor list, donor heart matching the patient tissue type, rejection of the heart.
2. Advantages: no need to wait for somebody to die.
   Disadvantages: can only be used for a short amount of time, requires batteries.

Lesson 18: Maths skills: Extracting and interpreting information

Lesson overview

Learning objectives

- To extract and interpret information from tables, charts and graphs.

Learning outcomes

- Able to interpret data in a table. [O1]
- Able to construct a graph when given a table of data. [O2]
- Able to interpret data in a graph. [O3]

Skills development

- WS 3.1 Presenting observations and other data using appropriate methods.
- WS 3.2 Translating data from one form to another.
- WS 3.3 Carrying out and representing mathematical and statistical analysis.
- WS 3.5 Interpreting observations and other data (presented in verbal, diagrammatic, graphical, symbolic or numerical form), including identifying patterns and trends, making inferences and drawing conclusions.

Maths focus

- 2c Construct and interpret frequency tables and diagrams, bar charts and histograms
- 2g Use a scatter diagram to identify a correlation between two variables
- 4a Translate information between graphical and numeric form
- 4c Plot two variables from experimental or other data
- 4d Determine the slope and intercept of a linear graph

Resources needed Worksheet 3.18.1 (low demand); Worksheet 3.18.2 (standard demand); Worksheet 3.18.3 (high demand)

Digital resources

Key vocabulary classify, compare, correlation, trend

Teaching and learning

Engage

- Ask students to look at Figure 3.51 on page 123 of the Student Book.
- In groups ask students to **write down** five different pieces of information from the graph. Explain that we are not just looking for the pattern but information about the axes and scales.
- Each group should then **report** back to the class; expected ideas may include the trend, the labels on each axis, the two different scales.
- Ask them if they can spot any missing information (there are no units or title).

Challenge and develop

- Use the starter activity to lead into a discussion about interpreting data from graphs and tables. **Explain** that when a set of results has been obtained from an investigation it is often important to draw a graph to identify any trends. However, sometimes this can be done directly from the table. [O1]
- Ask the students to **identify** the trend in the results table on page 122 of the Student Book. Explain that their answer must refer to the headings in both columns (often they will just say 'it goes up'). Encourage them to look for an 'er' 'er' statement – for example, 'the high**er** the light intensity, the high**er** the rate of photosynthesis'. [O1]

- Ask the students to **suggest** what sort of graph they would draw from the data, bar chart or line graph? Once you have determined it is a line graph, ask them to **suggest** some data that would need a bar chart (e.g. eye colour of a class). [O1, O2]

Explain

- Low demand students **answer** questions 1 and 2 on page 122 of the Student Book. [O1, O2, O3]
- Standard and high demand students **answer** questions 3 and 4 on page 123 of the Student Book. [O2, O3]

Consolidate and apply

- Low demand students then use Worksheet 3.18.1 to construct a table of results, draw a graph and interpret the data. The low demand students may also need help with the scale on their graph. [O1, O2, O3]
- Standard demand students use Worksheet 3.18.2 to construct a table of results, draw a graph and interpret the data. [O1, O2, O3]
- High demand students use Worksheet 3.18.3 to construct a table of results, draw a graph and interpret the data. [O1, O2, O3]

Extend

Ask students able to progress further to:

- **explain** the difference between a line graph and a scatter graph. (The line on a line graph connects each data point; whereas the line of a scatter graph – if drawn – just shows the trend.)

Plenary suggestions

Draw two axes on the board. Invite students to draw a line on the graph that represents different relationships. For example, zero correlation, negative correlation, higher temperature – faster reaction.

Answers to Worksheet 3.18.1

1. units of seconds and cm^3 added to headings, time in ascending order with matching volumes
2. Time should be on *x*-axis and volume on *y*-axis.
3. As time increases, so does the volume of oxygen; it then levels off at 120 seconds.

Answers to Worksheet 3.18.2

1. units of seconds and cm^3 added to headings, time in ascending order with matching volumes
2. headings of '3 g of liver' and '9 g of liver' added to the two columns under volume heading
3. axes as per Worksheet 3.18.1; two lines plotted separately and clear indication as to which is which
4. as per Worksheet 3.18.1 for 3 g of liver; for 9 g of liver: same pattern but the volume collected is higher for each time interval
5. 9 g of liver will conatin more enzyme so the reaction can go faster.

Answers to Worksheet 3.18.3

1. table drawn with correct headings including units of seconds and cm3 added, time in acending order with matching volumes; headings similar to that on Worksheet 3.18.2 (plus '15 g of liver')
2. axes as per Worksheet 3.18.2; three lines plotted separately and clear indication as to which is which
3. as per answer to Worksheet 3.18.2 (plus 9 g and 15 g are similar)
4. This will depend on line of best fit but approximately – 3 g = 0.13 cm^3/s, 9 g and 15 g = 0.3 cm^3/s
5. 9 g of liver will conatin more enzyme so the reaction can go faster, students may understand that there are more active sites available for the reaction to take place quicker. 9 g and 15 g are similar as the enzyme is not limiting the reaction, instead it could be limited by hydrogen peroxide concentration or temperature.

When and how to use these pages: Check your progress, Worked example and End of chapter test

Check your progress

Check your progress is a summary of what students should know and be able to do when they have completed the chapter. Check your progress is organised in three columns to show how ideas and skills progress in sophistication. Students aiming for top grades need to have mastered all the skills and ideas articulated in the final column (shaded pink in the Student book).

Check your progress can be used for individual or class revision using any combination of the suggestions below:

- Ask students to construct a mind map linking the points in Check your progress
- Work through Check your progress as a class and note the points that need further discussion
- Ask the students to tick the boxes on the Check your progress worksheet (Teacher Pack CD). Any points they have not been confident to tick they should revisit in the Student Book.
- Ask students to do further research on the different points listed in Check your progress
- Students work in pairs and ask each other what points they think they can do and why they think they can do those, and not others

Worked example

The worked example talks students through a series of exam-style questions. Sample student answers are provided, which are annotated to show how they could be improved.

- Give students the Worked example worksheet (Teacher Pack CD). The annotation boxes on this are blank. Ask students to discuss and write their own improvements before reviewing the annotated Worked example in the Student Book. This can be done as an individual, group or class activity.

End of chapter test

The End of chapter test gives students the opportunity to practice answering the different types of questions that they will encounter in their final exams. You can use the Marking grid provided in this Teacher Pack or on the CD Rom to analyse results. This shows the Assessment Objective for each question, so you can review trends and see individual student and class performance in answering questions for the different Assessment Objectives and to highlight areas for improvement.

- Questions could be used as a test once you have completed the chapter
- Questions could be worked through as part of a revision lesson
- Ask Students to mark each other's work and then talk through the mark scheme provided
- As a class, make a list of questions that most students did not get right. Work through these as a class.

Marking Grid for End of Chapter 3 Test

	Getting Started [Foundation Tier]				Going Further [Foundation and Higher Tiers]					More Challenging [Higher Tier]					Most demanding [Higher Tier]			Total marks	Percentage
Student Name	Q. 1 (A01) 2 marks	Q. 2 (A01) 2 marks	Q. 3 (A01) 4 marks	Q. 4 (A03) 2 marks	Q. 5 (A01) 1 mark	Q. 6 (A01) 1 mark	Q. 7 (A01) 2 marks	Q. 8 (A02) 4 marks	Q. 9 (A03) 2 marks	Q. 10 (A01) 2 marks	Q. 11 (A01) 2 marks	Q. 12 (A03) 2 marks	Q. 13 (A01) 1 marks	Q. 14 (A02) 4 marks	Q. 15 (A01) 2 marks	Q. 16 (A02) 4 marks	Q. 17 (A03) 4 marks		

Check your progress

You should be able to:

recall that osmosis describes water movement in and out of cells	→	explain osmosis as the movement of water through a partially permeable membrane	→	predict water movement during osmosis explain the words flaccid, plasmolysed and turgid
know that enzymes catalyse reactions in cells understand that substrate molecules fit into active sites of enzymes	→	describe how pH and temperature affect enzymes describe the lock-and-key theory	→	explain how pH and temperature affect enzyme activity use collision theory to explain enzyme action
describe the effect of SA:V on the diffusion of substances	→	describe the features of a range of exchange surfaces in plants and animals	→	explain the features of exchange surfaces
describe the functions of different parts of the circulatory system	→	describe how the circulatory system transports substances	→	explain how the circulatory system is adapted to its function
know that digested food is transported from the small intestine to body cells	→	describe the adaptations of the intestine as an exchange surface	→	explain how the small intestine is adapted for efficient food absorption

Worked examples

The diagram shows the human heart.

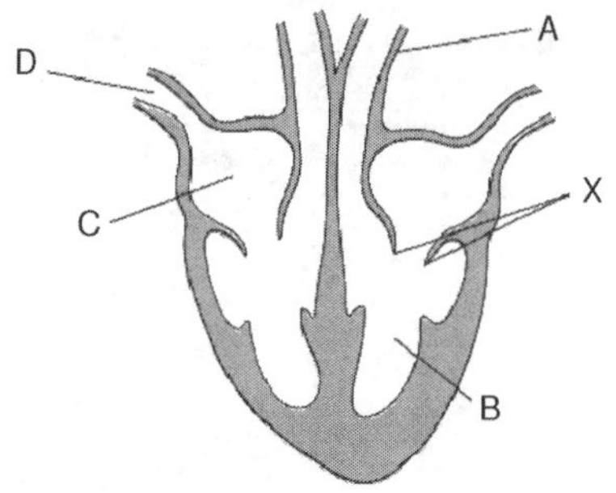

1 **Label parts A–D on the diagram.**

A Artery

B Left ventricle

C Right atrium

D Vena cava

> B–D correctly identified. Giving 'artery' for A is insufficient.
>
> 'Aorta' is a an alternative answer.

2 **a What is the name of part X?**

Valve

> Valve is correctly identified.

b Part X can become damaged or may leak. What effect will this have?

Person will have a lack of energy and breathlessness.

> Two symptoms correctly identified.

c How can a damaged or leaky part X be treated?

Transplant a valve from another person.

> One treatment has been identified.
>
> Artificial valves are another treatment.

d Give an advantage and a disadvantage of the treatment.

It will not be rejected.

> The student has correctly identified an advantage of one treatment but they have only answered half the question for this.

3 **The heart pumps blood to the lungs so that gas exchange can happen.**

Why are alveoli an efficient exchange surface?

They have a large surface area compared with volume to give a large area for gas exchange. Good blood supply to take oxygen to body cells where it is used for respiration for energy for growth.

> Two correct answers are identified. Then the student forgets the question being asked and moves away from the topic of alveoli.

> Overall, this student shows a good understanding of the topic but does not give complete answers. Careful reading of the questions, and consideration of exactly what is being asked for, in addition to more detailed responses, might be helpful.

Health matters: Introduction

When and how to use these pages

This chapter investigates communicable and non-communicable disease. It looks at ideas about pathogens and how the human body and plants defend themselves from infection. This chapter links Chapter 1 lesson 1.14 Growing microorganisms and lesson 1.15 Testing new antibiotics. There are also links to Chapter 3 lesson 3.17 Learning about coronary heart disease.

Overview of the chapter

In this chapter, students will learn about the differences between communicable and non-communicable disease. They will investigate risk factors linked to health and will then look at the different pathogens and some of the diseases they cause. Students will learn about the immune system and how vaccinations are used to prevent the spread of disease. They will then find out more about how drugs are developed and why clinical trials are needed. Finally, students will learn about monoclonal antibodies and their uses.

This chapter offers opportunities for the students to interpret data and look at the principles of sampling as applied to scientific data. They also carry out investigations into the effectiveness of antibiotics, which is similar to the Required practical in Chapter 1 – Investigating disinfectants.

Obstacles to learning

The students may need extra guidance with the following terms and concepts:

- Students struggle to explain the difference between health and disease; they should be given a clear definition of each.
- The terms 'communicable disease' and 'non-communicable disease' can easily be confusing for those students with a limited vocabulary. They should be given several examples of each to assist in their understanding.
- Analysing and evaluating data is an issue. Many students find it difficult to transfer skills they learn in maths lessons to science lessons. Practice is the best way to overcome this issue.
- Terminology used in the study of disease can be a problem for many students. Linking the different pathogens to different disease may cause problems.
- Students find the role of antibodies difficult to understand, especially when they look at immunity.
- Monoclonal antibodies is a concept new to GCSE, which can be very complicated at higher levels, so it is important that this topic is studied at a level appropriate to GCSE students.

Practicals in this chapter

In this chapter students will do the following practical work:

- Measure blood pressure
- Demonstrate the smoking machine
- Statistical study to test the hypothesis that taller people have a larger hand span
- Test the effectiveness of different antibiotics
- Extract chlorophyll from plant material

	Lesson title	Overarching objectives
1	Learning about health	Explain the differences between health and disease.
2	Key concept: Looking at risk factors	Explain the impact of lifestyle on non-communicable diseases.
3	Exploring non-communicable diseases	Identify risk factors for cancer and explain the differences between types of tumours.
4	Analysing and evaluating data	Translate information between graphical and numerical forms and evaluate the strength of evidence.
5	Studying pathogens	Recall the definition of a pathogen and explain how communicable diseases can be controlled.
6	Learning about viral diseases	Describe the transmission and control of some viral diseases.
7	Studying bacterial diseases	Describe the symptoms of some bacterial diseases and explain how they can be controlled.
8	Looking at fungal diseases	Describe the transmission and treatment of the fungal disease rose black spot.
9	Learning about malaria	Describe the transmission of malaria and evaluate control methods for the spread of malaria.
10	Protecting the body	Explain how the body protects itself from pathogens.
11	Exploring white blood cells	Explain how antibody production can lead to immunity.
12	Using antibiotics and painkillers	Explain how antibiotics and painkillers can be used to treat diseases, including their limitations.
13	Building immunity	Explain how mass vaccination programmes reduce the spread of a disease and evaluate the global use of vaccinations
14	Making new drugs	Recall some traditional drugs and their origins and describe how new drugs are developed.
15	Investigating monoclonal antibodies	Explain how monoclonal antibodies are produced and evaluate the use of monoclonal antibodies.
16	Looking at plant diseases	Describe the symptoms and identification methods of some plant diseases.
17	Learning about plant defences	Explain how plant defence systems help them survive and evaluate the use of clones in agriculture.
18	Maths skills: Sampling and scientific data	Understand why sampling is used in science and be able to explain different sampling techniques.

Lesson 1: Learning about health

Lesson overview

AQA Specification reference

AQA 4.2.2.6; AQA 4.2.2.7

Learning objectives

- Recall the difference between health and disease.
- Explain how some diseases interact.
- Evaluate data about lifestyle and health.

Learning outcomes

- Understand that health is complete physical and mental well-being and diseases are disorders that affect part or all of an organism. [O1]
- Describe examples of the interaction of different diseases. [O2]
- Evaluate data that links lifestyle choices to health. [O3]

Skills development

- WS 1.5 Evaluate risks both in practical science and the wider societal context, including perception of risk in relation to data and consequences.
- WS 3.2 Translating data from one form to another.
- WS 3.5 Interpreting observations and other data (presented in verbal, diagrammatic, graphical, symbolic or numerical form), including identifying patterns and trends, making inferences and drawing conclusions.
- WS 3.7 Being objective, evaluating data in terms of accuracy, precision, repeatability and reproducibility and identifying potential sources of random and systematic error.

Maths focus

- 4a Translate information between graphical and numeric form

Resources needed Worksheet 4.1.1 (low demand); Worksheet 4.1.2 (standard demand); Worksheet 4.1.3 (high demand); Practical sheet 4.1; Technician's notes 4.1

Digital resources

Key vocabulary communicable, depression, immune system, mental health, non-communicable

Teaching and learning

Engage

- **Write** the statement 'I am healthy' on the whiteboard. [O1]
- Students should **discuss** in groups what they believe the statement means. [O1]
- Groups then **report back** their ideas in a class discussion. [O1]

Challenge and develop

- Develop the **class discussion** to include the difference between ill health and disease. Ask students to **explain** what they think the difference is. [O1]
- **Introduce** the terms 'communicable disease' and 'non-communicable disease' using one example of each. [O1]
- Ask students to **list** some other examples of each of the two types of diseases. This can be done in groups using mini-whiteboards. The lists can then be compared and checked for accuracy. Low demand students can be given a list of diseases, which they then classify. [O1]

Explain

- Students **write a summary** about health and disease using the terms 'mental health', 'physical health', 'communicable disease' and 'non-communicable disease'. As an alternative, low demand students can complete the health and disease section (1) on Worksheet 4.1.1. [O1]
- Students **investigate** one aspect of health: blood pressure. They measure their own blood pressure following the instructions on Practical sheet 4.1. First **demonstrate** how to use the blood pressure monitors and then explain that systolic pressure is the maximum pressure caused by the blood being pushed out of the ventricles and diastolic pressure is the pressure between heartbeats. [O2]
- Students then **complete** section 2 of Worksheet 4.1.1 (low demand), Worksheet 4.1.2 (standard demand) or Worksheet 4.1.3 (high demand). [O2]

Consolidate and apply

- Low demand students **complete** section 3 Question 1 on Worksheet 4.1.1. [O3]
- Standard demand students **complete** section 3 questions 1 and 2 on Worksheet 4.1.2. [O3]
- High demand students **complete** section 3 on Worksheet 4.1.3. [O3]

Extend

Ask students able to progress further to:

- (high demand) **complete** the extension question on Worksheet 4.1.3. [O3]

Plenary suggestions

Call out the name of a disease and students should indicate if it is a communicable disease or non-communicable disease in some way. For example, this could be done using thumbs up for communicable disease and down for non-communicable disease.

Answers to Worksheet 4.1.3

NB. Low demand: answer up to Question 2 of section 3; standard demand: answer up to Question 3 of section 3; high demand: answer all questions and extension

Health and disease

mental and physical (either order); positive; organism; communicable; e.g. HIV;
non-communicable; e.g. cardiovascular disease

Interaction of diseases

1. Being overweight may cause high bold pressure, which increases the risk of a stroke.
2. Severe viral respiratory infections in early childhood can trigger asthma as children grow.
3. Cancer can lead to depression.
4. Infection of HPV can cause cervical cancer.

Health and lifestyle

1. e.g. more people with high blood pressure / diabetes / heart disease are obese
2. Yes, as the majority of people with high blood pressure are obese so there may be a link, but as some people with high blood pressure are not overweight, there might be other causes.
3. No, as there is no data that show the people with diabetes have high blood pressure.

Extension

1. Julie is correct, there is no data for females, only males or for people living outside of Canada. Other factors may also be involved, such as whether they smoke or not; these are not taken into account in the data.

Answers to Practical sheet 4.1

1. Systolic – maximum pressure; diastolic – pressure between heartbeats
2. e.g. age, gender, activity levels, weight, diet, stress levels, fitness levels
3. Change diet, increase exercise, stop smoking, reduce alcohol intake

Lesson 2: Key concept: Looking at risk factors

Lesson overview

AQA Specification reference

AQA 4.2.2.6

Learning objectives

- Recall the causes of some non-communicable diseases.
- Describe the impact of lifestyle on non-communicable diseases.
- Explain the impact of lifestyle on non-communicable diseases.

Learning outcomes

- Recall that different risk factors can increase the chance of having a non-communicable disease. [O1]
- Understand the financial implications of different risk factors. [O2]
- Understand the causal mechanisms for some risk factors of non-communicable diseases. [O3]

Skills development

- WS 1.3 Appreciate the power and limitations of science and consider any ethical issues which may arise.
- WS.1.4 Explain everyday and technological applications of science. Evaluate associated personal, social, economic and environmental implications, and make decisions based on the evaluation of evidence and arguments.
- WS 3.2 Translating data from one form to another.
- WS 3.5 Interpreting observations and other data (presented in verbal, diagrammatic, graphical, symbolic or numerical form), including identifying patterns and trends, making inferences and drawing conclusions.
- WS 3.7 Being objective, evaluating data in terms of accuracy, precision, repeatability and reproducibility and identifying potential sources of random and systematic error.

Maths focus

- 4a Translate information between graphical and numeric form

Resources needed Internet access; poster material. Worksheet 4.2.1 (low and standard demand); Worksheet 4.2.2 (high demand)

Digital resources

Key vocabulary causal mechanism, risk factor

Teaching and learning

Engage

- Ask the students to **look at** Figure 4.5 'Smoking and lung cancer' in the Student Book or put a copy of this onto the whiteboard. [O1]
- Students should **discuss** in groups what the graphs tell them about the link between smoking and lung cancer. [O1]
- Groups then **report back** their ideas in a class discussion. [O1]

Challenge and develop

- Develop the **class discussion**: introduce the term 'risk factor', explain that risk factors are linked to an increase in the rate of a disease. Apart from lung cancer, can they think of any other risks linked to smoking? [O1]
- Ask students to **list** some other risk factors and the diseases they are linked to. This can be done in groups using mini-whiteboards; they should be able to link back to the previous lesson on obesity. The lists can then be **compared** and checked for accuracy. [O1]

- Ask students to choose one of the risk factors, such as smoking or poor diet, and **list** the financial benefits of reducing this risk factor to both the individual and society as a whole. The lists can then be **compared** and checked for accuracy. [O2]
- As an alternative, low demand students could **calculate** the cost of smoking 20 cigarettes a day for one year and work out what they could buy instead. [O2]

Explain

- **Introduce** the term 'causal mechanism' and explain that this is the idea that the link between a risk factor and the disease can be identified. For example, we know how smoking is a risk factor for lung cancer and the causal mechanism is the chemicals in the cigarette smoke that cause the cancer. [O3]
- Low and standard demand students should **complete** Worksheet 4.2.1. [O1, O2]
- High demand students **complete** Worksheet 4.2.2. [O1, O2, O3]

Consolidate and apply

- Students **produce** a poster deterring teenagers from eating too much fast food. Instructions for the poster can be found on Worksheet 4.2.1 and 4.2.2 (same for both worksheets). If possible they could find further information on the Internet to include in their poster. [O1, O2, O3]

Extend

Ask students able to progress further to:

- **write a summary** that debates the statement 'The government should introduce a "sugar tax", which increases the tax on food containing high levels of sugar in the hope of reducing the incidence of type 2 diabetes'. [O3]

Plenary suggestions

Ask students to produce a mind map / spider diagram that links risk factors to the different diseases mentioned in the lesson. This can be done as a class activity or individually.

Answers to Worksheet 4.2.2

1 Risk factors

1. The following list is not all the possibilities but the more likely answers:
   high-fat diet – heart disease, obesity, type 2 diabetes
   drinking during pregnancy – possible side effects may include low birth weight, risk of miscarriage, premature birth, future behavioural problems of baby
   obesity – heart disease, type 2 diabetes, high blood pressure
   lack of exercise – heart disease
2. Cost to the individual – ideas could include the amount the cigarettes will cost to buy and the health risks, such as heart disease and cancer, which could lead to a loss of income.
   Cost to society – main idea is the cost of treatment on the NHS, which is linked to taxes that the individuals may have to pay, also money may not be available for other treatments if it is being spent on smokers.

2 Risk factors and heart disease

1. It increases – there is a genetic link or influenced by parents' diet.
2. 6 points
3. Joe – he has more points

3 Risk factors and causal mechanisms

1. There may be other factors involved; for example there may be a statistical link between smoking during pregnancy and low birthweights, but the mothers who smoke may also drink alcohol or have a poor diet, so it is difficult to determine which factor may be the cause or that it is a combination of risk factors.
2. They may already be addicted and find it difficult to give up, often they don't think it (lung cancer) will happen to them, plus many other reasons why they start, such as peer pressure.
3. Proving may be very difficult, but if statistics show there is a greater risk this might be enough to persuade them; also idea that actual link may be found in the future as it was with smoking and lung cancer.

Lesson 3: Exploring non-communicable diseases

Lesson overview

AQA Specification reference

AQA 4.2.2.6; AQA 4.2.2.7

Learning objectives

- Identify risk factors for cancer.
- Explain the differences between types of tumours.
- Explain the impact of non-communicable diseases.

Learning outcomes

- Recall the different risk factors for cancer. [O1]
- Understand the difference between benign and malignant tumours. [O2]
- Understand the human and financial cost of cancer. [O3]

Skills development

- WS.1.4 Explain everyday and technological applications of science. Evaluate associated personal, social, economic and environmental implications, and make decisions based on the evaluation of evidence and arguments.
- WS 1.5 Evaluate risks both in practical science and the wider societal context, including perception of risk in relation to data and consequences.
- WS 3.5 Interpreting observations and other data (presented in verbal, diagrammatic, graphical, symbolic or numerical form), including identifying patterns and trends, making inferences and drawing conclusions.
- WS 3.6 Presenting reasoned explanations including relating data to hypotheses.

Maths focus

- 4a Translate information between graphical and numeric form

Resources needed Internet access. Worksheet 4.3; Practical sheet 4.3; Technician's notes 4.3

Digital resources

Key vocabulary benign, carcinogen, malignant, tumour

Teaching and learning

Engage

- **Demonstrate** the smoking machine – see Practical sheet 4.3 – or if school policy on smoking does not allow this, show a video of the machine (search 'smoking machine demonstration'). [O1]
- Students should **complete** the questions on Practical sheet 4.3. [O1]

Challenge and develop

- Use the demonstration to **discuss** the effects of the chemicals in cigarette smoke and cancer. Ask students to **discuss** what they think cancer is. Do they know any other risk factors that may increase the chance of getting cancer? [O1]
- **Introduce** the term 'carcinogen' and discuss the different carcinogens associated with the different types of cancer. [O1]
- Students then **research** different types of cancer and **complete** causes of cancer on Worksheet 4.3. If the Internet is not available, then print the summary sheets about the different cancers from the website www.cancerresearchuk.org. [O1]

Explain

- **Introduce** the terms 'malignant' and 'benign' and explain what they mean, along with the idea of secondary tumours. Make it clear to students that benign tumours are not cancerous as they do not invade surrounding tissue; however, some can become cancerous, which is why they are removed by surgery. [O2]
- Low demand students should then **complete** the sentences about tumours on Worksheet 7.3. [O2]
- Standard demand and high demand students should **write a summary** that explains the difference between malignant and benign tumours. [O2]

Consolidate and apply

- Students **complete** the battle against cancer questions on Worksheet 4.3. [O1, O3]

Extend

Ask students able to progress further to:

- **discuss** the benefits of spending money on campaigns to stop people smoking. [O3]

Plenary suggestions

Provide students with two sticky notes. They should write on one a fact about risk factors they have learnt from this or the previous lesson. On the other they should write something they are not sure about. The sticky notes are then put on the whiteboard it two groups. Choose random ideas from the unsure section and see if other students can answer the query.

Answers to Worksheet 4.3

Causes of cancer

1. skin cancer – UV light; lung cancer – smoking; bowel cancer – lack of fibre in diet; liver cancer – alcohol
2. spend less time in the Sun, increase fibre in diet, reduce alcohol consumption, don't smoke

Tumours

slowly, harmless, surgery, quickly, spread, secondary, cancerous or cancer

Risk factors

1. melanoma, people spending more time in the Sun, more people holidaying abroad
2. lung cancer, greater awareness of the dangers of smoking
3. Ideas could include greater success rate of a cure if caught early, reduced cost of treatment, benefits are far greater than the cost of screening.

Answers to Practical 4.3

1. It will be yellow due to nicotine and tar.
2. lungs (and mouth)
3. damage tissue, cause cancer
4. Should turn yellow as the chemicals are acidic.

Lesson 4: Analysing and evaluating data

Lesson overview

AQA Specification reference

AQA 4.2.2.5; AQA 4.2.2.6

Learning objectives

- Translate information between graphical and numerical forms.
- Use scatter diagrams to identify correlations.
- Evaluate the strength of evidence.

Learning outcomes

- Identify trends and patterns in graphs. [O1]
- Understand the term correlation and uses it correctly. [O2]
- Evaluate data to determine if it is valid. [O3]

Skills development

- WS.1.4 Explain everyday and technological applications of science. Evaluate associated personal, social, economic and environmental implications, and make decisions based on the evaluation of evidence and arguments.
- WS 1.5 Evaluate risks both in practical science and the wider societal context, including perception of risk in relation to data and consequences.
- WS 3.1 Presenting observations and other data using appropriate methods.
- WS 3.2 Translating data from one form to another.
- WS 3.3 Carrying out and representing mathematical and statistical analysis.
- WS 3.6 Presenting reasoned explanations including relating data to hypotheses.
- WS 3.7 Being objective, evaluating data in terms of accuracy, precision, repeatability and reproducibility and identifying potential sources of random and systematic error.

Maths focus

- 4a Translate information between graphical and numeric form
- 4c Plot two variables from experimental or other data

Resources needed Graph paper. Worksheet 4.4.1 (low demand); Worksheet 4.4.2 (standard demand); Worksheet 4.4.3 (high demand); Practical sheet 4.4; Technician's notes 7.4

Digital resources

Key vocabulary bias, correlation, trend

Teaching and learning

Engage

- **Show** the students a graph that shows no correlation; there is an example on Technician's notes 4.4 that can be projected onto the whiteboard. [O1, O2]
- Students should **discuss** what the graph shows them. This is to introduce the idea that sometimes there is no relationship between two variables, something students find difficult in science as they always expect their results to show a pattern or trend. [O1, O2]

Challenge and develop

- Use the graph to **discuss** the idea of sample size; do they think increasing the sample size would make much difference to the graph? In this case maybe not, but in statistical analysis involving risk factors then yes. [O1]
- **Discuss** with students what information graphs and charts need to have in order to interpret them correctly. This should include ideas about the need for titles and correctly labelled axes. Ensure that they know how to

describe patterns and trends. Many students will simply state 'it goes up' and not use the axis labels in their description. [O1]

- Students then **complete** the bar chart questions on their respective worksheets (low demand, Worksheet 4.4.1; standard demand, Worksheet 4.4.2; high demand, Worksheet 4.4.3). [O1]

Explain

- **Introduce** the term 'correlation', the association between two sets of random data. **Explain** that with risk factors there may be a correlation seen in data, but this does not prove there is a link. Refer back to the example of smoking, where the correlation was seen long before the proof was found. [O2]
- Students then complete the scatter graph questions on Worksheet 4.4. (NB. The low demand worksheet misses out Question 4). [O2]
- Students **complete** the evaluation questions on Worksheet 4.4. Question 3 should be completed by high demand students only (it has been left off the other two worksheets). [O1, O2, O3]

Consolidate and apply

- Students **complete** their own study; see Practical sheet 4.4. [O1, O2, O3]

Extend

Ask students able to progress further to:

- **evaluate** their study further by answering the extension questions on Practical sheet 4.4. [O3]

Plenary suggestions

Students compare the findings of their study and discuss reasons why they found similar or different results.

Answers to Worksheet 4.4.3

Bar charts

1. Diagnosis increases with age until 60–66, then it drops and almost levels off.
2. Young people are less likely to be diagnosed.
3. There is no title, you do not know what they have been diagnosed with, is the information for one country or the whole world or even just a town, are they males or females or both?

Scatter diagrams

1. Density should be *x* axis percentage; *y* axis – check the units, line of best fit.
2. positive
3. The greater the density the higher the percentage of obese residents.
4. No, as it does not include information about the actual diet of the residents or how their diet would make them obese.

Evaluating the data

1. Percentage for area D is lower than expected. This may be due to the fact that the fast food restaurants could be serving healthy alternatives, which the majority of residents choose, or it may be that they don't use the restaurants that often; some students may refer to income, in that higher income areas are less likely to have higher proportions of obese residents – the suggestions just need to be sensible. They do not need to be true as students are not expected to know the actual reason.
2. There is very little data, only seven areas, other factors such as income and overall diet have not been taken into account.
3. genetic, amount of exercise, parental influence, overall diet, number of times a week the children eat junk food

Answers to Practical 4.4

1. Depends on results.
2. Depends on results.
3. no
4. need to increase sample size; they could take the whole class results

Extension

1. No, as the people were chosen at random by their proximity to the tester.
2. This will depend on results.
3. There could be a gender bias.

Lesson 5: Studying pathogens

Lesson overview

AQA Specification reference

AQA 4.3.1.1

Learning objectives

- Recall the definition of a pathogen.
- Explain how communicable diseases can be controlled.
- Distinguish between epidemics and pandemics.

Learning outcomes

- Recall that pathogens are microorganisms that cause infectious diseases. [O1]
- Understand how to prevent or reduce the spread of diseases. [O2]
- Understand the difference between epidemics and pandemics. [O3]

Skills development

- WS 3.5 Interpreting observations and other data (presented in verbal, diagrammatic, graphical, symbolic or numerical form), including identifying patterns and trends, making inferences and drawing conclusions.
- WS 3.8 Communicating the scientific rationale for investigations, methods used, findings and reasoned conclusions through paper-based and electronic reports and presentations using verbal, diagrammatic, graphical, numerical and symbolic forms.
- WS 4.1 Use scientific vocabulary, terminology and definitions.

Resources needed Poster material; Internet access. Worksheet 4.5.1 (low and standard demand); Worksheet 4.5.2 (high demand)

Digital resources

Key vocabulary Ebola, epidemic, pandemic, pathogen, protist

Teaching and learning

Engage

- **Show** the students a short video clip about an Ebola outbreak or similar. [O1]
- Students should **discuss** as a class what causes diseases and how they are spread from person to person. [O1, O2]

Challenge and develop

- **Explain** that communicable diseases are caused by pathogens, which are microorganisms such as bacteria, viruses, protists and fungi. **Ask** the students if they know which type of pathogens cause some common diseases, such as cold, flu, measles (viruses), food poisoning (bacteria). [O1]
- **Discuss** why pathogens make us ill. What do the students already know about this; how do they know when they are ill? Explain that high temperatures, nausea, headaches and rashes are the symptoms of disease, and not all disease have the same symptoms. Do they know any specific ones? [O1]
- Students then look at the image on Worksheet 4.5 and **identify** the ways in which the diseases are being spread. [O1]

Explain

- Use the image to **discuss** ways of preventing the spread of disease. [O2]
- Low demand students should complete the rest of section 1 of Worksheet 4.5.1. [O2]
- Standard and high demand students should **write a summary** on how to reduce the spread of diseases. [O2]
- **Explain** to high demand students the difference between pandemic and endemic. [O3]

Consolidate and apply

- Students **research** one example of a disease; see Worksheet 4.5.1 (low and standard demand) or Worksheet 4.5.2 (high demand). This can be done as a poster or PowerPoint and if there is no Internet then they can use the Student Book to present information about Ebola. [O1, O2, O3]
- High demand students should **complete** the extension, as detailed on Worksheet 4.5.2. [O3]

Extend

Ask students able to progress further to:

- **explain** why the Ebola outbreak was an epidemic but the 1918 influenza outbreak was a pandemic. [O3]

Plenary suggestions

Ask students to produce a flow chart to show what happens when a pathogen enters the body, causing a fever.

Answers to Worksheet 4.5.1

Diseases

a. Z
b. Y
c. X
d. W

Lesson 6: Learning about viral diseases

Lesson overview

AQA Specification reference

AQA 4.3.1.2

Learning objectives

- Describe the symptoms of some viral diseases.
- Describe the transmission and control of some viral diseases.
- Explain how some viral diseases are spread.

Learning outcomes

- Describe the symptoms of measles, HIV and TMV. [O1]
- Describe the transmission and control of measles, HIV and TMV. [O2]
- Understand why viral disease are harder to treat than bacterial diseases. [O3]

Skills development

- WS 1.2 Use a variety of models such as representational, spatial, descriptive, computational and mathematical to solve problems, make predictions and to develop scientific explanations and understanding of familiar and unfamiliar facts.
- WS 1.3 Appreciate the power and limitations of science and consider any ethical issues which may arise.
- WS 3.8 Communicating the scientific rationale for investigations, methods used, findings and reasoned conclusions through paper-based and electronic reports and presentations using verbal, diagrammatic, graphical, numerical and symbolic forms.
- WS 4.1 Use scientific vocabulary, terminology and definitions.

Resources needed Images of viruses including HIV; poster material; access to Internet. Worksheet 4.6.1 (low demand); Worksheet 4.6.2 (standard demand); Worksheet 4.6.3 (high demand)

Digital resources

Key vocabulary antiretroviral drug, vaccination

Teaching and learning

Engage

- **Show** the students an image of a virus. [O1]
- Ask the students if they know what it is. Tell them it is a virus and **discuss** what they know about viruses, can they name any diseases caused by a virus? [O1]

Challenge and develop

- **Explain** that HIV is a viral disease. Show the students an image of the virus. Ask the students to write down three facts they know about HIV. This could be done using sticky notes, which could then be placed on the board to see if there are any common misconceptions. [O1]
- Introduce measles and TMV and two other examples of viral diseases. Students should then read about the spread of these diseases in the Student Book. [O1, O2]
- Low demand students then **complete** the 'Spread and control of diseases' section on Worksheet 4.6.1. [O1, O2]
- Standard and high demand students write a summary of the information in the table on Student Book spread 4.6. [O1, O2]

Explain

- **Discuss** the idea that treating viruses is difficult because they reproduce inside the host cells. **Explain** that antiretroviral drugs are drugs that can prevent the virus reproducing. HIV can be treated with antiretroviral drugs in the early stages to suppress the HIV virus. However, if the virus enters the lymph nodes and attacks the immune system, then the person will develop AIDS. [O3]
- **Explain** that prevention by vaccination is often the best way to prevent the spread of viral diseases. [O3]
- (High demand students only) **Explain** how a virus reproduces inside the cells of the host using materials inside the host cell to copy its RNA and produce new proteins in order to multiply. Students may already know about mRNA, this will depend on whether they have covered Chapter 4. [O3]
- High demand students should **complete** the virus reproduction on Worksheet 4.6.3. [O3]

Consolidate and apply

- Students produce a poster encouraging parents to get their children vaccinated against measles (see Worksheet 4.6). [O1, O2, O3]

Extend

Ask students able to progress further to:

- **explain** the difference between HIV and AIDS. [O3]

Plenary suggestions

Refer back to the sticky notes from the start of the lesson. Can the students add to the facts they wrote down? Is there anything they wrote down that they now know is wrong?

Answers to Worksheet 4.6.2

Spread and control of viral diseases

Students should produce a table similar to the table in the Student Book.

Virus reproduction

Statements matched to picture number:

1. The virus attaches to a specific cell, called the host cell.
2. The genetic material from the virus is injected into the host cell where it enters the nucleus to make mRNA.
3. New viruses are made using material inside the host cell and the genetic material from the virus. This happens in the cytoplasm of the cell.
4. The cell splits open, allowing the viruses to leave but killing the host cell.

Lesson 7: Studying bacterial diseases

Lesson overview

AQA Specification reference

AQA 4.3.1.3

Learning objectives

- Describe the symptoms of some bacterial diseases.
- Explain how some bacterial diseases can be controlled.
- Compare and contrast bacterial and viral diseases.

Learning outcomes

- Describe the symptoms of *Salmonella* and gonorrhoea. [O1]
- Understand how *Salmonella* and gonorrhoea can be controlled. [O2]
- Understand the differences and similarities between bacterial and viral diseases. [O3]

Skills development

- WS 1.2 Use a variety of models such as representational, spatial, descriptive, computational and mathematical to solve problems, make predictions and to develop scientific explanations and understanding of familiar and unfamiliar facts.
- WS 3.2 Translating data from one form to another.
- WS 3.8 Communicating the scientific rationale for investigations, methods used, findings and reasoned conclusions through paper-based and electronic reports and presentations using verbal, diagrammatic, graphical, numerical and symbolic forms.
- WS 4.1 Use scientific vocabulary, terminology and definitions.

Maths focus

- 4a Translate information between graphical and numeric form

Resources needed Video of bacteria reproducing without the sound on; poster material; access to the Internet. Worksheet 4.7.1 (low demand); Worksheet 4.7.2 (standard demand); Worksheet 4.7.3 (high demand)

Digital resources

Key vocabulary diarrhoea, gonorrhoea, *Salmonella*

Teaching and learning

Engage

- **Ask** the students to **draw and label** a bacterial cell to see if they recall the structure from Chapter 1. [O1]
- Alternately, **show** the students a video of bacteria reproducing without the sound on. Ask the students to **describe** what is happening and why this could be a problem if it occurs inside the body. [O1]

Challenge and develop

- Ask the students to **read** the pages in the Student Book spread (pages 142–143); they must then **present** the information about *Salmonella* and gonorrhoea in some way. This could be as a spider diagram, flow chart, poster or written report. They must include the symptoms, and how the disease is spread and controlled. [O1, O2]
- Students should then pair up. One partner **tells** the other all they know about *Salmonella* in 1 minute. Their partner can then correct or add information. The second person then does the same for gonorrhoea. [O1, O2]

Explain

- **Explain** why people do not develop symptoms of a disease as soon as they are infected. [O2]
- **Ask** students to work in groups to list the differences and similarities between bacterial and viral diseases. [O3]
- Groups can then **report back** to the whole class on their discussion to identify the main differences and similarities between bacterial and viral diseases. [O3]
- Students then **write** their own summary of the discussion. [O3]

Consolidate and apply

- Low demand students complete safety rules on Worksheet 4.7.1. [O1, O2]
- Standard demand students complete *Salmonella* questions on Worksheet 4.7.2. [O1, O2]
- High demand students complete all the *Salmonella* questions on Worksheet 4.7.3. [O1, O2]

Extend

Ask students able to progress further to:

- **explain** the difference between the way bacteria and viruses reproduce using the terms 'binary fission' and 'RNA'. [O3]

Plenary suggestions

Put the names of the diseases mentioned in the last three lessons on the board and ask students to classify them into bacterial and viral diseases (bacterial – *Salmonella* and gonorrhoea; viral – HIV, TMV, measles, cold, flu).

Answers to Worksheet 4.7

Safety rules

Reasons could include some of the following:

- Wash hands before you start cooking – stops bacteria on your skin getting onto food.
- Keep raw and cooked food separate inside a fridge – stops bacteria from raw food getting into cooked food.
- Clean knives and chopping boards before use – bacteria can survive in scraps of food left on knives or chopping boards.
- Cook foods thoroughly – particularly chicken – cooking kills dangerous bacteria.

Salmonella

1. 2007, month 9
2. most cases in 2007, cases increase in month 6, peak in month 9 and then decrease to lowest in month 12
3. Warmer months see a rise in cases; bacterial growth increases in the warmth so food more likely to become contaminated; in winter, cold will reduce bacterial growth on food not kept in fridges.

   People more likely to take food out in warmer weather (picnics) where it may be left out in warmth for longer.

   Possible increase in barbecues where food is not cooked correctly.

Lesson 8: Looking at fungal diseases

Lesson overview

AQA Specification reference

AQA 4.3.1.4

Learning objectives

- Recall the name and symptoms of a fungal disease.
- Describe the transmission and treatment of rose black spot.
- Explain how rose black spot affects the growth of the plant.

Learning outcomes

- Describe the symptoms of rose black spot. [O1]
- Understand how rose black spot can be transmitted and treated. [O2]
- Understand how rose black spot affects the growth of the plant. [O3]

Skills development

- WS 1.2 Use a variety of models such as representational, spatial, descriptive, computational and mathematical to solve problems, make predictions and to develop scientific explanations and understanding of familiar and unfamiliar facts.
- WS 1.4 Explain everyday and technological applications of science. Evaluate associated personal, social, economic and environmental implications, and make decisions based on the evaluation of evidence and arguments.
- WS 3.8 Communicating the scientific rationale for investigations, methods used, findings and reasoned conclusions through paper-based and electronic reports and presentations using verbal, diagrammatic, graphical, numerical and symbolic forms.
- WS 4.1 Use scientific vocabulary, terminology and definitions.

Resources needed Video about black spot (e.g. search 'problem-solver rose-black spot' at www.yates.com.au). Worksheet 4.8

Digital resources

Key vocabulary mulch, pruning, spores

Teaching and learning

Engage

- **Show** the students a video about black spot. [O1]
- Ask the students to **recall** why the leaves of plants are so important for growth. [O3]

Challenge and develop

- Ask students to **recall** what they know about the way a fungus reproduces using spores. This may be very little so be prepared to explain that a fungus can produce spores that are a bit like seed. These spores are easily transmitted in the air, illustrate with a video clip (such as one from 'the private life of plants' at www.youtube.com). [O2]
- Ask the students to use the ideas about spore dispersal to **explain** why getting rid of rose black spot from a garden is very difficult. [O2]

Explain

- Students complete questions 1–4 about rose black spot in the Student Book (pages 144-145). [O1, O2]
- Discuss with high demand students why rose growth is greatly reduced by black spot in terms of photosynthesis rate. The high demand students should then answer Question 5 in the Student Book. [O3]

Consolidate and apply

- In groups, students **produce** a storyboard for a TV commercial advertising a fungicide used to treat black spot disease (see Worksheet 4.8). High demand students could do the alternative storyboard on a 3-minute clip for a gardening programme. [O1, O2, O3]

Extend

Ask students able to progress further to:

- **describe** the differences between the cells of black rot fungus and *Salmonella* bacteria cells; this is a bit of revision about eukaryotic and prokaryotic cells from Chapter 1.

Plenary suggestions

Volunteers could act out their TV clip.

Lesson 9: Learning about malaria

Lesson overview

AQA Specification reference

AQA 4.3.1.5

Learning objectives

- Recall that malaria is a protist disease.
- Describe the lifecycle of the malarial vector.
- Evaluate control methods for the spread of malaria.

Learning outcomes

- State that malaria is a protist disease. [O1]
- Describe the lifecycle of the mosquito and link this to transmission of malaria. [O2]
- Evaluate control methods for the spread of malaria and link them to the lifecycle of the mosquito. [O3]

Skills development

- WS 1.2 Use a variety of models such as representational, spatial, descriptive, computational and mathematical to solve problems, make predictions and to develop scientific explanations and understanding of familiar and unfamiliar facts.
- WS 1.4 Explain everyday and technological applications of science. Evaluate associated personal, social, economic and environmental implications, and make decisions based on the evaluation of evidence and arguments.
- WS 3.8 Communicating the scientific rationale for investigations, methods used, findings and reasoned conclusions through paper-based and electronic reports and presentations using verbal, diagrammatic, graphical, numerical and symbolic forms.
- WS 4.1 Use scientific vocabulary, terminology and definitions.

Resources needed Video about why malaria is such a problem (e.g. search for 'malaria' at http://video.nationalgeographic.com); video about malaria – search 'malaria human host' on the 'Biointeractive' pages at www.hhmi.org); Internet access or a fact sheet on malaria can be printed from the 'media centre' at www.who.int (high demand only). Worksheet 4.9.1 (low demand); Worksheet 4.9.2 (standard and high demand)

Digital resources

Key vocabulary protist, vector

Teaching and learning

Engage

- **Show** the students a video about why malaria is such a problem. [O1]
- Ask the students to **discuss** as a class what they know about malaria. Do they know what causes it, where in the world is it found and have any of them ever had to take malarial tablets? [O1]

Challenge and develop

- Use the introduction to **discuss** the cause and symptoms of malaria. Explain that it is caused by a protist known as *plasmodium*. The protist is spread by female mosquitoes, which feed on blood. **Introduce** the term 'vector', explaining that the mosquito is the vector. [O1]
- Low demand students can **complete** the sentences about transmission of malaria on Worksheet 4.9.1. [O1]
- Standard and high demand students **complete** the questions about malaria deaths on Worksheet 4.9.2. [O1]

Explain

- **Show** students a video clip about the transmission and symptoms of malaria (from www.hhmi.org). [O1, O2]
- **Discuss** with students ways to prevent the transmission of malaria; they may be able to make some suggestions themselves. [O2, O3]
- **Show** students an image of the lifecycle of the mosquito (see Worksheet 4.9). Ask them to **suggest** why removing stagnant water is one way of reducing the number of malaria cases in a community. [O2]
- Students **complete** the question related to the lifecycle of the mosquito on Worksheet 4.9.1 (low demand) or Worksheet 4.9.2 (standard and high demand). [O2]
- High demand students should also **complete** Question 6 in the Student Book (page 147) to evaluate the different methods of control. [O3]

Consolidate and apply

- In groups students **produce** a factsheet for somebody of their own age about travel to a possible malaria country. See Worksheet 4.9 for more information. The factsheet can be completed using IT or by hand if computers are not available. There is enough information in the Student Book, but a useful factsheet can be printed from the 'mediacentre' at www.who.int for use with high demand students. [O1, O2, O3]

Extend

Ask students able to progress further to:

- **explain** the social and economic benefits of producing a vaccine for malaria.

Plenary suggestions

Students work in pairs. One student picks a disease from the last few lessons and writes it down without the partner seeing it. The other student must identify the disease by asking their partner questions about symptoms, transmission, etc. The partner can only give yes or no answers. A rule could be set that the students cannot start guessing the name of the disease until they have asked at least three questions.

Answers to questions Worksheet 4.9

Malaria transmission

vectors, *Plasmodium*, protozoan, living, host, red

Malaria deaths

1. no
2. mostly around the equator, in Africa
3. tropical regions, so temperature may be a factor; mostly in less developed countries so economics may be a factor
4. 500 people

Mosquito lifecycle

draining stagnant pools of water – kills lava, no place to lay eggs
spraying pools of water with insecticides – kills adults
spraying pools of water with oil – prevents the larvae from breathing

Lesson 10: Protecting the body

Lesson overview

AQA Specification reference

AQA 4.3.1.6

Learning objectives

- Describe how the body protects itself from pathogens.
- Explain how the body protects itself from pathogens.
- Explain how communicable diseases can be spread.

Learning outcomes

- Recall ways the body protects itself from pathogens. [O1]
- Understand how simple actions can reduce the spread of communicable diseases. [O2]
- Understand in detail how the blood and respiratory system protects the body from pathogens. [O3]

Skills development

- WS 1.2 Use a variety of models such as representational, spatial, descriptive, computational and mathematical to solve problems, make predictions and to develop scientific explanations and understanding of familiar and unfamiliar facts.
- WS 3.8 Communicating the scientific rationale for investigations, methods used, findings and reasoned conclusions through paper-based and electronic reports and presentations using verbal, diagrammatic, graphical, numerical and symbolic forms.
- WS 4.1 Use scientific vocabulary, terminology and definitions.

Resources needed A video clip about how the blood clots; Internet access (higher tier only) or factsheet printed from the NHS website (click on 'Find conditions and treatments', then 'Haemophilia' at www.nhs.uk). Worksheet 4.10.1 (low demand); Worksheet 4.10.2 (standard demand); Worksheet 4.10.3 (high demand)

Digital resources

Key vocabulary cilia, goblet cells, sebaceous glands

Teaching and learning

Engage

- Ask the students to work in groups to **list** ways the body stops pathogens entering. [O1]
- Groups then **report back** their ideas in a class discussion to complete a final list. [O1]

Challenge and develop

- Students then **compare** their list with the one in the Student Book. [O1]
- Low demand students can **complete** the 'How the body protects itself' section on Worksheet 4.10.1. [O1]
- Standard and high demand students **write** a summary describing the way the body protects itself in their own words. [O1]
- Ask students to **suggest** why covering your mouth when you sneeze or cleaning a cut can help prevent the spread of diseases. [O2]

Explain

- **Explain** to students how the skin protects the body, including ideas about sebaceous glands and antimicrobial oils. The students should add this information to their diagram on Worksheet 4.10.1 (low demand) or their summary (if it is not already included). [O3]
- **Show** students a video clip about how the blood clots. [O3]
- Ask students to **suggest** reasons why it is important that blood clots. [O3]
- Students then **annotate** the diagram of a blood clot on Worksheet 4.10.1 (low demand), 4.10.2 (standard demand) or 4.10.3 (high demand). [O3]
- **Discuss** with high demand students how the respiratory system protects the body. [O3]
- High demand students then **annotate** the diagram of the respiratory system on worksheet 4.10.3. [O3]

Consolidate and apply

- Low and standard demand students could **produce** a poster encouraging people to cover their mouths when they sneeze. [O1, O2]
- High demand students should **research** haemophilia to find out what it is and how it is linked to the contents of the lesson. They should then **write** a report of what they find. If no Internet is available, information can be printed from the NHS website (search for 'conditions' then 'haemophilia' at www.nhs.uk). [O3]

Extend

Ask students able to progress further to:

- **suggest** reasons why, from the late 1970s to the mid-1980s, about half of all people with haemophilia became infected with HIV.

Plenary suggestions

Ask students why some of the diseases they have discussed over the last few lessons occur regularly in this country and why some do not. Can they identify which ones do and which ones don't and suggest reasons?

Answers to Worksheet 4.10.1

Low demand:

Skin acts as a barrier.

Trachea and bronchi secrete mucus, which traps pathogens.

Stomach makes hydrochloric acid that kills bacteria.

Platelets start the clotting process at wound sites.

Annotated diagrams should include all the information seen on the diagrams in the Student Book.

Lesson 11: Exploring white blood cells

Lesson overview

AQA Specification reference

AQA 4.3.1.6

Learning objectives

- Describe phagocytosis.
- Explain how antibody production can lead to immunity.
- Explain the specificity of immune system responses.

Learning outcomes

- Describe phagocytosis as the ingestion of pathogens by white blood cells. [O1]
- Understand how antibody production can lead to immunity. [O2]
- Understand the specificity of immune system responses. [O3]

Skills development

- WS 1.2 Use a variety of models such as representational, spatial, descriptive, computational and mathematical to solve problems, make predictions and to develop scientific explanations and understanding of familiar and unfamiliar facts.
- WS 3.2 Translating data from one form to another.
- WS 3.5 Interpreting observations and other data (presented in verbal, diagrammatic, graphical, symbolic or numerical form), including identifying patterns and trends, making inferences and drawing conclusions.
- WS 4.1 Use scientific vocabulary, terminology and definitions.

Maths focus

- 4a Translate information between graphical and numeric form

Resources needed A video of white blood cells in action without the sound; a video clip about the production of antibodies; image of a graph that shows primary and secondary infection. Worksheet 4.11

Digital resources

Key vocabulary antibody, antitoxin, immunity, lymphocyte, phagocyte

Teaching and learning

Engage

- **Show** students a video of white blood cells in action without the sound. [O1]
- In groups students **discuss** what the video is showing. [O1]
- Groups then **report back** their ideas in a class discussion. [O1]

Challenge and develop

- **Develop** the discussion about the video to **discuss** the role of white blood cells in protecting the body from pathogens. **Introduce** the terms 'phagocytosis', 'antibody' and 'antitoxin'. [O1]
- **Show** the video again with the sound (or provide your own commentary). [O1]
- Low demand students can **complete** the questions about white blood cells and phagocytosis on Worksheet 4.11.1. [O1]
- Standard and high demand students **write** a summary that explains the process of phagocytosis, including annotated diagrams. [O1]

Explain

- **Explain** to students how antitoxins and antibodies are used to protect the body from pathogens. [O2]
- **Show** students a video clip about the production of antibodies. [O2]
- Ask students to **suggest** reasons people become immune to diseases. [O2]
- **Explain** immunity in terms of lymphocyte cells recognising the pathogen and immediately making lots of antibodies. [O2]

Consolidate and apply

- Students then **complete** the questions about antibodies on Worksheet 4.11.1 (low demand), 4.11.2 (standard demand) and 4.11.3 (high demand). [O2]
- **Explain**, to high demand students only, why antibodies are specific to a pathogen. [O3]
- High demand students then **complete** the questions about specific antibodies on Worksheet 4.11.3. [O3]

Extend

Ask students able to progress further to:

- **suggest** reasons why babies that are being breastfed their mother's milk may be protected from some pathogens.

Plenary suggestions

Show students an image of a graph that shows primary and secondary infection. Ask the students to describe and explain (high demand) the shape of the graph.

Answers to Worksheet 4.11.3

White blood cells and phagocytosis

1. immune, ingesting, antibodies, antitoxins (the last two can be in either order)
2. White blood cell moves toward the bacteria.
   White blood cell ingests / engulfs bacteria.
   Bacteria is inside a vacuole where it is digested.

Antibodies

1. Sentence number in picture order 4, 2, 5, 3, 1
2. Antibodies bind to the pathogens; antitoxins destroy chemicals made by the pathogens.

Specific antibodies

Antibodies can only lock onto matching antigens (with complementary shape).

Lesson 12: Using antibiotics and painkillers

Lesson overview

AQA Specification reference

AQA 4.3.1.8

Learning objectives

- Describe the uses of antibiotics and painkillers.
- Explain how antibiotics and painkillers can be used to treat diseases.
- Explain the limitations of antibiotics.

Learning outcomes

- State the uses of antibiotics and painkillers. [O1]
- Understand how antibiotics and painkillers can be used to treat diseases. [O2]
- Understand the limitations of antibiotics. [O3]

Skills development

- WS 2.6 Make and record observations and measurements using a range of apparatus and methods.
- WS 3.1 Presenting observations and other data using appropriate methods.
- WS 3.5 Interpreting observations and other data (presented in verbal, diagrammatic, graphical, symbolic or numerical form), including identifying patterns and trends, making inferences and drawing conclusions.
- WS 4.1 Use scientific vocabulary, terminology and definitions.

Maths focus

- 2a Use an appropriate number of significant figures
- 5c Calculate areas of triangles and rectangles, surface areas and volumes of cubes

Resources needed Worksheet 4.12; Practical sheet 4.12; Technician's notes 4.12

Digital resources

Key vocabulary antiviral, aspirin, opiates, penicillin

Teaching and learning

Engage

- **Write** the words 'drugs' and 'medicines' onto the whiteboard. [O1]
- **Discuss** with the students what they understand by these two words. Do they mean the same thing? (The correct answer is that there is a lot of overlap; a medicine is any substance that is designed to prevent or treat diseases and a drug is designed to produce a specific reaction inside the body, such as a stimulant. To simplify this for students, just explain that medicines are mostly drugs that are used for treatment rather than drugs that are not used as treatments, such as caffeine and alcohol.) [O1]

Challenge and develop

- Ask students to **suggest** the difference between antibiotics and painkillers. [O1]
- Use their ideas to **discuss** uses of antibiotics and painkillers; include examples of both within the discussion. [O1, O2]
- Low demand students **complete** questions 1 and 2 on page 152 of the Student Book. [O1, O2]
- Standard demand students **complete** questions 1–4 on pages 152–153 of the Student Book. [O1, O2]
- High demand students **complete** questions 3–6 on page 153 of the Student Book. [O1, O2, O3]

Explain

- **Discuss** with the students why it is important to complete a course of antibiotics. [O3]
- Students **complete** the questions about MRSA on Worksheet 4.12. [O3]

Consolidate and apply

- Students then **investigate** antibiotics by following the instructions on Practical sheet 4.12. This needs to be left for two days before the students can analyse the results. Alternatively, they could be given a set of results to analyse. Low demand students could just measure the diameter and not calculate the cross-sectional area. [O3]

Extend

Ask students able to progress further to:

- **explain** the difference between antibiotics and antiviral drugs. [O3]

Plenary suggestions

Give one student a key word and ask them to describe the word or draw it for the class, who have to guess the word.

Answers to Worksheet 4.12

MRSA 'superbugs'

1. *Staphylococcus*
2. any strain of the *Staphylococcus* bacteria that is resistant to antibiotics
3. doubled
4. DNA
5. to ensure the more resistant strains are destroyed and cannot reproduce and pass on their DNA
6. Patients are already weak, and it is easily passed from patient to patient by doctors and nurses.
7. very high doses of antibiotics; new antibiotics
8. any two sensible suggestions, e.g. gloves, antiseptics

Answers to Practical 4.12

1–5 depends on results

6. as a control / comparison to show that the effect is not due to the paper disc but to the antibiotic in the paper disc

Lesson 13: Building immunity

Lesson overview

AQA Specification reference

AQA 4.3.1.7

Learning objectives

- Recall how vaccinations prevent infection.
- Explain how mass vaccination programmes reduce the spread of a disease.
- Evaluate the global use of vaccination.

Learning outcomes

- Recall that vaccinations are used to prevent infection of some diseases. [O1]
- Understand that vaccination involves introducing small quantities of dead or inactive forms of a pathogen into the body to stimulate the white blood cells to produce antibodies. [O2]
- Evaluate the global use of vaccination to eradicate polio. [O3]

Skills development

- WS 1.1 Understand how scientific methods and theories develop over time.
- WS 1.4 Explain everyday and technological applications of science. Evaluate associated personal, social, economic and environmental implications, and make decisions based on the evaluation of evidence and arguments.
- WS 3.2 Translating data from one form to another.
- WS 4.1 Use scientific vocabulary, terminology and definitions.

Maths focus

- 4a Translate information between graphical and numeric form

Resources needed Video about the first vaccinations (search videos 'history of smallpox vaccination'); video about the polio eradication programme (search 'polio video', adding UNICEF to the search should get you to a video about UNICEF's campaign to eradicate polio); timeline for smallpox; Worksheet 4.13.1 (low demand); Worksheet 4.13.2 (standard demand); Worksheet 4.13.3 (high demand)

Digital resources

Key vocabulary immunity, vaccination, vaccine

Teaching and learning

Engage

- Provide a list of ten different diseases and ask students to **identify** those you can be vaccinated against. [O1]
- **Discuss** with the students if they know what vaccinations they have had and why they had them. [O1]

Challenge and develop

- Ask students to **suggest** how vaccinations work; they should recall ideas about immunity and antibodies. [O1]
- **Show** students a short video about the first vaccinations (search videos 'history of smallpox vaccination'). [O1]
- Students **complete** the questions about Edward Jenner and smallpox on Worksheet 4.13 (tiered). [O1]
- High demand students have an extra question on their worksheet (Worksheet 4.13.3).

Explain

- Use the video to **explain** how vaccinations work and why different vaccines are needed for specific pathogens. [O2]
- Students **complete** the questions about vaccinations on Worksheet 4.13.1 (low demand), 4.13.2 (standard demand) or 4.13.3 (high demand). [O2]

Consolidate and apply

- **Introduce** the eradication programme for polio by showing a short video about the programme. [O3]
- Low demand students use the information about polio eradication on Worksheet 4.13.1 to **draw** a timeline for the eradication of polio. Some students may need an example to show them what a timeline is, use a timeline for smallpox. [O3]
- Standard and high demand students **draw** a timeline and **complete** the questions on Worksheet 4.13.2 and 4.13.3, respectively. [O3]

Extend

Ask students able to progress further to:

- **explain** why it is impossible to eradicate influenza. [O3]

Plenary suggestions

Students write the phrase 'immunity from disease' in the centre of a sheet of A4 paper. They then construct a mind map / spider diagram as a summary of what they know about immunity.

Answers to Worksheet 4.13

Edward Jenner and smallpox

1. virus
2. by infecting him with cowpox, which is similar but harmless
3. They were worried that they would develop cowpox.
4. Cowpox must have stimulated the body to make antibodies that could also destroy smallpox.

Vaccines

1. approximately 42
2. vaccine response – fewer antibodies are made, response is slower
3. Ideas include:
   - lymphocytes detect antigens on the dead or inactive pathogen; they produce a specific antibody for these antigens
   - antibodies lock onto the antigens
   - lymphocytes 'remember' the shape of the antigens
   - when there is a real infection due to a live pathogen entering the body, lymphocytes instantly recognise the pathogen because it has the same antigens as the vaccine
   - lymphocytes quickly make many specific antibodies
   - antibodies lock onto the pathogens and kill them before they have a chance to make you feel ill.

Polio

2. Idea that mass vaccination programmes increase the number of people who are immune to a pathogen, making it difficult for the pathogen to pass to people who are not immunised.

Lesson 14: Making new drugs

Lesson overview

AQA Specification reference

AQA 4.3.1.9

Learning objectives

- Recall some traditional drugs and their origins.
- Describe how new drugs are developed.
- Explain why 'double-blind' trials are conducted.

Learning outcomes

- Recall the origins of digitalis, aspirin and penicillin. [O1]
- Understand the stages in the development of drugs. [O2]
- Explain how 'double-blind' trials are conducted. [O3]

Skills development

- WS 1.1 Understand how scientific methods and theories develop over time.
- WS 1.6 Recognise the importance of peer review of results and of communicating results to a range of audiences.
- WS 2.3 Apply a knowledge of a range of techniques, instruments, apparatus, and materials to select those appropriate to the experiment.
- WS 2.4 Carry out experiments appropriately having due regard for the correct manipulation of apparatus, the accuracy of measurements, and health and safety considerations.

Resources needed A video of drugs being manufactured using a tablet press (e.g. type 'Advil Tablet Manufacturing Process' into www.youtube.com and watch the first 50 seconds of the film). Worksheet 4.14.1 (low demand); Worksheet 4.14.2 (standard and high demand); Practical sheet 4.14; Technician's notes 4.14

Digital resources

Key vocabulary dose, efficacy, placebo

Teaching and learning

Engage

- **Show** students a video of drugs being manufactured using a tablet press – play without the sound on. [O1]
- **Discuss** how they think new drugs are developed and what precautions there might be before new drugs can be prescribed. [O1, O2]

Challenge and develop

- **Explain** that some drugs contain chemicals that were extracted from plant material. [O1]
- Low demand students should **complete** the sentences in section 1 of Worksheet 4.14.1. Other students complete Question 1 on page 156 of the Student Book. [O1]
- Students then **extract** the chlorophyll from leaves to show how chemical extraction can be done. They follow the instructions on Practical sheet 4.14. [O1, O2]

Explain

- **Discuss** the importance of testing and trialling new medical drugs. [O2, O3]
- **Explain** how a double-blind trial works. Ask the students to **suggest** why the patients should not know if they are taking the placebo or not. [O3]

- Students **complete** the questions about developing new medical drugs on Worksheet 4.14.1 (low demand) or Worksheet 4.14.2 (standard and high demand). [O2]

Consolidate and apply

- Ask the students to **suggest** reasons why developing new medical drugs takes so long. [O2, O3]
- Students then **complete** the cost of developing new medical drugs flow chart on Worksheet 4.14.1 (low demand) or Worksheet 4.14.2 (standard and high demand). [O2, O3]

Extend

Ask students able to progress further to:

- **discuss** whether or not medical drugs should be tested on animals. [O2, O3]

Plenary suggestions

Ask students to suggest why saving the rainforests could be linked to the development of new medical drugs.

Answers to Worksheet 4.14

Traditional sources of drugs

foxgloves, willow, microorganisms, penicillin

Developing new medical drugs

NB. Low demand students answer questions 1 and 2 only.

1. effective – to find out if it is able to prevent or cure a disease, or make you feel better
   safe –to check it is not too toxic or with any undesirable side effects
   stable – idea of being able to store any new drug for a period of time
2. preclinical testing in laboratories (using cells, tissues and live animals) to find out side effects and efficacy; clinical trials use human volunteers
3. helps to prevent false claims of success
4. Double-blind trials are when patients are allocated randomly to groups, so that doctors and patients do not know, until the trial is complete, if they are taking the new drug or a placebo.
5. prevents patients thinking they are getting better just because they are taking the drug / 'placebo effect'

The cost of developing new medical drugs

1d, 2e, 3a, 4c, 5b

Lesson 15: Investigating monoclonal antibodies

Lesson overview

AQA Specification reference

AQA 4.3.2.1; AQA 4.3.2.2

Learning objectives

- Describe uses of monoclonal antibodies.
- Explain how monoclonal antibodies are produced.
- Evaluate the use of monoclonal antibodies.

Learning outcomes

- Understand how monoclonal antibodies are used in laboratories, research and treatment of disease. [O1]
- Understand how monoclonal antibodies are produced. [O2]
- Evaluate uses of monoclonal antibodies. [O3]

Skills development

- WS 1.3 Appreciate the power and limitations of science and consider any ethical issues which may arise.
- WS 1.4 Explain everyday and technological applications of science. Evaluate associated personal, social, economic and environmental implications, and make decisions based on the evaluation of evidence and arguments.
- WS 1.5 Evaluate risks both in practical science and the wider societal context, including perception of risk in relation to data and consequences.
- WS 3.8 Communicating the scientific rationale for investigations, methods used, findings and reasoned conclusions through paper-based and electronic reports and presentations using verbal, diagrammatic, graphical, numerical and symbolic forms.
- WS 4.1 Use scientific vocabulary, terminology and definitions.

Resources needed A video that illustrates how monoclonal antibodies are made (e.g. 'Monoclonal Antibody by John Munro' at www.youtube.com); poster material; access to the Internet. Worksheet 4.15

Digital resources

Key vocabulary hybridoma, monoclonal antibodies

Note this lesson is for high demand students only.

Teaching and learning

Engage

- Ask students to write down five facts they remember about antibodies and antigens. This can be done in small groups using mini-whiteboards. [O1]
- Each group reports back to the class to build up a picture of what they remember about antibodies and antigens. [O1]

Challenge and develop

- Use the introduction to **discuss** the idea of antibodies being specific. **Introduce** the term 'monoclonal antibody', explaining that they are specific to one binding site on one protein antigen and so are able to target a specific chemical or specific cells in the body. [O1]
- **Show** students a video that illustrates how monoclonal antibodies are made. [O2]
- Students then **complete** the questions on monoclonal antibodies on Worksheet 4.15. [O1, O2]

Explain

- **Discuss** some of the uses of monoclonal antibodies with students; explain that they are not expected to recall any specific tests or treatments but need to understand how they work in a given situation. In each example the basic idea is that they bind to a specific substance. [O3]

Consolidate and apply

- Students **research** the uses of monoclonal antibodies **using** Worksheet 4.15. This can be done as a poster or PowerPoint. If there is no Internet they can use the information in the Student Book. [O1, O2, O3]

Extend

For students able to progress further:

- Monoclonal antibodies are being developed for use with cancer patients and people with diabetes. Using monoclonal antibodies in humans has risks, **suggest** why a cancer patient may be more willing to take the risk than someone with diabetes. [O3]

Plenary suggestions

Ask students to show their presentations to others in the class.

Answers to Worksheet 4.15

Diagram should be similar to that in the Student Book (page 159).

Lesson 16: Looking at plant diseases

Lesson overview

AQA Specification reference

AQA 4.1.2.3; AQA 4.3.3.1

Learning objectives

- Recall the causes of plant diseases.
- Describe the symptoms and identification methods of some plant diseases.
- Explain the use of monoclonal antibodies in identifying plant pathogens.

Learning outcomes

- Recall the causes of plant diseases as bacteria, viruses, fungus and insects. [O1]
- Describe the symptoms of infection by pathogens and damage by ion deficiency. [O2]
- Understand how plant disease can be identified using monoclonal antibodies. [O3]

Skills development

- WS 1.4 Explain everyday and technological applications of science. Evaluate associated personal, social, economic and environmental implications, and make decisions based on the evaluation of evidence and arguments.
- WS 3.8 Communicating the scientific rationale for investigations, methods used, findings and reasoned conclusions through paper-based and electronic reports and presentations using verbal, diagrammatic, graphical, numerical and symbolic forms.
- WS 4.1 Use scientific vocabulary, terminology and definitions.

Resources needed Image of plant with rose black spot; image of a plant with tobacco mosaic virus; video of aphids feeding; poster material; access to Internet. Worksheet 4.16.1 (low demand); Worksheet 4.16.2 (standard demand); Worksheet 4.16.3 (high demand)

Digital resources

Key vocabulary chlorosis

Teaching and learning

Engage

- **Show** students an image of a plant with rose black spot. [O1]
- Ask students to **write down** what is wrong with the plant, what might have caused it and why it is a problem for the plant. This can be done in small groups using mini-whiteboards and should be revision from the lesson on fungal diseases. [O1]
- Each group **reports back** to the class to build up a picture of what they remember about fungal plant diseases. [O1]

Challenge and develop

- Use the introduction to **discuss** causes of disease in plants; explain that as well as bacteria, fungi and viruses, insects will also damage plants. [O1]
- **Show** students a video of aphids feeding. Ask the students to **suggest** what they are feeding on and what damage that will do to the plant, including the spread of diseases. [O2]
- **Show** students an image of a plant with tobacco mosaic virus. Ask them to describe the symptoms and **suggest** why it will affect the rate of photosynthesis. [O1, O2]
- Ask students to **recall** what they remember about mineral deficiency in plants. Remind them of the symptoms of nitrate deficiency and **introduce** the term 'chlorosis', caused by magnesium deficiency. [O1, O2]

Explain

- Students should **complete** the table on Worksheet 4.16.1 (low demand), Worksheet 4.16.2 (standard demand) or Worksheet 4.16.3 (high demand). Low demand students have been given a help sheet, which contains the information needed to complete the table. [O1, O2]
- **Discuss** with high demand students the different ways plant diseases can be detected; they will already be familiar with some of these from the discussions about the different diseases. [O2]
- **Explain** to high demand students how diseases are identified. Students should then use the Internet to **identify** some different diseases listed on Worksheet 4.16.3. The information is sourced from www.planetnatural.com (go to 'Learning centre', then 'Plant diseases') and could be printed if the Internet is not available. [O2, O3]

Consolidate and apply

- Low and standard demand students **research** one of the plant diseases mentioned and produce a poster (see Worksheet 4.16.1 or 4.16.2). If there is no Internet they can use the information in the Student Book. [O1, O2]
- High demand students should **produce** a fact sheet for inexperienced gardeners about plant diseases. The fact sheet should include indicators to look out for and advice on how to get the diseases identified. [O2, O3]

Extend

Ask students able to progress further to:

- **discuss** why using chemical insecticides to kill aphids may not be the best solution. [O3]

Plenary suggestions

Read out the statements on the Technician's notes – some are true and others are false. Ask the students to identify if they are true or false – this can be done with thumbs up or thumbs down.

Once a false statement is identified, can they change it to make it true?

Answers to Worksheet 4.16

Sample answers for table

Plant disease or deficiency	Symptoms	Cause	Treatment or control
tobacco mosaic virus	mottling or discoloured leaves curled leaves stunted growth yellow streaks or spots on leaves	infection by a virus	use insecticide to kill insects that spread the disease destroy any infected plants clean infected tools
aphid infection	decreased growth rate mottled or yellow leaves wilting low yields	aphids feeding on the sap	insecticides or removing the aphids by hand or using natural predators
nitrate deficiency	stunted growth	not enough nitrates in soil	add minerals to soil
magnesium deficiency	chlorosis (yellow leaves)	not enough magnesium in soil	add minerals to soil

Identifying plant diseases

1. brown rot, 2. apple scab, 3. club root, 4. blossom end rot

Lesson 17: Learning about plant defences

Lesson overview

AQA Specification reference

AQA 4.3.3.2

Learning objectives

- Recall some physical plant defence responses.
- Explain how mechanical plant defence systems help them survive.
- Explain how chemical plant defence systems help them survive.

Learning outcomes

- Recall some physical plant defence responses to include cellulose cell walls, waxy cuticles and layers of dead cells. [O1]
- Understand how mechanical plant defence systems help them survive. [O2]
- Understand how chemical plant defence systems help them survive. [O3]

Skills development

- WS 1.4 Explain everyday and technological applications of science. Evaluate associated personal, social, economic and environmental implications, and make decisions based on the evaluation of evidence and arguments.
- WS 3.8 Communicating the scientific rationale for investigations, methods used, findings and reasoned conclusions through paper-based and electronic reports and presentations using verbal, diagrammatic, graphical, numerical and symbolic forms.
- WS 4.1 Use scientific vocabulary, terminology and definitions.

Resources needed Poster material, access to Internet, *Mimosa pudica* plant or a video showing the plant folding its leaves; image of leaf structure; a video about the uses of witch hazel; Worksheet 4.17

Digital resources

Key vocabulary antibacterial chemicals, mimicry, rust

Teaching and learning

Engage

- Ask students to **write down** five facts they remember about the way the body protects itself from disease. This can be done in small groups using mini-whiteboards. [O1]
- Each group **reports back** to the class to build up a picture of what they remember then ask them to suggest if plants might have any similar defence mechanisms. [O1]

Challenge and develop

- Use the introduction to **discuss** physical defences that plants have; link this to the idea of the human body having a skin to protect against infection. [O1]
- Show students an image of leaf structure to remind them where the waxy cuticle is. Ask students to **recall** another reason why the waxy cuticle is important to the plant. [O1]
- Ask students to **suggest** why roses have thorns. Then discuss the other mechanical adaptations, such as drooping leaves and mimicry. If a *Mimosa pudica* plant is available (they can be purchased online) then get the students to touch the leaves to see the response, or show them a video. [O2]
- **Show** students a video about the uses of witch hazel. Ask the students to **suggest** why witch hazel is used in health/beauty products; you may need to explain to some students that it is an antiseptic and ask them to **recall** what that means. [O3]

- Ask the students to **suggest** why the witch hazel plant makes the chemicals that we can extract and use in health/beauty products. [O3]

Explain

- Students should **complete** a mind map that shows all they know about plant defences. They should include pictures on their mind map. Low demand students may need more help so provide them with Worksheet 4.17. [O1, O2, O3]

Consolidate and apply

- Ask students to **explain** how a chemical used as defence in a plant can become part of a new health product by producing an advert for a new 'health or beauty product'. The students should work in groups and use Worksheet 4.17 to help them. [O3]

Extend

Ask students able to progress further to:

- **debate** the cloning and genetic modifications of disease-resistant plants. [O3]

Plenary suggestions

Ask students to try and 'sell' their new 'beauty product' to another group.

Lesson 18: Maths skills: Sampling and scientific data

Lesson overview

Learning objectives

- Understand why sampling is used in science.
- Be able to explain different sampling techniques.

Learning outcomes

- Suggest appropriate sampling techniques. [O1]
- Explain why sampling is used in science. [O2]
- Understand how to minimise bias. [O3]

Skills development

- WS 2.5 Recognise when to apply a knowledge of sampling techniques to ensure any samples collected are representative.
- WS 2.6 Make and record observations and measurements using a range of apparatus and methods.
- WS 3.3 Carrying out and representing mathematical and statistical analysis.
- WS 3.5 Interpreting observations and other data (presented in verbal, diagrammatic, graphical, symbolic or numerical form), including identifying patterns and trends, making inferences and drawing conclusions.

Maths focus

- 1a Recognise and use expressions in decimal form
- 1c Use ratios, fractions and percentages
- 1d Make estimates of the results of simple calculations
- 2b Find arithmetic means
- 2d Understand the principles of sampling as applied to scientific data
- 2f Understand the terms mean, mode and median

Resources needed Worksheet 4.18; Practical sheet 4.18; Technician's notes 4.18

Digital resources

Key vocabulary bias, systematic

Teaching and learning

Engage

- Ask students to **suggest** how many students there are in the school. Could they find out without counting them (or looking at the actual number in documents)?
- **Discuss** how they could use the number of students in the class to estimate the total number of students in the school. How accurate would their estimate be? How might they improve the accuracy again without counting them (or looking at the actual number in documents)?

Challenge and develop

- Use the starter to **discuss** the idea of sampling a population when it is difficult to measure the whole population. They may have completed Chapter 6 and know about different sampling methods used in ecology. [O1, O2]
- **Discuss** why samples have to be chosen randomly; again, you could refer back to ecology and the use of quadrats. Ask them to explain why you cannot just choose where to put a quadrat. [O1, O2]

Explain

- Introduce the practical investigation; students can work in pairs and follow the method on Practical sheet 4.18. [O1, O2, O3]

Consolidate and apply

- All students then answer the questions on Worksheet 4.18. [O1, O2, O3]
- Low demand students may need help with the calculations. This could be done by asking each group to record one set of results and then complete the calculations together with a class set of results. [O1, O2, O3]

Extend

Ask students able to progress further to:

- (high demand) **complete** questions 5 and 6 on page 165 of the Student Book. [O3]

Plenary suggestions

Describe this scenario to the students.

'In a class of 20 students, two girls smoked and no boys. The class belonged to a school with 2000 students.' Ask the students to **estimate** the number of boys (zero) and girls (10% or 200) that smoke in the school, would this be a true estimate? How could you improve the estimate?

Answers to Worksheet 4.18

1. Answer depends on their results.
2. Answer depends on their results.
3. This will depend on the number of beads the students were given.
4. Depends on their results – but it is likely to be one of the larger values.
5. For an antibiotic to be effective it should cure 95–100% so their answer will depend on the proportion of blue beads they were given.
6. Idea of using placebo in a blind trial where neither the patients nor the doctors can tell which are the test or the placebo drugs, so they don't know which patients get the test drug.

When and how to use these pages: Check your progress, Worked example and End of chapter test

Check your progress

Check your progress is a summary of what students should know and be able to do when they have completed the chapter. Check your progress is organised in three columns to show how ideas and skills progress in sophistication. Students aiming for top grades need to have mastered all the skills and ideas articulated in the final column (shaded pink in the Student book).

Check your progress can be used for individual or class revision using any combination of the suggestions below:

- Ask students to construct a mind map linking the points in Check your progress
- Work through Check your progress as a class and note the points that need further discussion
- Ask the students to tick the boxes on the Check your progress worksheet (Teacher Pack CD). Any points they have not been confident to tick they should revisit in the Student Book.
- Ask students to do further research on the different points listed in Check your progress
- Students work in pairs and ask each other what points they think they can do and why they think they can do those, and not others

Worked example

The worked example talks students through a series of exam-style questions. Sample student answers are provided, which are annotated to show how they could be improved.

- Give students the Worked example worksheet (Teacher Pack CD). The annotation boxes on this are blank. Ask students to discuss and write their own improvements before reviewing the annotated Worked example in the Student Book. This can be done as an individual, group or class activity.

End of chapter test

The End of chapter test gives students the opportunity to practice answering the different types of questions that they will encounter in their final exams. You can use the Marking grid provided in this Teacher Pack or on the CD Rom to analyse results. This shows the Assessment Objective for each question, so you can review trends and see individual student and class performance in answering questions for the different Assessment Objectives and to highlight areas for improvement.

- Questions could be used as a test once you have completed the chapter
- Questions could be worked through as part of a revision lesson
- Ask Students to mark each other's work and then talk through the mark scheme provided
- As a class, make a list of questions that most students did not get right. Work through these as a class.

Marking Grid for End of Chapter 4 Test

	Getting Started [Foundation Tier]					Going Further [Foundation and Higher Tiers]					More Challenging [Higher Tier]		Most demanding [Higher Tier]			Total marks	Percentage
Student Name	Q. 1 (A01) 1 mark	Q. 2 (A01) 1 mark	Q. 3 (A02) 2 marks	Q. 4 (A04) 3 marks	Q. 5 (A03) 2 marks	Q. 6 (A01) 2 marks	Q. 7 (A01) 1 mark	Q. 8 (A02) 3 marks	Q. 9 (A03) 2 marks	Q. 10 (A01) 2 marks	Q. 11 (A01) 2 marks	Q. 12 (A02) 2 marks	Q. 13 (A01) 2 marks	Q. 14 (A02) 4 marks	Q. 15 (A03) 4 marks		

Check your progress

You should be able to:

describe the major causes of ill health	→	explain how diseases and lifestyle factors affect health	→	evaluate graphical data about lifestyle and health
recall a number of interacting factors which cause different diseases	→	explain how risk factors are linked to an increased rate of some non-communicable diseases	→	evaluate evidence linking risk factors and increased rates of disease
describe different types of tumours	→	describe lifestyle changes to reduce cancer risk	→	evaluate evidence linking smoking and cancer
describe the symptoms of some viral, bacterial, fungal and protist diseases	→	describe the transmission and control of different diseases	→	compare and contrast bacterial and viral diseases evaluate control measures for malaria explain the specificity of antibodies
describe how the body protects itself from pathogens	→	explain the role of the immune system	→	explain the impact of antibiotic-resistant bacteria
recall why vaccinations are used	→	explain how vaccinations trigger an immune response	→	evaluate the global use of vaccinations
describe the use of antibiotics and painkillers	→	explain how antibiotics and painkillers treat disease	→	explain the limitations of antibiotics
recall some plant diseases and plant defences	→	explain the identification and symptoms of some plant diseases explain some plant defences	→	explain the use of monoclonal antibodies in identifying plant pathogens

Worked example

Viruses are pathogens that cause many diseases. Babies and children are vaccinated to protect them from many viruses.

1 a Why are antibiotics not used to kill viruses in the body?

They cannot kill them.

> The correct answer is that viruses live inside cells, where they are protected from the action of antibiotics.

b What is in a vaccine?

A weak form of the virus.

> Correct answer identified, although 'weak form' could be improved by replacing it with 'inactive or dead form'.

c What does the vaccine do in the child's body?

It makes antibodies.

> This is inaccurate. White blood cells/lymphocytes make antibodies.

d Why do pathogens make us feel ill?

They produce poisons.

> This is partially correct, but the answer should also include the fact that pathogens damage cells.

2 a One type of tumour is known as benign. Name the other type of tumour and explain how it is different from a benign tumour.

Cancer tumour spreads through the body.

> Incorrect tumour type identified. Correct answer is malignant tumour, but the difference is correct.

b Name a risk factor for lung cancer.

Smoking

> Correct answer identified

3 a Describe how malaria is spread by mosquitoes.

Mosquitoes suck the blood of an infected person.

> Answer should also include how the mosquito then passes the protest pathogen/*Plasmodium* to an uninfected person.

b Explain how malaria is controlled by insecticide.

Insecticide kills eggs.

> Answer should mention mosquito eggs, and also include that mosquitoes lay their eggs in water.

Coordination and control: Introduction

When and how to use these pages

This chapter builds on ideas relating to organ systems, introducing students to the function and detailed structure of organs within the nervous and endocrine systems. Students will also learn how these two systems work together to bring about control and homeostasis within the body. The role of hormones in plants will also be explored.

Overview of the chapter

In this chapter, students will consider how the brain works and outline its basic structure. Students will learn about the structure of different neurones, explaining how they work within different reflex actions. Students will explain the structure of different parts of the eye in detail, and how they are adapted to enable sight of near and far objects. Students will learn about the endocrine system, how this is controlled and the functions of the main hormones involved. The role of hormones in maintaining the balance of water, glucose and temperature in the body will be examined in detail. Students will explain the process of negative feedback in different contexts, explaining how this helps to maintain balance. They will compare and contrast hormonal and nervous responses. In developing ideas about the role of hormones in the reproductive system, students will learn to evaluate the use of IVF to promote fertility and the use of contraceptives to control fertility. They will question the social, spiritual and moral aspects of associated technologies. Students will learn how plants are controlled by hormones and their responses to different tropisms. Finally, students will explore the way we use plant hormones to improve food production and control the growth of plants.

This chapter offers a number of opportunities for the students to research and select appropriate information. They will analyse data in a number of different contexts, including social contexts relating to the problems of diabetes and kidney failure. Students will draw and interpret graphical information to draw conclusions, using these to evaluate various applications of scientific technology, including the use of contraceptives and kidney donation. Students will make models of the neurones and kidney, using these to explain the processes occurring within them. Students will make many links in this topic, identifying how the endocrine and nervous system work together to bring about control.

Obstacles to learning

Students may need extra guidance with the following terms and concepts:

- Students may think that all neurones are the same, and they have the same structure and function.
- Students may confuse neurotransmitters with hormones.
- Students will think that the brain is involved in all reactions in the body, and not realise that some reflexes occur without the involvement of the brain.
- Students often confuse long- and short-sightedness and the adjustments needed to correct these defects.
- The reasons for Type 1 and Type 2 diabetes are not always clearly understood.
- In the menstrual cycle, students often interchange ovulation and fertilisation.
- Students can confuse the way auxins work in the shoot and the root.

Practicals in this chapter

In this chapter students will do the following practical work.

- Investigate different reflex actions
- Plan and carry out an investigation to determine the effect of caffeine on reaction time
- Carry out a survey to investigate long- and short-sightedness
- Use a model to investigate how sweating cools down the body
- Investigate the effect of gravity on plant seedlings
- Investigate the effect of auxins on plant growth
- Required Practical: Investigate the effect of light intensity on the growth of seedling

	Lesson title	Overarching objectives
1	Homeostasis	Explain the importance of homeostasis in regulating internal conditions in the body. Recall that these control systems involve nervous or chemical responses. Describe how control systems involve receptors, coordination centres and effectors.
2	The nervous system	Describe the structure and function of the nervous system. Explain how the nervous system is adapted to its functions. Describe the structure of sensory, motor and relay neurones.
3	Reflex actions	Explain the importance of reflex actions. Describe the path of a reflex arc. Explain how the structures in the reflex arc relate to their function.
4	The brain	Identify the three main regions of the brain and describe their functions. Describe how the regions of the brain are mapped. Explain why it is difficult to treat brain disorders.
5	Required Practical: Investigating reaction times	Select appropriate apparatus and techniques for the measurement of biological processes. Carry out physiological experiments safely. Use appropriate techniques in problem-solving contexts.
6	The eye	Relate the structures of the eye to their functions. Explain how the eye is adapted to seeing in colour and in dim light.
7	Seeing in focus	Relate the structures of the eye to their functions. Understand how the eye is able to focus on near or distant objects.
8	Eye defects	Describe why short-sightedness (myopia) occurs. Describe why long-sightedness (hyperopia) occurs. Demonstrate how techniques are used to correct eye defects.
9	Controlling body temperature	Describe the mechanisms by which body temperature is controlled when too hot or cold. Explain how body temperature can be controlled in a specific context.
10	The endocrine system	Recall that the endocrine system is made up of glands that secrete hormones into the blood. Know the location of the major endocrine glands. Understand why the pituitary gland is the 'master gland'.
11	Controlling blood glucose	Recall that blood glucose is monitored and controlled by the pancreas. Understand how insulin controls blood glucose levels. Understand how insulin works with another hormone – glucagon – to control blood sugar levels.
12	Diabetes	Understand the causes of Type 1 and Type 2 diabetes. Compare Type 1 and Type 2 diabetes. Evaluate information on the relationship between obesity and diabetes, and make appropriate recommendations.
13	Diabetes recommendations	Understand the causes of Type 1 and Type 2 diabetes. Compare Type 1 and Type 2 diabetes.

		Evaluate information on the relationship between obesity and diabetes, and make appropriate recommendations.
14	Water balance	Recall the ways in which the body loses water. Explain why cells do not function efficiently if they lose or gain too much water. Explain how excess protein is converted to urea for excretion.
15	The kidneys	Recall that excess water, ions and urea are removed from the body by the kidneys in urine. Describe how the kidneys produce urine. Explain how the hormone ADH regulates the amount of water in the urine and, therefore, in the body.
16	Negative feedback (Higher Tier)	Explain the role of thyroxine in the body. Understand the principles of negative feedback, as applied to thyroxine.
17	Kidney failure	Recall that people who suffer from kidney failure can be treated by dialysis or kidney transplant. Understand the principles of dialysis. Evaluate the advantages and disadvantages of treating organ failure using a mechanical device or transplant.
18	Dialysis or transplant?	Describe different forms of dialysis. Use data to draw conclusions. Use data to evaluate the advantages and disadvantages of dialysis and kidney donation.
19	Human reproduction	Describe the roles of hormones in sexual reproduction. Explain how hormones interact in the menstrual cycle.
20	IVF (Higher Tier)	Describe the technique of *in-vitro* fertilisation. Explain the use of hormones in technologies to treat infertility.
21	IVF evaluation (Higher Tier)	Evaluate data regarding *in-vitro* fertilisation and use this to draw conclusions. Evaluate the scientific, emotional, social and ethical issues of *in-vitro* fertilisation.
22	Systems working together (Higher Tier)	Describe the effects of adrenaline. Understand that automatic control systems may involve nervous responses and chemical responses. Understand that combinations of hormones work to produce a response.
23	Contraception	Define the purpose of contraception Describe hormonal methods and hormonal methods of contraception Explain how these methods are effective at preventing pregnancy
24	Which contraceptive?	Describe the advantages and disadvantages of different contraceptive methods. Use date to evaluate the effectiveness of different contraceptive methods.
25	Auxins	Recall that plants produce hormones to coordinate and control growth, and responses to light and gravity. Describe how unequal distributions of auxins cause unequal growth rates in plant shoots and roots.

26	Applications of auxins	Explain how auxins coordinate and control responses to light and gravity. Explain that auxins act on 'stem cells' in plants called meristems. Describe some applications of auxins.
27	Required Practical: The effect of light and gravity on the growth of germinating seeds	Select appropriate apparatus and techniques for the observation and measurement of biological processes. Carry out experiments safely to measure response to the environment. Use appropriate techniques to make observations of biological specimens and produce labelled scientific drawings.
28	Other plant hormones (Higher Tier)	Recall that gibberellins are important in seed germination, and ethene in cell division and ripening of fruit Explain the application of the plant hormones gibberellins and ethene.
29	Maths skills: The spread of scientific data	Be able to use range bars on graphs. Understand how box and whisker plots can be used to show the spread of data. Understand how to use percentiles.

Lesson 1: Homeostasis

Lesson overview

AQA specification reference

AQA 4.5.1

Learning objectives

- Explain the importance of homeostasis in regulating internal conditions in the body.
- Recall that these control systems involve nervous or chemical responses.
- Describe how control systems involve receptors, coordination centers and effectors.

Learning outcomes

- Identify factors which need to be controlled in the body. [O1]
- Explain why specific factors need to be controlled in the body. [O2]
- Explain how factors may be automatically controlled by the nervous or endocrine system. [O3]

Skills development

- WS 1.2 Use a variety of models such as representational, spatial, descriptive, computational and mathematical to solve problems, make predictions and to develop scientific explanations and understanding of familiar and unfamiliar facts.
- WS 3.5 Interpret observations and other data (presented in verbal, diagrammatic, graphical, symbolic or numerical form), including identifying patterns and trends, making inferences and drawing conclusions.
- WS 4.1 Use scientific vocabulary, terminology and definitions.

Resources needed Worksheet 5.1.1 (low and standard demand); Worksheet 5.1.2 (high demand)

Digital resources PowerPoint slides

Key vocabulary endocrine system, homeostasis, hormone, nervous system, target organ

Teaching and learning

Engage

- Show students the images on the PowerPoint of people in different situations. Ask them to **explain** why the people might have become ill. [O1]
- Read out loud the passage from the Student Book about Eve the marathon runner on page 172, or provide students with the story on section 1 of the worksheet. Ask students to **sketch a graph** of how the temperature, water levels and sugar levels change as Eve runs her race, as they listen to the story. [O1]
- **Discuss** students' responses and why it is important for the levels of water, temperature and glucose to remain constant. Revert back to the PowerPoint slide and ask students to **relate each image** to one of these factors. Ask students to **complete** task 1 on Worksheet 5.1. [O1, O2]

Challenge and develop

- Introduce the idea of homeostasis and the role of receptors, coordination centres and effectors in bringing about control. Use the PowerPoint slides and the Student Book. [O1]
- Low and standard demand students should **identify** the three components of the automatic control system in the body and **explain** their roles (section 2 on Worksheet 5.1.1). High demand students should **relate** each of the three parts of the control system to the scenario given in section 2 of Worksheet 5.1.2. [O3]
- Introduce the two different control systems of the body using the Student Book and PowerPoint slides. Students should complete section 3 of their worksheet, where they **connect** each statement to a particular system. Show the last PowerPoint slide to check the answers. [O3]

Explain

- Students work in small groups. They should return to the story of Eve, the marathon runner. Students should **discuss** their sketches and **produce** a poster-sized sketch of their agreed ideas for water, temperature and glucose control. Students should **annotate** each part of the sketch, **explaining** what is happening in Eve's body. Ask students to **share** their posters and **discuss** their ideas. [O1, O2, O3]

Consolidate and apply

- Students should **analyse** the situations provided in section 4 of the worksheet. They should **write** some advice for the people in each situation, to help them maintain homeostasis. [O1, O2, O3]

Extend

Ask students able to progress further to:

- **find out** about how different organisms, such as organisms that live in extreme environments, control their internal organs, e.g. deep sea creatures. They could **produce** a PowerPoint of their ideas. [O1, O2]

Plenary suggestions

Students could **draw annotated** diagrams in a concept map of homeostasis to explain what they have learned in the lesson about what needs to be controlled and the systems in our body that are involved. [O1, O2, O3]

Answers to Worksheet 5.1

1 What needs to be controlled

1. Temperature – starts at 37 °C. This increases but reaches a stable level as her sweating increases, then decreases as she drinks water, back to 37 °C. Water level – high to start with but decreases as sweating increases. Increases to normal levels as she drinks water. Glucose level – high to start with, but decreases throughout the race as she is respiring. Levels increase as she drinks from the glucose drink. But towards the end of the race glucose levels fall significantly.
2. 1B, 2C, 3A

2 Automatic control

(Low and standard demand):

1. **Receptors** – detect changes to the internal conditions in our body
2. **Coordination centre** (brain, spinal cord, pancreas) – receives messages about our internal condition and take action
3. **Effectors** – muscles or glands that bring about changes to maintain optimal levels

(High demand):

Receptor cells: cells in skin and in the hypothalamus; Coordination centre: hypothalamus in the brain; Effector: sweat glands, muscles on arms and muscles on hair cells

3 Coordination centres

	Nervous system	**Endocrine system**
Response	Rapid and precise	Slower but acts for longer
Nature of message	Nerve impulse – electrical	A **hormone** – chemical
Action	Carried in nerves to specific location, e.g. muscle	Carried in blood to all organs, but affects the **target organ** only

4 Keeping control

Image	**Advice**	**Receptor cells**	**Coordination centre**	**Effector cells**
A	Do not eat so much sugar. Remove the sugar from the spoon.	Blood sugar detector cells in the blood	Brain and endocrine system	Hormone insulin released by the pancreas
B	Take shelter and drink plenty of water. Stay out of the sun.	Temperature detector cells in the skin and hypothalamus; water detector cells in the blood	Brain and endocrine system	Sweat glands, muscles on hair cells on skin, glands to induce thirst, kidney to control water

Lesson 2: The nervous system

Lesson overview

AQA specification reference

AQA 4.5.2.1

Learning objectives

- Describe the structure and function of the nervous system.
- Explain how the nervous system is adapted to its functions.
- Describe the structure of sensory, motor and relay neurones.

Learning outcomes

- Describe the organisation of the nervous system and its function. [O1]
- Describe the stimulus–response pathway. [O2]
- Use models to explain the role of sensory, motor and relay neurones. [O3]

Skills development

- WS 1.2 Use a variety of models such as representational, spatial, descriptive, computational and mathematical to solve problems, make predictions and to develop scientific explanations and understanding of familiar and unfamiliar facts.
- WS 4.1 Use scientific vocabulary, terminology and definitions.
- WS 4.4 Use prefixes and powers of ten for orders of magnitude (e.g. tera, giga, mega, kilo, centi, milli, micro and nano).

Maths focus

- 1b Recognise and use expressions in standard form
- 2h Make order of magnitude calculations

Resources needed Head injury symptoms from www.nhs.uk (search conditions for 'severe head injury'); video of what happens at the synapse (from Bitesize at www.bbc.co.uk); materials for making models of neurones, e.g. cardboard, rubber tubing, cling film, jelly, string, sticky tape, modelling clay. Worksheet 5.2.1 (low demand); Worksheet 5.2.2 (standard and high demand)

Digital resources PowerPoint slides

Key vocabulary central nervous system, myelin sheath, neurone, peripheral nervous system, receptor

Teaching and learning

Engage

- Show students the list of symptoms produced after a severe head injury from www.nhs.uk. **Discuss** the function of the brain by considering what happens when something goes wrong with it. [O1]
- Ask students to work in pairs and **identify** five things they already know about the brain and spinal cord. Ask students to **share** their ideas with another pair. Take responses from groups and from a class list, which can be referred to at the end of the lesson. [O1]

Challenge and develop

- Provide students with the list from section 1 of Worksheet 5.2.1 (low demand) or 5.2.2 (standard and high demand) and ask them to **pick out** all the functions of the nervous system. Check their answers. [O1]
- Show students the PowerPoint slide of the peripheral and central nervous systems. **Discuss** the attributes of each and the reasons for having two parts to the nervous system. Ask students to **compare** and **contrast** each system using section 2 of their worksheet. [O1]
- Introduce the idea that the CNS responds to information received by sensory cells. Use the PowerPoint slide to **discuss** the connections with how information is received, processed and acted upon. [O2]

- Show students the PowerPoint slide of the structure of a sensory neurone, relay neurone and motor neurone, and a video of what happens at the synapse. **Discuss** the structure and function of each type of neurone.
- Low demand students **use** section 3 of Worksheet 5.2.1 to **make** a scaled model of each neurone. [O3]
- Standard and high demand students should **work out** a suitable scale from the information given in section 3 of Worksheet 5.2.2 and use this to **make** their own models. [O3]

Explain

- Ask low demand students to **use** all the cut-out images and words in section 4 of Worksheet 5.2.1 **to create an annotated** poster **explaining** how information passes to and from the CNS. [O1, O2, O3]
- Standard and high demand students should **use** their models of the neurones they have produced and **create** their own role play to **explain** how messages are passed to and from the CNS. Students should **explain** how the nervous system is adapted to carry out its functions. [O1, O2, O3]

Consolidate and apply

- Ask students to **complete** the table in section 5 of their worksheet. They should **identify** which part of the nervous system is involved and whether or not the condition is treatable. [O1, O2, O3]

Extend

Ask students able to progress further to:

- **find out** about the nervous system of one other organism, e.g. squids which have very long, highly adapted neurones, and **compare** and **contrast** their system to ours. [O1, O2, O3]

Plenary suggestions

Return to the class list. Ask students to work in pairs and **write down** additional information from the lesson which can be added to the list. Take responses from selected pairs and add to the list. [O1, O2, O3]

Students could compare and contrast the nervous system with a computer. [O1, O2, O3]

Answers to Worksheet 5.2

1 Functions of the nervous system

1, 3, 4, 6, 7 and 10

2 Peripheral and central nervous systems

	Peripheral nervous system	Central nervous system
Which organ(s) of the body are involved in the systems?	Sensory organs, neurones, muscles and glands	Brain and spinal cord
What are their main functions?	To receive information and send it to the central nervous system. To take messages from the central nervous system and send to the appropriate effector (muscle or gland).	Process information from the peripheral nervous system, and send the appropriate responses to the body.
What happens if they are damaged?	Limited ability to repair damage	Unable to repair damage

5 Problems that can be encountered

Situation	Peripheral nervous system / central nervous system or both affected?	Long term or short term damage?	Can the problems be treated? Explain.
Taking too much alcohol and too many drugs over a lifetime	Central	Long term	No – synapses destroyed and neural pathways affected.
Drinking coffee or cola every day	Central	Short term	Yes – no long term damage to synapses.
Injury to spinal cord causing paralysis	Central	Long term	No – damaged nerve cells cannot be repaired or replaced.
Touching a hot object	Peripheral	Short term	Yes – as soon as stimulus is removed, the nerve cells recover.

Lesson 3: Reflex actions

Lesson overview

AQA specification reference

AQA 4.5.2.1

Learning objectives

- Explain the importance of reflex actions.
- Describe the path of a reflex arc.
- Explain how the structures in the reflex arc relate to their function.

Learning outcomes

- Describe some reflex actions and recognise their importance. [O1]
- Explain the path of the reflex arc. [O2]
- Explain how the structures in the reflex arc are adapted to carry out their function. [O3]

Skills development

- WS 1.2 Use a variety of models such as representational, spatial, descriptive, computational and mathematical to solve problems, make predictions and to develop scientific explanations and understanding of familiar and unfamiliar facts.
- WS 2.4 Carry out experiments appropriately having due regard for the correct manipulation of apparatus, the accuracy of measurements, and health and safety considerations.
- WS 4.1 Use scientific vocabulary, terminology and definitions.

Resources needed Simulation from www.sumanasinc.com (click 'Animation library', 'General biology', 'Reflex arcs'). Worksheet 5.3.1 (low demand); Worksheet 5.3.2 (standard demand); Worksheet 5.3.3 (high demand); Practical sheet 5.3; Technician's notes 5.3

Digital resources PowerPoint slides

Key vocabulary reflex action, reflex arc, relay neurone, synapse

Teaching and learning

Engage

- Ask students to work in pairs. Show students the images on the PowerPoint slide and ask them to **decide** what the images have in common. Take responses and **discuss** their ideas. [O1]
- Demonstrate the knee jerk reaction, by asking a student to cross their legs and tapping the knee just below the patella. Ask students to **try this out** on themselves to experience the reflex action. [O1]
- **Discuss** the importance of reflex actions and why we need them, comparing them to voluntary actions. Use the PowerPoint and Student Book to support this. Ask students to **use the table** from section 1 of their worksheet and **sort** the given examples into reflex or voluntary actions. Students then **add** two more examples of each. [O1]

Challenge and develop

- Introduce the reflex arc by showing the simulation of a person pricking their finger on a rose bush from www.sumanasinc.com. [O1]
- **Discuss** the reflex arc using the PowerPoint slide and the Student Book. [O2]
- Remind students of the synapse using the PowerPoint slide and/or Student Book. **Discuss why** this may be an important part of the reflex arc, and how it affects the speed of transmission of the messages. [O3]
- Ask students to **investigate** the activities outlined on Practical sheet 5.3 and **carry out** different reflex actions. They should **record** their observations on the practical sheet and answer the questions. [O1, O2]
- Students then complete section 2 of the worksheet, which is differentiated: Worksheet 5.3.1, low demand; Worksheet 5.3.2, standard demand; Worksheet 5.3.3, high demand. [O2]

Explain

- Ask students **to consider** all the parts of the reflex arc and **explain** how each is adapted to carry out its job using the table from section 3 of the worksheet. [O3]

Consolidate and apply

- Ask students to work in small groups to **produce** a short presentation of why the knee jerk reaction is important and **explain** the reflex arc. They should **explain how** each part is adapted to carry out its function in their presentation. [O1, O2, O3]

Extend

Ask students able to progress further to:

- **find out** more about the synapse and how different drugs affect its function. They could **classify** the substances they research into those that help to speed up the message and those that slow it down. [O3]

Plenary suggestions

Students could **correct** the mistakes in the reflex arc in section 4 of the worksheet. [O1, O2]

Students could **analyse and explain** what would happen if the particular parts of the reflex arc given in section 5 of the worksheet did not work effectively. [O1, O2, O3]

Answers to Worksheet 5.3

1 Reflex or voluntary

Reflex actions: blinking, coughing, sneezing, baby gripping, removing hand from hot object, knee jerk, falling; Voluntary actions: laughing, running, talking, driving, eating

2 The reflex arc

Low demand: 3, 6, 4, 1, 7, 2, 5; Standard demand: see diagram on page 176 of the Student Book.

High demand: students should draw a diagram similar to the one below to describe the pupil reflex.

The pathway of impulses during the pupil reflex

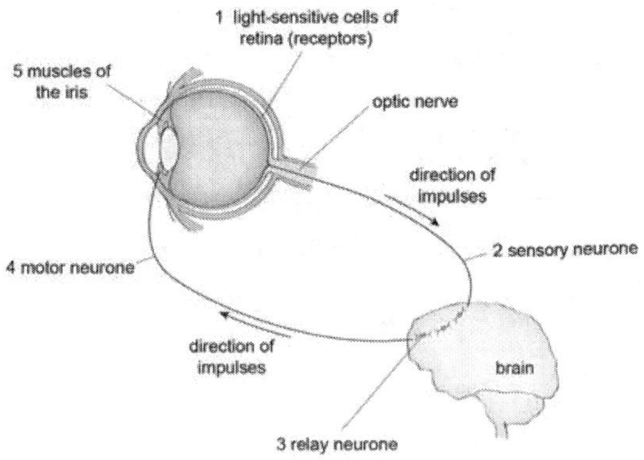

3 Adaptations to perform their function

Part of the reflex	Adaptation with example where appropriate
Receptor cell	e.g. Retina cells – highly specialised to receive and respond to different wavelengths of light. Impulses produced from light stimulus is passed onto sensory neurone.
Sensory neurone	Lots of dendrites to gather information from different and many receptor cells.
Synapse	Contains neurotransmitters, which are released when an electrical signal is passed.
Relay neurone	Short axon so impulses can be carried quickly.
Central nervous system	Millions of neural networks present for quick responses. Different parts of the brain carry out specific functions so messages relating to these are directed appropriately.
Motor neurone	Long axon to transmit messages to effectors a long way from CNS.
Effector	Muscle has lots of long fibres, which can expand and contract; contains many mitochondria. Glands – have specialised cells that make and secrete hormones or enzymes.

4 Correct the mistakes

See diagram on page 176 of the Student Book.

5 What happens when the reflex arc goes wrong?

Part	Impact
Receptor cell	Stimulus would not be detected and part of body could be damaged.
Sensory neurone	Message that the stimulus might be dangerous would not be passed to the brain and the part of body could be damaged.
Motor neurone	Brain would register stimulus but be unable to quickly respond. This delay could lead to the part of the body being damaged.

Lesson 4: The brain

Lesson overview

AQA specification reference

AQA 4.5.2.2

Learning objectives

- Recall that the brain controls complex behaviour using billions of interconnected neurones.
- Identify the three main regions of the brain and describe their functions.
- Describe how the regions of the brain are mapped.

Learning outcomes

- Identify the cerebral cortex, cerebellum and medulla and describe their functions. [O1]
- Explain how mapping and scanning techniques can link different parts of the brain to particular functions. [O2]
- Evaluate the benefits and risks of carrying out procedures on the brain and nervous system. [O3]

Skills development

- WS 1.4 Explain everyday and technological applications of science. Evaluate associated personal, social, economic and environmental implications, and make decisions based on the evaluation of evidence and arguments.
- WS 1.5 Evaluate risks both in practical science and the wider societal context, including perception of risk in relation to data and consequences.
- WS 4.1 Use scientific vocabulary, terminology and definitions.

Resources needed Video about brain wiring from www.bbc.co.uk (search BBC News for 'scans reveal intricate brain wiring'); video from www.med.nyu.edu (search 'the courage to cut'); Internet access. Worksheet 5.4.1 (low and standard demand); Worksheet 5.4.2 (high demand)

Digital resources

Key vocabulary cerebellum, cerebral cortex, magnetic resonance imaging, medulla, non-invasive

Teaching and learning

Engage

- Ask students work in pairs to **analyse** the different activities in section 1 of the worksheet and **decide** which do not directly involve the brain. **Discuss** the complexity of the activities co-ordinated by the brain. [O1]

Challenge and develop

- Show students the video about brain wiring from www.bbc.co.uk. **Discuss** how many billions of interconnected neurones are involved in sending messages to different parts of the brain. Recap and **discuss** ideas about how neurones interconnect and the role of the synapse. [O1]
- Ask students to use the Internet to **research** key facts about the brain and **find out** about the structure and functions of the cerebral cortex, cerebellum and medulla. They should **use** the top trumps card templates provided in section 2 of the worksheet. [O1]
- Low and standard demand students then **label** the diagram of the brain provided on Worksheet 5.4.1. They should **identify** which pieces of information about the brain are true and which are false. [O1]
- Low and standard demand students then **read** the section 'The regions of the brain' on page 178 of the Student Book. Students should **answer** questions 1 and 2 from the Student Book. [O2]
- High demand students should **read** the information about brain mapping and scanning techniques in the Student Book and Worksheet 5.4.2 then **summarise** each method in bullet points. [O2]
- High demand students then **list** some disorders they know of that can go wrong in the brain. They should **identify** methods used to treat these problems. Introduce them to the idea that historically electric shocks were used regularly and are still used today. **Discuss** the damage that could be done with invasive techniques. [O3]

- Students should **listen** to the thoughts of different neurosurgeons about the job they undertake in treating brain cancer (from www.med.nyu.edu). Ask students to work in pairs and **present** some ideas about why the brain is so difficult to treat once it goes wrong. **Discuss** these ideas with the class. [O3]

Explain

- Students should work in small groups and **create** an information leaflet for neurological patients. They should **explain** why the brain is not like any other organ of the body. They should **outline** what the dangers are in interfering with the brain. They should **explain** why non-invasive techniques are used to identify problems. [O1, O2, O3]

Consolidate and apply

- Low and standard demand students should **identify** which of the activities would be affected if the cerebrum, cerebellum or medulla were to be damaged (section 3 of Worksheet 5.4.1).
- High demand students should **compare** and **contrast** different scanning and imaging techniques by completing section 3 of Worksheet 5.4.2. [O1, O2, O3]

Extend

Ask students able to progress further to:

- **find out** about the history of discovery about different parts of the brain and **produce a timeline** of discovery to present to the class. [O1, O2, O3]

Plenary suggestions

Students could **debate** the statement 'The brain is the most important organ in the body'. They should **identify** the evidence for and against this statement. [O1, O2, O3]

Answers to Worksheet 5.4

1 Functions of the brain

5 and 7

2 Fact finding

2. Three structures correctly labelled
3. True satements about the brain: a, c, d, f, g, h, j

3 Damage to brain regions

Activity	Damage to cerebral cortex	Damage to cerebellum	Damage to medulla
1	no	yes	yes
2	yes	no	no
3	yes	yes	no
4	no	yes	yes
5	yes	no	no

3 Imaging and scanning techniques

MRI

- very safe
- very detailed image produced
- cancer cells, clots and abnormal blood flow can be detected
- can be used to map brain activity

CT scans

- some risk from ionising radiation and injection of dye
- detailed image produced
- cancer cells, clots and abnormal blood flow can be detected
- can be used to map brain activity

EEGs

- very safe
- used to detect abnormal changes to brain activity
- can't be used to detect cancer but used for epilepsy and mental problems
- can be used to map brain activity

Lesson 5: Required practical: Investigating reaction time

Lesson overview

AQA specification reference

AQA 4.5.2.1

Learning objectives

- Select appropriate apparatus and techniques for the measurement of biological processes.
- Carry out physiological experiments safely.
- Use appropriate techniques in problem-solving contexts.

Learning outcomes

- Revise the structure of a neurone and plan an investigation to examine reaction time safely. [O1]
- Record and analyse data selecting suitable methods to display data. [O2]
- Use data and interpret data to draw conclusions. [O3]

Skills development

- WS 2.2 Plan experiments or devise procedures to make observations, produce or characterise a substance, test hypotheses, check data or explore phenomena.
- WS 2.4 Carry out experiments appropriately having due regard for the correct manipulation of apparatus, the accuracy of measurements, and health and safety considerations.
- WS 3.5 Interpret observations and other data (presented in verbal, diagrammatic, graphical, symbolic or numerical form), including identifying patterns and trends, making inferences and drawing conclusions.

Maths focus

- 3d Solve simple algebraic equations
- 4a Translate information between numerical and graphical forms
- 4c Plot two variables from experimental or other data

Resources needed Video 'How synapses work' from BBC Bitesize; information about caffeine from www.sciencemuseum.org.uk (search 'how does caffeine affect you'). Worksheet 5.5.1 (low demand); Worksheet 5.5.2 (standard demand); Worksheet 5.5.3 (high demand); Practical sheet 5.5; Technician's notes 5.5

Digital resources

Key vocabulary dominant hand, reaction time, valid

Teaching and learning

Engage

- Place students in pairs. Remind students of the structure of a neurone. Ask them to **relate** five things they already know about how the neurone works. [O1]
- Remind students of the synapse using a video (e.g. 'How synapses work' from www.bbc.co.uk). [O1]
- **Discuss** the possible factors affecting the transmission of messages from one neurone to another and **make a class list** on the board. [O1]

Challenge and develop

- Ask students to **measure** their reaction time using a simulation, such as the one from BBC Science (search www.bbc.co.uk for 'choop dach'). [O1]
- Ask students to work in pairs and **devise** a list of factors that might affect reaction time. Ask students to **make connections** between these factors and those that affect the speed of transmission of neurones on the board list discussed earlier. [O1]

- Students then **research** the effect of caffeine on reaction time (they could use **Error! Hyperlink reference not valid.**www.sciencemuseum.org.uk).
- Place students in small groups. Provide them with a ruler and a timer and some caffeinated drinks, e.g. cola. Students should **read** the Student Book page 180 and **answer** the questions on page 181.
- Ask students to use what they have found out to **develop** a hypothesis for the effect of caffeine on reaction time. They should **plan** an investigation using Practical sheet 5.5 to **test** this hypothesis. They should **decide** how many times the investigation should be carried out with each person to produce reliable data. [O1]
- Students should **decide** in their groups how to present the data recorded and **draw** suitable graphs of their data. [O2]

Explain

- Students should **explain** their results and **draw conclusions** from their data. They should **explain** whether the quality of their data was reliable, reproducible and accurate and **suggest** further improvements that could be made to improve the quality of their results. [O2, O3]

Consolidate and apply

- Ask students to **analyse** the data given in the worksheet.
- Low demand students should **explain** the relationship between reaction time and amount of substance taken (Worksheet 5.5.1). [O2]
- Standard demand students should **explain** the conclusions for the results and whether the quality of the data is good and if not how it could be improved (Worksheet 5.5.2). [O2, O3]
- High demand students should **use** the equation given to **calculate** the actual reaction time for their own data (Worksheet 5.5.3). [O2, O3]

Extend

Ask students able to progress further to:

- **find out** more about how different chemicals such as alcohol affect the synapse and reaction time. They should use the information to **analyse** the alcohol limit for driving, and **explain** whether it is steep enough or the value should be lower. [O1]

Plenary suggestions

Students should work in pairs and **suggest** how the investigation might be adapted to be carried out as a double-blind trial. [O1]

Ask students to **compare** and **contrast** the two methods of measuring reaction time: the simulation vs catching the ruler. They should suggest the advantages and disadvantages of each. [O1]

Answers to Worksheet 5.5

Low demand:

1. Average reaction times
2.

	Subject 1, day 1	Subject 1, day 2	Subject 2, day1	Subject 2, day 2
Average reaction times	12.94	17.7	14.06	19.02

3. The substance added increased the reaction times for both subjects.

Standard demand:

1. Without the substance, the reactions times are faster than with the substance.
2. More subjects should be tested (a large sample size of people, e.g.1000, of a similar age, equal numbers of the same gender and the same ethnic backgrounds); a range in concentration of the substance should be tested. The tests should be carried out with sufficient time inbetween to ensure there is no residual test chemical left in the system of any of the subjects.

Lesson 6: The eye

Lesson overview

AQA specification reference

AQA 4.5.2.3

Learning objectives

- Relate the structures of the eye to their functions.
- Explain how the eye is adapted to seeing in colour and in dim light.

Learning outcomes

- Identify different parts of the eye. [O1]
- Explain the roles of each of the main parts of the eye. [O2]
- Explain how the eye is adapted to see in colour and dim light. [O3]

Skills development

- WS 1.2 Use a variety of models such as representational, spatial, descriptive, computational and mathematical to solve problems, make predictions and to develop scientific explanations and understanding of familiar and unfamiliar facts.
- WS 3.5 Interpret observations and other data (presented in verbal, diagrammatic, graphical, symbolic or numerical form), including identifying patterns and trends, making inferences and drawing conclusions.
- WS 4.1 Use scientific vocabulary, terminology and definitions.

Resources needed Internet access; websites with optical illusions for students to test; video on how the eye works; strong elastic bands, sticky tape, card, scissors, string. Worksheet 5.6.1 (low demand); Worksheet 5.6.2 (standard demand); Worksheet 5.6.3 (high demand); Practical sheet 5.6

Digital resources PowerPoint slides

Key vocabulary cones, iris, retina, rods

Teaching and learning

Engage

- Ask students to work in pairs and **describe** how we see objects. **Discuss** their responses, and challenge their ideas where necessary. [O1]
- **Discuss** the fact that our eyesight is very unique and that sometimes what we see can also be deceiving. [O1]
- Show students an optical illusion (available online), and ask them to **carry out** the instructions. **Discuss** how the brain is involved in interpreting what our eyes see, and that this information is not always correct. [O1]

Challenge and develop

- Ask students to **complete** activity 1 on Practical sheet 5.6; they **use** images to try and find their blind spot.
- Show students the PowerPoint image of the eye and its structure. Use the PowerPoint to **discuss** the various parts of the eye and the job they do. [O1, O2]
- Show students a video on how the eye works and how it actually sees, to introduce rods and cones. [O3]
- Ask students to **read** the section about rods and cones in the Student Book and **answer** the questions. They should then **answer** the true/false questions in section 1 of Worksheet 5.6. [O1, O2, O3]
- Remind students of the investigation where they watched the eye reflex. **Discuss** the roles of the radial and circular muscles in controlling the amount of light entering the eye using the PowerPoint. [O3]

Explain

- Ask students to complete section 2 of Worksheet 5.6.1 (low demand) or Worksheet 5.6.2 (standard demand), where they **label** parts of the eye.
- High demand students should **make** a model to **explain** in detail how the eye controls the amount of light entering it using the radial and circular muscles. They can use large elastic bands, string and sticky tape to construct the model (Worksheet 5.6.3). [O1, O2, O3]

Consolidate and apply

- Ask students to **write an explanation** of how we see objects, in section 3 of the worksheet. Low demand students **put** the sentences in order; standard and high demand students use the keyword list to **construct** their own written explanation. [O1, O2, O3]

Extend

Ask students able to progress further to:

- **find out** more about colour blindness and why people might go blind in general. They should discover which parts of the eye are not working and why, then **write** a short information leaflet on the defects of the eye. [O3]

Plenary suggestions

Ask students to complete the table in section 4 of the worksheet and explain what would happen if we did not have each part of the eye given. [O1, O2, O3]

Ask students to return back to the task they did on the blind spot. Ask them to **explain why** we have a blind spot using ideas about the structure of the eye. [O1, O2, O3]

Answers to Worksheet 5.6

1 True or false

F, T ,T ,F, F, T, F, T, F, T

2 The structure of the eye

Low demand:1c, 2i, 3f, 4b, 5h, 6d, 7a, 8e, 9g

Standard demand:

Retina: contains rod and cone cells which are connected to neurones. The rods and cones receive information about the light and pass them onto the neurones.

Fovea: highest concentration of cone cells, as most of the light is focused here.

Sclera: thick, strong outer coating to protect and hold the eye in place.

Cornea: transparent covering over the front of the eye which is curved so that most of the light can be refracted into the pupil.

Optic nerve: all the information from the neurones is collected here. Lots of dendrites to receive information and send to the brain.

Iris: contains radial and circular muscles which are antagonistic. They relax and contract to control the shape of the pupil.

Ciliary muscles and suspensory ligaments: the suspensory ligaments can shorten or lengthen, as the ciliary muscles expands and contracts to alter the shape of the lens.

Lens: made of a material that is transparent and highly refractive so light can be refracted onto the retina.

High demand: the circular and radial muscles are antagonistic. As the circular muscles contract, the radial muscles relax and the pupil gets smaller. As the radial muscles contract, the circular muscles relax and the pupil gets bigger.

3 Explaining how we see

Low demand: 5, 2, 8, 1, 6, 3, 9, 4, 7

4 Eye parts not working

Lens: light would not be focused on the retina causing vision to be blurred, as in short- and long-sightedness.

Optic nerve: no impulses would be sent to the brain, so no image would be interpreted. Blindness would result.

Iris: too much light could enter the eye, causing damage to the receptor cells on the retina. Alternatively, insufficient light would reach the retina, meaning everything would look very dark.

Cone cells: colour vision would be impaired.

Lesson 7: Seeing in focus

Lesson overview

AQA specification reference

AQA 4.5.2.3

Learning objectives

- Relate the structures of the eye to their functions.
- Understand how the eye is able to focus on near or distant objects.

Learning outcomes

- Identify the parts of the eye related to refraction. [O1]
- Explain how these parts are adapted to their function. [O2]
- Explain how the shape of the lens affects the extent of refraction. [O3]

Skills development

- WS 1.2 Use a variety of models such as representational, spatial, descriptive, computational and mathematical to solve problems, make predictions and to develop scientific explanations and understanding of familiar and unfamiliar facts.
- WS 3.5 Interpret observations and other data (presented in verbal, diagrammatic, graphical, symbolic or numerical form), including identifying patterns and trends, making inferences and drawing conclusions.
- WS 3.8 Communicate the scientific rationale for investigations, methods used, findings and reasoned conclusions through written and electronic reports and presentations using verbal, diagrammatic, graphical, numerical and symbolic forms.

Resources needed Simulation 'Lens Accommodation in the Human Eye' from http://demonstrations.wolfram.com; string, elastic bands, balloon, ray box, fat convex lens, thin convex lens and screen. Worksheet 5.7.1 (low demand); Worksheet 5.7.2 (standard demand); Worksheet 5.7.3 (high demand)

Digital resources PowerPoint slides

Key vocabulary accommodation, ciliary muscles, refraction, suspensory ligaments

Teaching and learning

Engage

- Ask students to **carry out** the quick quiz from the PowerPoint to recap the structure of the eye and the functions of its parts. [O1]

Challenge and develop

- Ask students to **look** at an object far in the distance and one nearby. **Discuss** what might be happening within the eye to see these different objects. [O2]
- Use a ray box, a lens and a screen to demonstrate what happens to the image when the object is close to the lens and when it is far way. **Discuss** the way light rays are bent by the lens. Repeat the demonstration with a fatter lens and then with a thinner lens. Students should **discuss** in pairs what they see and **describe** the differences between the lenses. Students should **discuss** what would need to happen in order to see the object at its correct size at near and far distances. [O1]
- Introduce the role of the aqueous and vitreous humour in providing shape to the eye and helping with refraction, using the PowerPoint. Ask students to **recall** all the parts of the eye that help to focus an image onto the retina. [O1]
- **Discuss** how the shape of the lens affects where the image is most likely to fall. Use the PowerPoint and the information in the Student Book to **explain** how and why the lens changes shape. [O3]

- Show students the simulation 'Lens Accommodation in the Human Eye' from http://demonstrations.wolfram.com and **discuss** how the shape of the lens alters with the distance from the object. [O3]

Explain

- Low demand students should **label** the parts of the eye that are involved in refraction and **explain** to what extent they refract the light (Worksheet 5.7.1). [O1, O2]
- Standard demand students should **draw** the light rays to show how the light is bent through the eye. They should **annotate** their diagram to **explain** the role of each part of the eye involved (Worksheet 5.7.2). [O1, O2, O3]
- High demand students should **draw** and **annotate** their own diagrams to **explain** how the lens changes when it sees near and far objects. [O1, O2, O3]

Consolidate and apply

- Students should **answer** the questions from the Student Book (pages 184-185). [O1, O2, O3]
- Ask students to **make a model** of the ciliary body, suspensory ligaments and lens, using string, elastic bands and a balloon. Students should **use** this model to produce a demonstration of what happens when a near and far object is seen. [O1, O2, O3]

Extend

Ask students able to progress further to:

- **find out** how compound eyes work in arthropods and **compare** and **contrast** these with human eyes.

Plenary suggestions

Ask students to **imagine** they are a light ray. They should **write** a short account to **explain** what happens to them as they enter the eye and how each part of the eye affects them. [O1, O2, O3]

Answers to Worksheet 5.7

Low demand:

Students should label the cornea, aqueous humour, lens and vitreous humour. They should explain that 70% of the light is first refracted by the cornea, then by a small amount through the aqueous humour, all the light falling on the lens is refracted onto the fovea, and there is some further refraction by the vitreous humour.

Standard and high demand:

See Figure 5.19 from the Student Book.

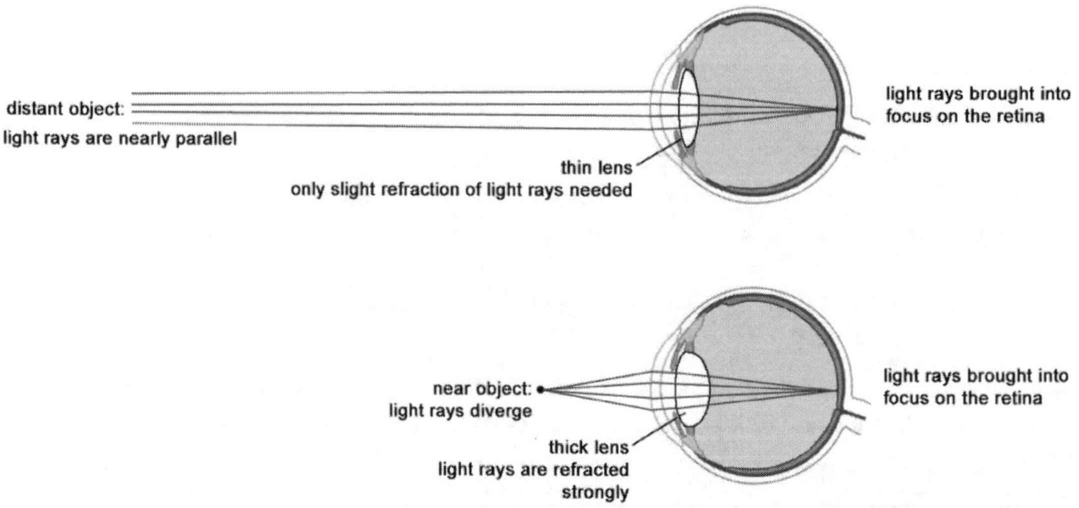

Lesson 8: Eye defects

Lesson overview

AQA specification reference

AQA 4.5.2.3

Learning objectives

- Describe and understand why short-sightedness (myopia) occurs.
- Describe and understand why long-sightedness (hyperopia) occurs.
- Demonstrate how techniques are used to correct eye defects.

Learning outcomes

- Identify the consequences of being short- or long-sighted. [O1]
- Explain the causes of short- and long-sightedness. [O2]
- Explain how corrective lenses and new technologies can correct these eye defects. [O3]

Skills development

- WS 3.1 Present observations and other data using appropriate methods.
- WS 3.5 Interpret observations and other data (presented in verbal, diagrammatic, graphical, symbolic or numerical form), including identifying patterns and trends, making inferences and drawing conclusions.
- WS 4.1 Use scientific vocabulary, terminology and definitions.

Maths focus

- 2c Construct and interpret frequency tables and diagrams, bar charts and histograms

Resources needed Vision simulations (search 'eye clinic' pages at www.gcu.ac.uk). Worksheet 5.8.1 (low demand); Worksheet 5.8.2 (standard demand); Worksheet 5.8.3 (high demand); Practical sheet 5.8

Digital resources

Key vocabulary gel, hyperopia, laser surgery, myopia

Teaching and learning

Engage

- Ask students to **devise and carry out** a class survey of the number of people who are short- and long-sighted on Practical sheet 5.8. Ask students to include members of their families. Students should **record** the data, **draw a graph** of the tally-chart and **make conclusions** from this data. **Discuss** their findings. [O1]
- Show students the statistics about short- and long-sightedness on the practical sheet. Ask students to **compare** the national figures with their own and account for the differences. [O1]

Challenge and develop

- Ask students to **look out** to the horizon as far as they can see, and **compare** this with a partner. Now they should hold the same object up to their eyes and **compare** how close they can move the objects, while still seeing it in focus. **Discuss** the findings of the class. [O1]
- Show students and **discuss** with them simulations of normal, myopic and hyperopic vision from www.gcu.ac.uk.
- Ask students to **read** the Student Book (pages 186 and 187) on the causes of myopia and hyperopia and how they can be corrected using lenses. **Discuss** the ideas further and ask students to **answer** the questions from the Student Book. [O2]
- Students should **research** hard and soft contact lenses, laser eye treatment and lens replacement using the Student Book and the Internet and complete section 1 of their worksheet. [O3]

Explain

- Ask students to **compare** and **contrast** normal vision, myopia and hyperopia, and to **explain** the differences between all three in section 2 of the worksheet. [O1]

Consolidate and apply

- Ask students to **write** an information leaflet to **explain** what is meant by long- and short-sightedness and how each can be corrected. They should **outline** the new technologies that are available.
- Low demand students should **write** an explanation of how to correct sight defects (Worksheet 5.8.1).
- Standard demand students should **use annotated diagrams** of the eye in their explanations of myopia and hyperopia, and how they are caused (Worksheet 5.8.2).
- High demand students should **use annotated diagrams** and ray diagrams to **explain how** myopia and hyperopia are caused and corrected (Worksheet 5.8.3). [O1, O2, O3]

Extend

Ask students able to progress further to:

- use ray diagrams to **compare** and **contrast** the vision of a short-sighted young person with someone over 40 years old. They should **explain** what corrective measures could be used with each. [O1, O2, O3]

Plenary suggestions

Students could match the linking sentences in section 4 of the worksheet. [O1, O2, O3]

Students could **explain** which type of eye correction they would prefer if they were short- or long-sighted and **justify** their choice. [O1, O2, O3]

Answers to Worksheet 5.8

1 Researching eye treatments

1.

	Hard contact lenses	**Soft contact lenses**
Materials they are made from	Used to be glass, and plastics, PMMA plastic, now a range of materials incorporating silicone.	Silicone hydrogels
Benefits over alternative contact lens	More durable and last 6–12 months. Better at retaining shape so sharper vision	Thinner, more comfortable, oxygen can pass easily to the cornea, easily disposable

2. In laser eye treatment, the shape of the cornea is altered to improve refraction, using a laser.
3. People who can afford it, and only people with mild disorders.
4. No need for spectacles or contact lenses.
5. There are complications and side effects, e.g. thinning of the eye wall, dry eyes, glare of halo effects with night driving.
6. The whole lens is replaced with an artificial one.
7. Generally older people.

2 Eye defects

	Normal vision	**Myopia**	**Hyperopia**
Where is the image formed?	On the back of the retina, mostly at the fovea	In front of the retina	Behind the retina
Why does this happen?	Light is correctly refracted by the cornea, lens and humours in the eye.	The lens is too strong, or the eyeball is too long.	The lens is too weak or the eyeball is too short.
What can be done to correct this, if necessary?		Use a concave lens	Use a convex lens

4 Matching the sentences

1c, 2f, 3e, 4g, 5a, 6b, 7d

Lesson 9: Controlling body temperature

Lesson overview

AQA specification reference

AQA 4.5.2.4

Learning objectives

- Describe the mechanisms by which body temperature is controlled when too hot or cold.
- Explain how body temperature can be controlled in a specific context.

Learning outcomes

- Recognise the parts of the body used to detect and control temperature. [O1]
- Describe the different methods used to reduce or increase body temperature. [O2]
- Explain how these methods work to produce the desired effect. [O3]

Skills development

- WS 1.2 Use a variety of models such as representational, spatial, descriptive, computational and mathematical to solve problems, make predictions and to develop scientific explanations and understanding of familiar and unfamiliar contexts.
- WS 3.1 Present observations and other data using appropriate methods.
- WS 3.6 Present reasoned explanations including relating data to hypotheses.

Maths focus

- 4a Translate information between graphical and numeric form
- 4c Plot two variables from experimental or other data

Resources needed Video clip of vasodilation from www.davidbarlowarchive.com (search for video number 5009). Worksheet 5.9; Practical sheet 5.9; Technician's notes 5.9

Digital resources PowerPoint slides

Key vocabulary cooling, evaporation, negative feedback, thermoregulatory centre, vasoconstriction, vasodilation

Teaching and learning

Engage

- Ask students to work in pairs and **look at** the cartoon of the person in the cold and in the heat from the PowerPoint. Ask them to **write down** all the things they can think of that the body does to control temperature in each situation. [O1, O2]
- **Discuss** the reasons for keeping the body temperature at 37.5 °C. Ask students to **consider** and **explain** what happens when the blood temperature increases above 41 °C or decreases below 35 °C. [O1, O2]

Challenge and develop

- Show students the thermal image of the man on the PowerPoint. **Discuss** how the thermal image was created and introduce the idea that radiation is given out by all bodies with energy. **Discuss** where the hottest parts of the body are and why. [O1]
- Use the Student Book and/or PowerPoint to **discuss** how changes in temperature are detected and what responses are made. [O1, O2, O3]
- Show students the clip of vasodilation from www.davidbarlowarchive.com. Ask students to **identify** what is happening to the blood vessels near the surface of the skin, **how** this action helps the body to cool down, and how the thermal image of the body might change when the blood vessels are dilated. [O1, O2, O3]

- Ask students to set up and **carry out** the investigation on Practical sheet 5.9 to model the effect of sweating on the body. Students should **develop** their own hypothesis about what will happen.
- High demand students should **use ideas** about evaporation and energy to **explain** what is happening and how sweating provides a cooling effect. [O1]
- Show students the image on the PowerPoint of a person who is very cold. **Discuss** what has happened to the blood flow to the surface of the skin and why this helps to control body temperature.
- Use the PowerPoint and Student Book to **discuss** how shivering is used to increase temperature.
- High demand students should use ideas about respiration to **explain** how this action produces heat. [O1, O2, O3]

Explain

- Ask students to **annotate** the image in section 1 of Worksheet 5.9, to **highlight** all the parts of the body that are involved in temperature regulation. They should **explain** how the body detects and changes temperature.
- High demand students should **explain in detail** how each of the actions alters the temperature. [O1, O2, O3]

Consolidate and apply

- Ask students to **predict** the shape of the graph of a triathlon runner (Worksheet 5.9). Ask students to **explain** the shape of the graph and **identify** where temperature control measures might take place. They should **include** a stimulus response pathway for shivering. [O1, O2, O3]

Extend

Ask students able to progress further to:

- **research** one mammal that lives in an extremely cold environment and one in an extremely hot environment. They should **find out** what further methods these organisms have developed to regulate and maintain a constant body temperature. Students should **explain** how these methods work. [O1, O2, O3]

Plenary suggestions

Ask students to work in small groups to **carry out** a short role play to **explain** how the body detects changes in temperature and what responses occur. They should **explain** how these responses work to bring about desired changes to the temperature. [O1, O2, O3]

Answers to Worksheet 5.9

1 How we detect and control temperature

Body too hot – annotations to the skin: thermal receptor cells send a message to the thermoregulatory centre. Response is vasodilation of the blood vessels in the skin, and sweat production.
High demand: vasodilation increases heat loss by thermal radiation; sweating produces water which removes heat from the body by evaporation. The heat leaves the body with the water vapour.
Annotations to main organs: reduced blood flow to main organs – this is why we feel tired. Less activity to reduce amount of heat produced by respiration.
Body too cold – annotations to the skin: thermal receptor cells send a message to the thermoregulatory centre. Response is vasoconstriction of the blood vessels in the skin, and complete reduction in sweat production.
Annotations to main organs: increased blood flow to main organs to ensure temperature decrease does not affect their function.
Annotation to muscles: message from thermoregulatory centre to increase muscle movement in the form of shivering.
High demand: vasoconstriction reduces heat loss by radiation; no sweat produced, which reduces heat loss by surface water; shivering increases respiration of muscles so releasing heat.

2 Participating in a triathlon

1. Initial drop in temperature as person enters the water – vasoconstriction and shivering may occur. However, temperature soon increases with movement and respiration in muscles.
After about 20 minutes, temperature may increase due to high level of exercise, vasodilation and sweating may occur.
At start of bike ride, reduction in temperature as water from swim evaporates and reduces body temperature.
Vasoconstriction and shivering may result. After a time temperature increases to about 30 minutes. Then body may be too hot – so vasodilation and sweating may increase. Towards the end, sweating may decrease as the person gets thirsty. So body temperature may increase.
At the start of the run, the person may need to drink more.
When running, the body may get too hot again at the start so increased sweating and vasodilation occur.

4. Stimulus (temperature decrease) – receptor cells (thermal receptors in skin) – sensory neurone – thermoregulatory centre in brain (hypothalamus) – relay neurone – motor neurone – effector (muscle cells) – response – shivering.

Lesson 10: The endocrine system

Lesson overview

AQA specification reference

AQA 4.5.3.1

Learning objectives

- Recall that the endocrine system is made up of glands that secrete hormones into the blood.
- Know the location of the major endocrine glands.
- Understand why the pituitary gland is the 'master gland'.

Learning outcomes

- Describe the endocrine system. [O1]
- Explain the function of the major endocrine glands and their hormones. [O2]
- Explain how the endocrine system works to bring about changes to the body. [O3]

Skills development

- WS 3.5 Interpret observations and other data (presented in verbal, diagrammatic, graphical, symbolic or numerical form), including identifying patterns and trends, making inferences and drawing conclusions.
- WS 3.8 Communicate the scientific rationale for investigations, methods used, findings and reasoned conclusions through written and electronic reports and presentations using verbal, diagrammatic, graphical, numerical and symbolic forms.
- WS 4.1 Use scientific vocabulary, terminology and definitions.

Resources needed Video of endocrine glands from Medline Plus at www.nlm.nih.gov; poster paper; Internet access. Worksheet 5.10.1 (low and standard demand); Worksheet 5.10.2 (high demand)

Digital resources

Key vocabulary endocrine gland, endocrine system, hormone

Teaching and learning

Engage

- Show students the image of Robert Wadlow on page 190 of the Student Book. **Discuss** why he grew so tall and died so young. Introduce the idea that hormones are chemicals produced by different organs that have a profound effect on our body. Ask students to **list** any hormones they have heard of and what their roles might be. [O1]

Challenge and develop

- Show students the video about endocrine glands from www.nlm.nih.gov, which introduces the endocrine system and its control. **Discuss** how the pituitary gland acts as the 'master gland'. Ask students to **think** of an analogy to compare the role of the pituitary with the other endocrine glands. [O1, O2]
- Ask students to work in small groups and use the Internet to **research** the main endocrine glands, their hormones and the effects on the body. Students should **complete** section 1 of Worksheet 5.10. [O1, O2]
- Ask students to **use** the information they have collected and **produce** a PowerPoint presentation on the endocrine system. They should **explain** the role of the main endocrine organs, pituitary, thyroid, adrenal gland, pancreas, ovary and testes. [O1, O2, O3]

Explain

- Ask students to **annotate** the diagram of the endocrine system in section 2 of Worksheet 5.10.1 and **explain** the roles of all the main hormones, and how this is brought about.
- High demand students (Worksheet 5.10.2) should **compare** the length of time each hormone is likely to act over, and **rank** each hormone according to how much it affects the body. [O1, O2, O3]

Consolidate and apply

- Ask students to **complete** section 3 of their worksheet, where they **match** the scenarios given to the hormones that are most likely to be involved. [O1, O2]
- Students should work in small groups. They should use one of the scenarios and **produce a poster** to **explain** how the signals are picked up, and how the pituitary gland controls the release of the perspective hormone. They should **explain** what impact the hormone has on the body. [O1, O2, O3]

Extend

Ask students able to progress further to:

- **find out** more about the role of human growth hormone and how it is used today. They could discover why people with large amounts of growth hormone do not live very long. [O1, O2]

Plenary suggestions

Play 'What am I?' Think of a part of the endocrine system, without telling the students. Ask them to **think** of 'yes or no' questions. Students must **guess** the part with the fewest number of questions.

Students **write** their own 'true or false' statements about the endocrine system, and test the class. [O1, O2, O3]

Answers to Worksheet 5.10

1 Roles of the endocrine glands and their hormones

Name of gland	Name of hormone(s) produced
pituitary	growth hormone, prolactin, TSH, ACTH, ADH, oxytocin
thyroid	thyroxine
pancreas	insulin, glucagon
adrenal	adrenaline, corticosteroids,
ovary	oestrogen, progesterone
testes	testosterone

Name of hormone	Effects on the body
growth hormone	stimulates growth of cells and tissues
ADH	regulates water balance in the body
thyroxine	regulates the rate of metabolism
insulin	reduces the levels of glucose in the blood
glucagon	increases the levels of glucose in the blood
adrenaline	increases heart rate, breathing rate, affects the release of insulin and glucagon, prepares the body for emergency action
oestrogen and progesterone	control the menstrual cycle,pregnancy, puberty and secondary female characteristics
testosterone	controls puberty and secondary male characteristics

2 Where are the endocrine glands?

Diagram annotated as per Figure 5.28 in the Student Book.

High demand:

Hormones likely to have a very long-term effect – oestrogen, progesterone, testosterone, growth hormone; hormones likely to have a more immediate effect (days or weeks) – insulin, glucagon, thyroxine, ADH; hormones likely to have a very quick response – adrenaline; most profound effects on the body – oestrogen, progesterone, testosterone, growth hormone; weaker effects – adrenaline, insulin, glucoagon, thyroxine

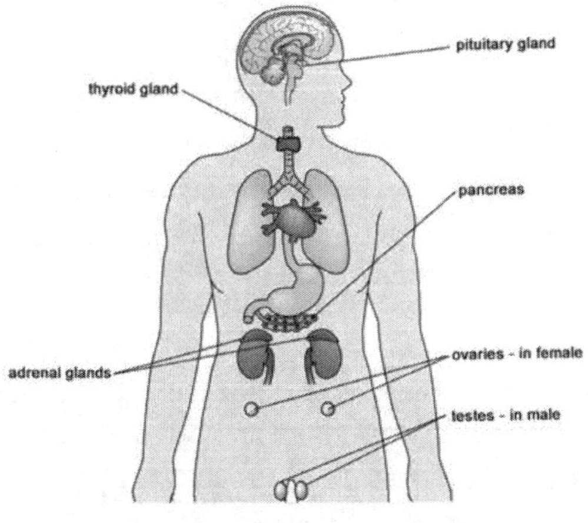

3 When do hormones act?

1. insulin, glucagon, thyroxine, pituitary
2. thyroxine, insulin, glucagon, pituitary
3. oestrogen, progesterone, testosterone
4. adrenaline, thyroxine, pituitary, insulin, glucagon
5. pituitary, thryoxine, insulin, and glucoagon
6. adrenaline, pituitary, thyroxine, insulin, glucagon

Lesson 11: Controlling blood glucose

Lesson overview

AQA specification reference

AQA 4.5.3.2; AQA 4.5.3.7

Learning objectives

- Recall that blood glucose is monitored and controlled by the pancreas.
- Understand how insulin controls blood glucose levels.
- Understand how insulin works with another hormone – glucagon – to control blood sugar levels.

Learning outcomes

- Identify the organs involved in blood glucose control. [O1]
- Explain how insulin reduces blood glucose levels. [O2]
- Explain how glucagon increases blood glucose levels. [O3]

Skills development

- WS 1.2 Use a variety of models such as representational, spatial, descriptive, computational and mathematical to solve problems, make predictions and to develop scientific explanations and understanding of familiar and unfamiliar facts.
- WS 3.5 Interpret observations and other data (presented in verbal, diagrammatic, graphical, symbolic or numerical form), including identifying patterns and trends, making inferences and drawing conclusions.
- WS 4.1 Use scientific vocabulary, terminology and definitions.

Resources needed Simulation 'What happens when you have diabetes' (www.diabetes.org.uk); coloured card, scissors, pens, rulers. Worksheet 5.11.1 (low demand); Worksheet 5.11.2 (standard demand); Worksheet 5.11.3 (high demand)

Digital resources

Key vocabulary glucagon, insulin

Teaching and learning

Engage

- Show students the graph of what happens to blood glucose levels after a meal. Ask students to work in pairs and **describe** the pattern in the data. They should try and **explain** what they think happens to bring glucose levels in balance. [O2]
- **Discuss** the role of glucose in our cells and what the impact might be if glucose stayed in our blood. [O1]

Challenge and develop

- Remind students of the endocrine organs and ask them to **recall** the organ and hormones responsible for blood sugar control. [O1]
- Ask students to **watch** the first 2 minutes of the simulation 'What happens when you have diabetes' (www.diabetes.org.uk), which shows how insulin regulates the amount of glucose in the blood by sending it into our cells. **Discuss** the reasons for needing to keep blood glucose levels between certain limits. [O2]
- Show and **discuss** the further effects of insulin control using an online simulation. [O1, O2, O3]
- Show and **discuss** with high demand students the effects of both insulin and glucagon control. [O1, O2, O3]
- Ask students to complete section 1 of Worksheet 5.11.1 (low demand), 5.11.2 (standard demand) or 5.11.3 (high demand). Students **produce a flow chart** of how the body detects high and low glucose levels and how the pituitary gland and pancreas work to bring the levels down or up. High demand students should include the effects of both insulin and glucagon in their flow charts. [O1, O2, O3]

- Ask students to **read** pages 192–193 in the Student Book and to **answer** the questions. Low and standard demand students should answer questions 1–4 only. High demand students only should read the section on glucagon and also answer questions 5 and 6. [O1, O2, O3]

Explain

- Low demand students should **complete** the task in section 2 of Worksheet 5.11.1 to explain how insulin controls blood glucose levels. [O1, O2]
- Standard demand students should **use** the images in section 2 of Worksheet 5.11.2 and **write** an explanation of how the body reduces blood glucose levels. [O1, O2]
- High demand students should **use** the images in section 2 of Worksheet 5.11.3 and **write** a full explanation of how the body regulates blood glucose levels, including information about glucagon in their answer. [O1, O2, O3]

Consolidate and apply

- Ask students to work in small groups. They should **make large models** of the pancreas, liver, insulin molecules, glucose molecules, glucagon molecules (high demand only), glycogen molecules and the bloodstream, using coloured cards. Students should **use** their models to **produce a simulation** of what happens in the body after a large meal (all students) and when running a marathon (high demand only). [O1, O2, O3]

Extend

Ask students able to progress further to:

- **find out** why high glucose levels are bad for the body and what impact they have over a long period of time. They could **write a short piece** for a scientific journal, including some of the evidence cited. [O1]

Plenary suggestions

Ask students to look at the graph in the section 3 of the worksheet. They should **predict and draw** how the levels of insulin and glucagon (high demand only) might change according to the changes in glucose concentration. [O2, O3]

Answers to Worksheet 5.11

1 Detecting and controlling blood glucose levels

Students should produce a flow chart similar to the one on page 193 of the Student Book. Only high demand need include information about glucagon.

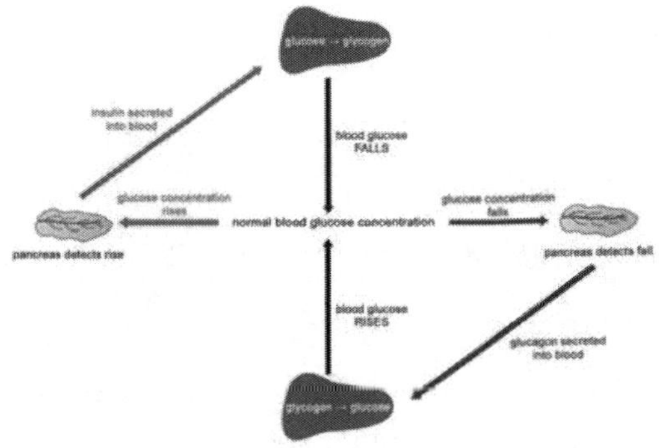

2 Explaining blood glucose control

Low demand: high, pancreas, cells, key, reduces, respiration, glycogen, liver, muscle

3 Predicting changes to hormone levels

Insulin levels will follow the shape of the glucose levels; glucagon levels will be high when the glucose levels are low, and low when the glucose levels are high.

Lesson 12: Diabetes

Lesson overview

AQA specification reference

AQA 4.5.3.2; AQA 4.5.3.7

Learning objectives

- Understand the causes of Type 1 and Type 2 diabetes.
- Compare Type 1 and Type 2 diabetes.
- Evaluate information on the relationship between obesity and diabetes, and make appropriate recommendations.

Learning outcomes

- Describe the causes of Types 1 and Type 2 diabetes. [O1]
- Explain the differences between Type 1 and Type 2 diabetes. [O2]
- Use data to explain the connections between diabetes and obesity and what can be done to improve health. [O3]

Skills development

- WS 1.2 Use a variety of models such as representational, spatial, descriptive, computational and mathematical to solve problems, make predictions and to develop scientific explanations and understanding of familiar and unfamiliar facts.
- WS 1.4 Explain everyday and technological applications of science. Evaluate associated personal, social, economic and environmental implications, and make decisions based on the evaluation of evidence and arguments.
- WS 3.5 Interpret observations and other data (presented in verbal, diagrammatic, graphical, symbolic or numerical form), including identifying patterns and trends, making inferences and drawing conclusions.

Resources needed Videos 'What happens when you have diabetes' (from 2 min 25 sec to 5 min 15 sec) and 'Type 2 diabetes' (from 5 min 15 sec to the end) from www.diabetes.org.uk; props for role play – coloured card and scissors. Worksheet 5.12.1 (low demand); Worksheet 5.12.2 (standard and high demand)

Digital resources PowerPoint slides

Key vocabulary body mass index, glucose tolerance test, Type 1 diabetes, Type 2 diabetes

Teaching and learning

Engage

- Show students some images on the PowerPoint of people with Type 1 diabetes. Ask students to **discuss** the images in pairs and **identify** one thing they all have in common. [O1]
- Show students a graph of the blood sugar concentrations of people with Type 1 diabetes. Ask them to **find a connection** between this and insulin production. [O1]

Challenge and develop

- Ask students to **watch** the video from www.diabetes.org.uk on the causes and effects of Type 1 diabetes.
- Low demand students should **identify** the causes and symptoms of Type 1 diabetes in section 1 of Worksheet 5.12.1. [O1]
- Standard and high demand students should **complete** the table in section 1 of Worksheet 5.12.2 to **explain** why the symptoms occur. [O1]
- Show students the images of people with Type 2 diabetes. Ask them to **find** two things these people have in common. [O1]
- Show students the video from www.diabetes.org.uk that describes Type 2 diabetes.

- Ask students to **summarise** the differences between Type 1 and Type 2 diabetes on their worksheet. [O2]
- Show students the graph on linking diabetes with obesity in section 3 of the worksheet and ask them to **use** the data to **justify** the correlation. [O2]

Explain

- Ask students to use role play to **model** and **explain** what happens to people with Type 1 diabetes and Type 2 diabetes after a meal. They should **show** what happens to the glucose and insulin in both cases and **compare** the two types of diabetes. [O1, O2, O3]

Consolidate and apply

- Ask students to **devise** five questions about what it is like to live with Type 1 diabetes. Students could **watch** a video of a young teenager with Type 1 diabetes to see if their questions are answered. [O1]
- Ask students to **debate** whether it is more dangerous to have Type 1 or Type 2 diabetes. Ask students to **consider** any preventative measures and what the risk to life is in each case. [O1, O2, O3]

Extend

Ask students able to progress further to:

- **find out** more about how the drugs to treat Type 2 diabetes work. They could **compare** their effectiveness with injecting insulin, and **explain why** insulin injections are not used for treating Type 2 diabetes. [O1]

Plenary suggestions

Provide students with a list of key words from the PowerPoint. Ask them to **select** those that belong to Type 1 diabetes and those that belong to Type 2. Students should **use** the key word list to **make up** five sentences about each, using as many words as possible. [O1, O2, O3]

Answers to Worksheet 5.12

1 Type 1 diabetes

Low demand:

1. genetics and unknown
2. younger people
3. quite rare
4. thirst, frequent urination, tiredness, blurred vision, weight loss
5. Yes, glucose in the blood can lead to many problems, including amputation of limbs, retinal damage and sometimes death.

Symptom	Reason
Tiredness	Glucose cannot enter cells as there is no insulin to unlock the cells.
Frequent urination	Gluose is passed out of the body by the kidney.
Blurred vision	Glucose builds up in the eye and affects vision.
Loss of weight	As cells cannot respire they try to burn off fat.
Thirst	Lots of water is also passed out of the body while trying to get rid of glucose.

2 Comparing Type 1 and Type 2 diabetes

Type 1 diabetes	Type 2 diabetes
Genetic and most common in childhood	Linked to obesity and middle age – influenced by diet and lifestyle.
Caused by complete lack of insulin	Caused by cells becoming resistant to insulin and it being no longer able to function.
Treated using injections of insulin	Treated by controlling lifestyle – improved diet and more exercise, or tablets.
Similar symptoms in both types of diabetes	

3 Diabetes and obesity

In both men and women, as the body mass index increases, the percentage of diabetes increases, with steep increases at the most obese end. Women seem to be more prone to Type 2 diabetes than men.

Lesson 13: Diabetes recommendations

Lesson overview

AQA specification reference

AQA 4.5.3.2; AQA 4.5.3.7

Learning objectives

- Understand the causes of Type 1 and Type 2 diabetes.
- Compare Type 1 and Type 2 diabetes.
- Evaluate information on the relationship between obesity and diabetes, and make appropriate recommendations.

Learning outcomes

- Describe the comparative treatments of Type 1 and Type 2 diabetes. [O1]
- Identify people most at risk to Type 2 diabetes using data. [O2]
- Evaluate information and debate how society can help to reduce Type 2 diabetes. [O3]

Skills development

- WS 1.4 Explain everyday and technological applications of science. Evaluate associated personal, social, economic and environmental implications, and make decisions based on the evaluation of evidence and argument.
- WS 1.5 Evaluate risks both in practical science and the wider societal context, including perception of risk in relation to data and consequences.
- WS 3.5 Interpret observations and other data (presented in verbal, diagrammatic, graphical, symbolic or numerical form), including identifying patterns and trends, making inferences and drawing conclusions.

Maths focus

- 2c Construct and interpret frequency tables and diagrams, bar charts and histograms

Resources needed Worksheet 5.13

Digital resources

Key vocabulary Type 1 diabetes, Type 2 diabetes

Teaching and learning

Engage

- Students should **review** the main differences between Type 1 and Type 2 diabetes and **answer** the true or false questions in section 1 of Worksheet 5.13. Ask students to work in pairs and **reflect on** how Type 1 diabetes is treated and how this is different from the treatments of Type 2 diabetes. [O1]
- Ask students to **read** the passage from section 2 of the worksheet. They should **identify** the main differences in treating Type 1 and Type 2 diabetes, and **explain why** there are more choices for the treatment of Type 2 diabetes. [O1]

Challenge and develop

- Ask students to **analyse** the graph on Type 2 diabetes from section 3 of the worksheet, **draw conclusions** and **explain** their findings. **Discuss** what students have found out, and whether the quality of the data is reproducible. Ask students to **think about** how the data was collected and what further experiments could be done to improve the quality of the data. [O2]
- Ask students to **analyse** the graph of the connection of diabetes with society. Ask students to **draw conclusions** from this set of data. Students should **identify** other factors in society that might put some groups of people more at risk than others and **write a list** of groups along with explanations, in the worksheet. [O3]
- Ask students to **read** the information from the Student Book and **answer** the questions. [O1, O2, O3]

Explain

- Ask students to **write** an advice sheet for the treatment of Type 2 diabetes.
- Low demand students should **describe** in simple terms what can be done to prevent Type 2 diabetes. [O1, O2, O3]
- Standard demand should **explain** which people are most likely to be at risk from Type 2 diabetes and why. They should **outline** the main things that people can do to avoid getting diabetes and what they can do to reduce their symptoms if they already have it. [O1, O2, O3]
- High demand should **explain** in detail the differences between Type 1 and Type 2 diabetes. They should **explain why** some ethnic and social groups may be more vulnerable to the disease than others. They should **explain why** various treatment options work. [O1, O2, O3]

Consolidate and apply

- Ask students to work in small groups. They should **imagine** they are part of a health committee whose aim is to reduce the spread of diabetes. They should **devise** a proposal of what society can do to raise awareness of the disease and prevent it from becoming more widespread. Students should **consider** what can be done for children as well as adults. [O1, O2, O3]

Extend

Ask students able to progress further to:

- **research** further into how common Type 1 diabetes is in different populations and whether there are any underlying social factors. [O3]

Plenary suggestions

Ask students to complete section 5 of the worksheet. They **draw a graph** to predict how many people are likely to have diabetes in the countries given on the worksheet. [O3]

Answers to Worksheet 5.13

1 Type 1 and Type 2 diabetes

F, F, T, T, T, T

2 Treating diabetes

1. Type 1 diabetes uses insulin injections; Type 2 diabetes can be controlled by eating less carbohydrate, exercising, and drugs.
2. It is caused by an unhealthy lifestyle and strongly linked to obesity.

3 Analysing ethnic backgrounds

1. The greatest incidence of diabetes is in the Pakistani or Bangaleshi communities, followed by south Asian and African. The white community has the lowest incidence.
2. White communities are five times less likely to be diabetic compared with Pakistani communities. There may be some genetic links that increase the liklihood of diabetes; different communities eat different types of food – Pakistani communities may use more oil in their cooking and therefore be more likely to develop Type 2 diabetes.
3. It is unclear what the sample size is; people from different communities were identified to see if they had Type 2 diabetes or not; at least 100 people were sampled.
4. A greater sample size could be used of native populations and those that had migrated to this country. This would determine whether genetic factors are dominant.

4 Analysing social backgrounds

1. Men are more prone to diabetes than women. The more deprived you are the more likely you are to get diabetes.
2. Men in deprived areas are more likely to do less exercise and drink more; the more deprived you are, the more likely you are to eat cheaper foods that contain a lot of sugar.
3. whether it is Type 1 or Type 2 diabetes

5 Making predictions about diabetes

USA should have the highest bar, followed by UK, then India and Uganda.

Lesson 14: Water balance

Lesson overview

AQA specification reference

AQA 4.5.3.3; AQA 4.5.3.7

Learning objectives

- Recall the ways in which the body loses water.
- Explain why cells do not function efficiently if they lose or gain too much water.
- Explain how excess protein is converted to urea for excretion.

Learning outcomes

- Identify ways that water is gained and lost by the body. [O1]
- Explain the impact of too much or too little water on cells. [O2]
- Explain the process of deamination and its importance. [O3]

Skills development

- WS 1.2 Use a variety of models such as representational, spatial, descriptive, computational and mathematical to solve problems, make predictions and to develop scientific explanations and understanding of familiar and unfamiliar facts.
- WS 3.5 Interpret observations and other data (presented in verbal, diagrammatic, graphical, symbolic or numerical form), including identifying patterns and trends, making inferences and drawing conclusions.
- WS 4.1 Use scientific vocabulary, terminology and definitions.

Maths focus

2c Construct and interpret frequency tables and diagrams, bar charts and histograms

Resources needed A small mirror, a piece of cobalt chloride paper. Worksheet 5.14.1 (low demand); Worksheet 5.14.2 (standard demand); Worksheet 5.14.3 (high demand)

Digital resources PowerPoint slides

Key vocabulary deamination, dehydration, urea

Teaching and learning

Engage

- Ask students to work in pairs to **list** all the different ways we take in water and lose water from our bodies. Students should **rank** these processes in order of most to least water added or lost. [O1]
- Show students the table in the PowerPoint and **discuss** which processes they had missed or added. [O1]

Challenge and develop

- Provide students with a small mirror and a piece of cobalt chloride paper. Ask them to **breathe** on the mirror and **test** the substance on the mirror with the cobalt chloride paper. This shows the presence of water. **Discuss** where the water in our lungs come from and why there is more water in exhaled air compared with inhaled air. [O1]
- Remind students of osmosis using the PowerPoint, and of the investigation on the effect of different solute concentrations on plant cells carried out in Chapter 2. Ask students to **describe** what has happened to the blood cells in the different solutions in section 1 of their worksheet
- Low demand students should **describe** the red blood cells in each diagram (Worksheet 5.14.1).
- Standard demand students should **explain why** the cells are different (Worksheet 5.14.2).
- High demand students should **predict** the concentration of each solution (Worksheet 5.14.3).

- **Discuss** the importance of keeping cells in the right concentration environment. [O2]
- Introduce high demand students to the process of deamination using the Student Book and/or the PowerPoint. **Discuss** the role of the liver and the kidneys in this process. [O3]

Explain

- Low and standard demand students should **write** a short summary to explain why water is important in the body, the processes through which we gain and lose water, and **explain** what happens to the cells when there is too much or too little water. [O1, O2]
- High demand students should **use** an annotated diagram to **explain** the process of deamination in section 2 of Worksheet 5.14.3. [O3]

Consolidate and apply

- Ask students to **imagine** a person taking part in a race. Ask students to **draw a graph** of their water levels and **explain** at each stage what is happening in the body (on their worksheets). They should **identify** when the water levels might get dangerously low and what the person should do. [O1, O2]
- Ask students to **read** pages 198–199 in the Student Book and to **answer** questions 1–4 (low and standard demand) or questions 1–6 (high demand). [O1, O2, O3]

Extend

Ask students able to progress further to:

- **find out** the extreme measures that different organisms go to in order to conserve water in their bodies. They could also **find out** how organisms are adapted to survive in salty conditions, e.g. organisms living in the sea. [O1, O2]

Plenary suggestions

Students should complete the last section of their worksheet. They **draw** a concept map and use this to **explain how** water is lost and gained from the body and why it needs to be kept in balance. High demand students should **show how** the process of deamination links in to water loss. [O1, O2, O3]

Answers to Worksheet 5.14

1 How does water affect cells?

Low demand: in Diagram 2, the cell has expanded and increased in size; in Diagram 3, the cell has shrunk.

Standard demand: in Diagram 2, water has entered the cell as it is placed in a concentration that is lower than the inside of the cell; in Diagram 3, water has left the cell, as it is placed in a concentration that is higher than the inside of the cell.

High demand: Diagram 2 – below 0.85%; Diagram 3 – above 0.85%. Diagram 2 – they would burst; Diagram 3 – they would shrink further.

2 Explaining deamination (high demand)

Diagram showing a chain of amino acids in the liver being broken down, with the amino groups being removed and turned into ammonia. Diagram to show that ammonia is toxic. The ammonia is converted in the liver into urea, which is less toxic. Urea enters the blood and is taken to the kidneys. Ureas is expelled from the body as urine.

2 / 3 Predicting changes in water levels

Levels start at 2.9. Water levels increase to start with to a maximum of 3.3, due to increased respiration. High amounts of water are exhaled to balance the extra water being produced. Levels should return to 2.9. As the run proceeds, the body will become hot and produce sweat. More water is lost as sweat, but some continues to be produced through respiration. So water levels should not decrease below about 2.4. At this point the person should drink some more water, and so levels will rise high. They will continue to lose water through exhaled air and sweat, until it comes back into balance, and may decrease. At the end of the race, water levels are likely to be below normal.

Lesson 15: The kidneys

Lesson overview

AQA Specification reference

AQA 4.5.3.3

Learning objectives

- Recall that excess water, ions and urea are removed from the body by the kidneys in urine.
- Describe how the kidneys produce urine.
- Explain how the hormone ADH regulates the amount of water in the urine and, therefore, in the body.

Learning outcomes

- Identify the position and function of the kidneys in removing excess water, ions and urea from the body. [O1]
- Describe the processes of filtration and selective reabsorption of glucose, ions and water. [O2]
- Explain how ADH is controlled by negative feedback. [O3]

Skills development

- WS 1.2 Use a variety of models such as representational, spatial, descriptive, computational and mathematical to solve problems, make predictions and to develop scientific explanations and understanding of familiar and unfamiliar facts.
- WS 3.5 Interpret observations and other data (presented in verbal, diagrammatic, graphical, symbolic or numerical form), including identifying patterns and trends, making inferences and drawing conclusions.
- WS 4.1 Use scientific vocabulary, terminology and definitions.

Resources needed A real kidney; paper, coloured card, pipe cleaners to make a filter, sticky tape, cardboard boxes, rubber tubing and glue. Worksheet 5.15.1 (low and standard demand); Worksheet 5.15.2 (high demand)

Digital resources PowerPoint slides

Key vocabulary ADH, aorta, filtrate, negative feedback, selective reabsorption

Teaching and learning

Engage

- Show the students a real kidney. Ask them to **devise** five questions they could ask about the kidney and see if they can answer these at the end of the lesson. [O1]
- Ask students to **identify** the position of the kidneys, pituitary gland and adrenal glands on the outline of the body in section 1 of the worksheet. [O1]
- Show the overview and **discuss** the way the kidney works using the PowerPoint slides; students do not need to know the details of the structure of the kidney, but just focus on the processes of filtration and reabsorption.

Challenge and develop

- Ask students to work in small groups. Provide students with a description of how the kidney works from the Student Book. Provide them with some raw materials – paper, coloured card, pipe cleaners to make a filter, sticky tape, cardboard boxes, rubber tubing and glue. Ask students to **use** the information and **build a model** of how the kidney works and what it does.
- High demand students should **use** the information from the table on page 200 of the Student Book and **reflect** the values in their model. [O1, O2]
- Ask students to **show their** model of a kidney working normally and **discuss** the strengths and limitations of their models. [O1]
- Introduce the hormone ADH to high demand students only. Show a simulation of how ADH is used to control the amount of water reabsorbed by the kidney. **Discuss** the way ADH works. [O3]
- (High demand) **Discuss** what is meant by the term 'negative feedback' and how ADH controls the balance of water in the blood. [O3]

Explain

- Provide students with the flow diagram in Worksheet 5.15.1 (low and standard demand). Ask students to **fill in** the process and steps to **explain** the function of the kidney. [O1. O2]
- High demand students should use Worksheet 5.15.2 and **annotate** the flow diagram to include the function of ADH. They should **explain how** the control of water is an example of negative feedback. [O3]

Consolidate and apply

- Ask students to work in ability groups. They should **choose** one model to work with and **adapt** their models to show what would happen if the body was: doing a lot of exercise and sweating a lot (low demand); eating a very salty meal (standard demand); eating a meal with lots of protein (high demand) – students should **include** the process of deamination in their explanation. [O1, O2, O3]
- Ask students to **demonstrate** each scenario and **include** the role of ADH (high demand) and the pituitary gland in the process. [O1, O2, O3]

Extend

Ask students able to progress further to:

- **find out** more about the detailed structure of the kidney and how ultrafiltration and reabsorption take place. [O1, O2]

Plenary suggestions

Ask students to **return** to the questions they devised at the start of the lesson and see if they can **answer** them. Students **devise** five harder questions to ask another group and **swap** questions with this group. [O1, O2, O3]

Ask students to **match** the sentences in section 3 of the worksheet. [O1, O2, O3]

Answers to Worksheet 5.15

1 Locating the main organs associated with the kidney

Refer to Figure 5.28 from the Student Book.

2 Explaining how the kidneys work

First box – filtration with proteins and blood cells going into the box and glucose, urea, water and ions going to the next box.

Second box – glucose going into the blood being selectively reabsorbed, with urea, water and ions going to the next box.

Third box – most ions going to the blood with some ions, urea and water going to the next box.

Fourth box – water going into the blood and being reabsorbed – urea, some water and ions go to the next box.

1.
    a. Too much water: message picked up by body and sent to pituitary gland. Small amount of ADH released or no ADH released. Less water is reabsorbed back into the body from the kidney, so dilute urine produced.
    b. Too little water: message picked up by body and sent to pituitary gland. Increase in ADH released so more water is reabsrobed back into the blood from the kidney, so more concentrated urine is formed.
2. Negative feedback: when the body makes changes to restore the amount to normal. With an increase in water levels in the blood, less ADH is produced, so more water is lost in urine to restore the balance. With a decrease in water levels in the blood, more ADH is released, so more water is reabsorbed back into the blood, so less is lost, to restore the balance.

3 Summarising the function of the kidney

The first step	is ultrafiltration.
Large proteins and blood cells	are too large to pass through the kidney and remain in the blood.
Small molecules like glucose, ions and water	pass through into the kidney.
Once inside the kidney, the glucose and ions are first	selectively reabsorbed back into the blood.
Water is then also	selectively reabsorbed.
The only thing that is not reabsorbed is	urea.
Urea, excess salts and excess water are stored	in the bladder and pass out of the body as urine.

Lesson 16: Negative feedback

Lesson overview

Note: the content of this lesson is all Higher Tier.

AQA specification reference

AQA 4.5.3.7

Learning objectives

- Explain the role of thyroxine in the body.
- Understand the principles of negative feedback, as applied to thyroxine.

Learning outcomes

- Describe the effects of thyroxine and adrenaline in the body. [O1]
- Explain what happens when there is too much or too little thyroxine or adrenaline. [O2]
- Explain how thyroxine and adrenaline levels are controlled by negative feedback mechanisms. [O3]

Skills development

- WS 1.2 Use a variety of models such as representational, spatial, descriptive, computational and mathematical to solve problems, make predictions and to develop scientific explanations and understanding of familiar and unfamiliar.
- WS 4.1 Use scientific vocabulary, terminology and definitions.

Resources needed Internet access; website about thyroxine (www.yourhormones.info – search hormones for 'thyroxine'). Worksheet 5.16 (high demand)

Digital resources PowerPoint slides

Key vocabulary basal metabolic rate, negative feedback, pituitary gland, thyroxine

Teaching and learning

Engage

- Ask students to work in pairs and recall as many hormones from the endocrine system as they can, and what roles they have in the body. [O1]

Challenge and develop

- Show the PowerPoint image of where the thyroid and adrenal glands are situated. Introduce the hormone adrenaline and use the simulation to **explain** how adrenaline works. **Discuss** the role of adrenaline as the 'fight, flight and fright' hormone and the way the body is prepared for action. [O1, O2]
- Ask students to research the function of thyroxine and its effects in the body. Students should **summarise** these as a list in the worksheet. They should use the Student Book and the Internet (www.yourhormones.info – search hormones for 'thyroxine'). [O1]
- Ask students to **identify** other organs and glands in the body that are affected by each hormone. They should **create** a mind map of the function of each hormone on their worksheet, **showing the links** to different parts of the body. [O1, O2]
- Show students the negative feedback system for thyroxine from the PowerPoint. **Discuss** how the body responds when there is too much or too little of the hormone and how the negative feedback system helps to bring the levels back in line. [O2, O3]

Explain

- Students should **compare and contrast** this negative feedback system with a thermostat given in the worksheet and show where the similarities lie. [O3]

Consolidate and apply

- Ask students to work in small groups and **role play** or **produce a presentation** of the effects of adrenaline and thyroxine on the body. They should **demonstrate** what happens when there is too much or too little of each hormone and how the body responds. [O1, O2, O3]

Extend

Ask students able to progress further to:

- **research** other negative feedback mechanisms in the body. They could **evaluate** the strengths and weaknesses of using negative feedback mechanisms as a way of controlling levels. [O3]

Plenary suggestions

Students should **correct** the mistakes on the roles of adrenaline and thyroxine and negative feedback control in the passage on the worksheet. [O1, O2, O3]

Answers to Worksheet 5.16

1 Functions of adrenaline and thyroxine

Thyroxine	Adrenaline
Controls metabolic rate, i.e. how fast chemical processes like respiration occur in cells. Regulates breathing, heart rate, muscle strength, bone development, body weight, body temperature	Fight, flight, fright hormone Increases heart rate, increases breathing rate, sends more blood to the muscles and brain, increases sugar levels (by converting glycogen to glucose), dilates the pupils so more light enters the eye

2 How the hormones act

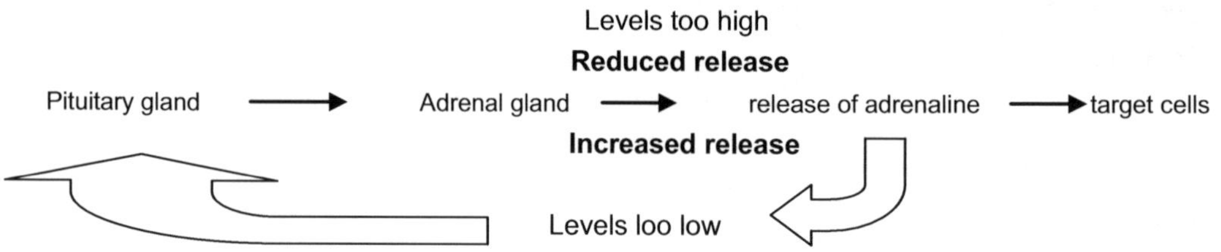

3 Negative feedback

The thermostat works in the same way as the feedback system for adrenaline. When the temperature is too high, the thermostat detects this and the heating is switched off. The levels of adrenaline are detected in the blood and messages sent to the pituitary gland. If the levels are too high, a message is sent from the pituitary gland to the adrenal glands to slow or stop the release of adrenaline. The difference here is that there is a control centre, the pituitary gland, and the feedback may be slower, whereas with the thermostat the connection is direct and more immediate. The opposite is true when the levels fall.

4 Find and correct the mistakes

Thyroxine is made in the thyroid gland. Its job is to ~~increases~~ **control** heart rate and breathing rate **at all times** ~~in times of emergency~~. The hormone adrenaline controls the heart rate and breathing rate ~~at all times~~ **in times of emergency** and **thyroxine** controls the rate of metabolism. When there is too much adrenaline, a message is sent from the blood to the ~~adrenal gland~~ **pituitary gland** and the amount of adrenaline is ~~increased~~ **decreased**.

Lesson 17: Kidney failure

Lesson overview

AQA specification reference

AQA 4.5.3.3

Learning objectives

- Recall that people who suffer from kidney failure can be treated by dialysis or kidney transplant.
- Understand the principles of dialysis.
- Evaluate the advantages and disadvantages of treating organ failure using a mechanical device or transplant.

Learning outcomes

- Describe the causes of kidney failure. [O1]
- Explain how dialysis and organ transplants work to treat kidney failure. [O2]
- Evaluate and compare the advantages and disadvantages of kidney transplants or dialysis. [O3]

Skills development

- WS 1.2 Use a variety of models such as representational, spatial, descriptive, computational and mathematical to solve problems, make predictions and to develop scientific explanations and understanding of familiar and unfamiliar facts.
- WS 1.4 Explain everyday and technological applications of science. Evaluate associated personal, social, economic and environmental implications, and make decisions based on the evaluation of evidence and arguments.
- WS 4.1 Use scientific vocabulary, terminology and definitions.

Resources needed Materials to make dialysis models – tubing, plastic box with holes at either end for tubing to pass through, fluid. Worksheet 5.17

Digital resources

Key vocabulary dialysis, donor

Teaching and learning

Engage

- Ask students to work in pairs and **identify** all the functions of the kidney from the previous lesson. Students should **predict** what would happen to the body if the kidneys no longer worked properly. [O1]
- Ask students to select all symptoms on the worksheet that might occur in kidney failure. [O1]

Challenge and develop

- Ask students to **carry out** their own research on kidney failure and methods to treat this, including dialysis and organ transplants. They could use the information from the Student Book and the Internet (try www.nhs.uk). [O1, O2]
- Students should work in small groups and **make** their own model to **explain** how dialysis works using the materials given. [O2]
- Place 'for' and 'against' posters around the room for both dialysis and organ transplants. Ask students to **write** statements on sticky notes, to **explain** the reasons 'for' and 'against' transplant and dialysis respectively, from their research. [O3]

Explain

- Low demand students should **complete** the cloze exercises on Worksheet 5.17 to **explain** how dialysis and kidney transplants works. [O2, O3]

- Standard demand students should **create** a brief news bulletin to **explain** why kidney donors are needed and why this is a better option than dialysis. [O2, O3]
- High demand students should **prepare** a **debate** to **compare** and **contrast** dialysis with kidney donation. They should **explain** the benefits and drawbacks of each, using data from their research. [O1, O2, O3]

Consolidate and apply

- Students should **write** an information leaflet to patients in a kidney treatment centre. They should **explain** what the kidneys do, why they might fail and what treatment options there are. They should **advise** people on how to avoid kidney disease, and **outline** the benefits and drawbacks of each treatment option. [O1, O2, O3]
- Students should **answer** all the questions on pages 204–205 of the Student Book. [O1, O2, O3]

Extend

Ask students able to progress further to:

- **find out** more about the history of the development of dialysis and how the techniques have changed over time. They could **evaluate** the improvements. [O2, O3]

Plenary suggestions

Students should **return** to the lists they prepared at the start of the lesson. They should **extend** or **correct** the list to include the new learning from the lesson. [O1, O2]

Students could **role play** being in a doctors' surgery and **model** what might happen with someone with kidney disease, discussing the treatment options available. [O1, O2, O3]

Answers to Worksheet 5.17

1 Kidney failure

protein in urine   increased need to urinate   water retention   weight loss

2 How does dialysis work?

fail, dialysis machine, glucose, ions, membrane, opposite, urea, salts, body, four ,three

3 What is a kidney transplant?

diseased, donor, operation, tissue, blood, relative, rejected, immunosuppressant drugs, reduced, increased

Lesson 18: Dialysis or transplant?

Lesson overview

AQA specification reference

AQA 4.5.3.3

Learning objectives

- Describe different forms of dialysis.
- Use data to draw conclusions.
- Use data to evaluate the advantages and disadvantages of dialysis and kidney donation.

Learning outcomes

- Describe the differences between haemodialysis and peritoneal dialysis. [O1]
- Use data to draw and interpret graphs. [O2]
- Use data to justify the advantages and disadvantages of different forms of treatment of kidney disease. [O3]

Skills development

- WS 1.4 Explain everyday and technological applications of science. Evaluate associated personal, social, economic and environmental implications; and make decisions based on the evaluation of evidence and arguments.
- WS 3.1 Present observations and other data using appropriate methods.
- WS 3.5 Interpret observations and other data (presented in verbal, diagrammatic, graphical, symbolic or numerical form), including identifying patterns and trends, making inferences and drawing conclusions.

Maths focus

- 2c Construct and interpret frequency tables and diagrams, bar charts and histograms

Resources needed Worksheet 5.18.1 (low demand); Worksheet 5.18.2 (standard demand); Worksheet 5.18.3 (high demand)

Key vocabulary haemodialysis, peritoneal dialysis

Teaching and learning

Engage

- Show students a video that compares haemodialysis and peritoneal dialysis.
- **Discuss** the differences between peritoneal dialysis and haemodialysis, using ideas from the Student Book. Ask students to **complete** section 1 of their worksheet to **compare** the two methods. [O1]

Challenge and develop

- Ask students to **look at** the data in section 2 of Worksheet 5.18.1 (low demand), 5.18.2 (standard demand) or 5.18.3 (high demand), which shows the prevalence of kidney disease in people of different ages and ethnic backgrounds. [O2]
- Low and standard demand: Ask students to **draw a graph** to show the way kidney disease affects people of different ages. Students should **describe** the link between kidney disease and age.
- High demand: Students should **explain** which people are most at risk and which are least at risk, using data from the table. They should **draw** a suitable graph to provide evidence for their ideas and provide possible **reasons** for these differences.
- All students should **explain** whether the quality of the data is valid and precise. [O2]
- Students should work in pairs. They should **read** the sections of the Student Book, and **use** their understanding from the previous lesson. They should **identify** some reasons for why people of a particular age and ethnic background might suffer more from kidney disease than others.They should **consider** which treatment options are likely to suit people most at risk. **Discuss** their responses. [O2]

Explain

- Students should work in pairs and **produce a poster** to **summarise** all the problems with kidney donation, haemodialysis and peritoneal dialysis. They should **explain** which treatment options would suit people from particular backgrounds and why. [O1, O3]

Consolidate and apply

- Students should **answer** the questions on kidney transplants in section 3 of their worksheet.
- Students should work in small groups and **devise** a list of strategies to improve the situation for people waiting for a new kidney. [O2, O3]

Extend

Ask students able to progress further to:

- **find out** what strategies the government has put into place to improve the situation of kidney donation. They should **identify** how successful these measures have been and what further steps could be taken. [O2, O3]

Plenary suggestions

Ask students to **complete** the card sort in section 4 of their worksheet. [O2]

Ask students to **make up** three sentences about each heading using the words from the list. [O2]

Answers to Worksheet 5.18

1 Peritoneal dialysis and haemodialysis

Haemodialysis	Peritoneal dialysis
Normally done in hospital	Normally done at home during the night
Takes about 4 hours, three times a week	People who are still working opt for this treatment
People who are too weak for a transplant will undergo this treatment	People who are too weak for a transplant will undergo this treatment
It is a life-long treatment	It is a life-long treatment
Complications of infection are common and it can reduce life expectancy in the elderly	Complications of infection are common and it can reduce life expectancy in the elderly
	Some people may progress to donation, which improves independent living
	Problems can occur if there are power failures at home

2 Who is most at risk of kidney disease?

Low and standard demand:

1. Bar chart should be drawn, with age along the *x* axis and percentage of people affected along the *y* axis. Students can choose any or all of the ethnic groups to show the data.
2. As age increases from 40 to 60, the percentage of kidney disease increases slowy, until over 60 years where the increase is rapid and even more so over 70 years.

High demand:

1. Graph of age along the *x* axis and two sets of data on the *y* axis – one for male and one for female. Alernatively, bars drawn in different colours for male and female and for different ethnic backgrounds.
2. The people most at risk are the male Bangaledeshi over 70 years old. The least at risk are the white males and females 40–49 years old. Reasons: could be genetic – linked to diet and exercise, white people of that age group eat less fatty foods and exercise more.
3. There is a large sample size for all ethnic backgrouds, so the data is likely to be valid. There are good records in this country for kidney disease, so the data is likely to be precise.

3 Data on kidney transplants

Low demand: no, it is no a good option treatment because 1000 people die each year because unsuitable kidneys are found.

Standard and high demand: there are many people (90%) that want to be donors but haven't signed up to be (only 30%). This has caused a big decrease in the potential number of donors available, causing many people to die (1000 per year).

4 Using kidney terminology

Normal kidney: filtration, urea, glucose, water, ions, selective reabsorption; Haemodialysis: counterflow, dialysis fluid, partially pereable membrane; Kidney transplant: immunosuppressant drugs, donor

Lesson 19: Human reproduction

Lesson overview

AQA specification reference

AQA 4.5.3.4

Learning objectives

- Describe the roles of hormones in sexual reproduction.
- Explain how hormones interact in the menstrual cycle.

Learning outcomes

- Name the hormones involved in sexual secondary characteristics and describe their roles. [O1]
- Describe the roles of all the hormones involved in the menstrual cycle. [O2]
- Explain how the hormones interact to control the menstrual cycle. [O3]

Skills development

- WS 1.2 Use a variety of models such as representational, spatial, descriptive, computational and mathematical to solve problems, make predictions and to develop scientific explanations and understanding of familiar and unfamiliar facts.
- WS 3.5 Interpret observations and other data (presented in verbal, diagrammatic, graphical, symbolic or numerical form), including identifying patterns and trends, making inferences and drawing conclusions.
- WS 4.1 Use scientific vocabulary, terminology and definitions.

Resources needed Video of the menstrual cycle (search 'menstrual cycle animation' at www.nhs.uk); video of hormone levels (search 'Sex hormones – the menstrual cycle' at www.abpischools.org.uk); large sheet of paper to draw a timeline of the menstrual cycle – poster paper and four different colours. Worksheet 5.19.1 (low and standard demand); Worksheet 5.19.2 (high demand)

Digital resources

Key vocabulary follicle, FSH, LH, oestrogen, progesterone, secondary sex characteristics, testosterone

Teaching and learning

Engage

- Ask students to work in pairs and identify as many secondary sexual characteristics as they can. Ask pairs to join together to make fours and share ideas. Finally, ask each four to join with another four and combine their lists. Discuss their ideas, and why these characteristics occur during puberty. [O1]
- Using the Student Book, discuss the names and roles of the hormones involved in producing secondary sexual characteristics. [O1]

Challenge and develop

- Play the video from www.nhs.uk showing what happens overall in the menstrual cycle. [O2, O3]
- Introduce the names of the main four hormones in the menstrual cycle and which parts of the body they are made in. Use the Student Book to explain how the menstrual cycle works and the role of each of the four hormones within the cycle.
- Ask low and standard demand students to **sort** the cards from section 1 of Worksheet 5.19.1 and **place the cards** under their respective hormone. [O2]
- High demand students should **use** the image of the menstrual cycle on Worksheet 5.19.2. They should **annotate** the diagram to include the role of all the hormones at the correct points in the cycle. [O2, O3]

Explain

- Ask students to work in groups of four. They should adopt one of the hormones. Show students the animation of the menstrual cycle from www.abpischools.org.uk.
- Ask students to follow their hormone on the animated graph and **write** an explanation of how and why the levels change in the worksheet, using the graphs provided. They should explain what **impact** these changes have on the menstrual cycle. [O2]

Consolidate and apply

- Ask students to work in groups of four. They should **draw** a timeline of the menstrual cycle on a large sheet of paper with images to **show** what is happening at key points. Students should **produce** a role play about one hormone to **explain** where it comes from and its role in the menstrual cycle, They should stand at the correct place on their menstrual cycle timeline while giving their performance. [O2, O3]
- High demand students should also **explain** how the hormones interact to control the menstrual cycle.

Extend

Ask students able to progress further to:

- **find out** about the menstrual cycle of other mammals, and **compare** and **contrast** their similarities and differences. They could find out if similar hormones are used. [O2, O3]

Plenary suggestions

Ask students to **spot** the odd one out in section 3 of the worksheet and **explain** why. [O1, O2, O3]

Students should **draw** a flow chart on poster paper to **explain** how the hormones work together within the menstrual cycle. They should use different colours for each of the four different hormones. [O2, O3]

Answers to Worksheet 4.19

1 Hormones in the menstrual cycle

Low and standard demand: FSH – 1, 3, 6; LH – 6, 5; Oestrogen – 4, 7, 10, 11,12; Progesterone – 2, 8, 9, 12

High demand: Correctly annotated diagram including ideas such as (clockwise): FSH secreted by pituitary gland and causes eggs to mature in ovary; stimulates release of oestrogen; oestrogen builds up the lining of the uterus after menstruation; as oestrogen levels increase to a maximum, it stimulates the production of LH and inhibits the production of FSH, so no more follicles mature; LH causes ovulation to occur on day 14; progesterone is produced by the empty follicle after ovulation and maintains the lining of the uterus; If the egg is not fertilised, the empty follicle decays and the levels of progesterone drop so the lining of the uterus falls away and is passed out of the body during menstruation.

2 Interpreting graphs

LH – low level to start with. As oestrogen levels rise to a maximum, levels of LH begin to increase sharply. When they are at a maximum, ovulation occurs, after which LH levels fall sharply, coinciding with a drop of oestrogen levels. They remain low for the remainder of the menstrual cycle.

Progesterone – low levels until after ovulation, where levels begin to increase.They reach a manimum at about day 21 and then fall rapidly. They do not coincide with any other hormone increases. However, high levels of progesterone coincide with low levels of FSH and LH. The lining of the uterus is maintained while the levels of progesterone remain high. When the levels decrease, the lining of the uterus is removed from the body.

Oestrogen – levels begin to increase afer day 1 of the cycle and reach a manximum just before ovulation. As oestrogen levels increase, LH levels also increase, and FSH levels begin to fall. The lining of the uterus increases as the oestrogen levels increase.

FSH – levels start relatively high and increase. As they do so, oestrogen levels also increase. FSH stimulates oestrogen production. Just before ovulation, FSH levels fall sharply and continues to drop. They increase again after progesterone levels drop.

3 Odd one out

Breasts develop – not controlled by testosterone

Oestrogen – not produced by the pituitary gland

LH – does not affect the lining of the uterus

Day 21 – after ovulation

Lesson 20: IVF

Lesson overview

Note: the content of this lesson is Higher Tier only.

AQA specification reference

AQA 4.5.3.6

Learning objectives

- Explain the use of hormones in technologies to treat infertility.
- Describe the technique of *in-vitro* fertilisation.
- Evaluate the scientific, emotional, social and ethical issues of *in-vitro* fertilisation.

Learning outcomes

- Describe what *in-vitro* fertilisation is and why it is needed. [O1]
- Describe the steps involved in *in-vitro* fertilisation. [O2]
- Explain how hormones are used to carry out the process of *in-vitro* fertilisation. [O3]

Skills development

- WS 1.2 Use a variety of models such as representational, spatial, descriptive, computational and mathematical to solve problems, make predictions and to develop scientific explanations and understanding of familiar and unfamiliar facts.
- WS 1.4 Explain everyday and technological applications of science. Evaluate associated personal, social, economic and environmental implications, and make decisions based on the evaluation of evidence and arguments.
- WS 4.1 Use scientific vocabulary, terminology and definitions.

Resources needed Worksheet 5.20

Digital resources PowerPoint slides

Key vocabulary fertility drug, *in-vitro* fertilisation, IVF cycle

Teaching and learning

Engage

- Show the PowerPoint slide of the first IVF baby in 1978. Discuss why *in-vitro* fertilisation might be needed. Show students a list of all the common reasons for infertility. Ask students to **identify** which reasons might be related to hormones. [O1]

Challenge and develop

- Show students a video that shows all the stages of IVF. Ask students to **compare** and **contrast** IVF with normal fertilisation in section 1 of Worksheet 5.20. [O1, O2]
- Students should **draw** a flow chart of all the steps of *in-vitro* fertilisation on their worksheet. [O1, O2]
- Ask students to **read** the section about *in-vitro* fertilisation on pages 210–211 of the Student Book and **research** other sources (e.g. www.nhs.uk – search conditions for 'IVF'), which explain the process in more depth. Students should **identify** the hormones involved in the process and **suggest** reasons for their use. Students should also **identify** any drawbacks with the process. [O3]

Explain

- Ask students to complete the table in section 4 of the worksheet and explain the advantages and disadvantages of IVF, including issues with the potential success rate. [O2, O3]

Consolidate and apply

- Students should **write** an information leaflet for an IVF clinic to **explain** what will happen to the couple during the process. They should **explain** any side effects they might encounter and **provide** an estimate of the chances of success. Students should **use** the Student Book to help them prepare their leaflet. [O1, O2, O3]

Extend

Ask students able to progress further to:

- **find out** about the historical development of IVF and how scientists came to perfect the techniques. They could **research** which key developments made the technique possible. [O1, O2, O3]

Plenary suggestions

Ask students to **prepare some advice** for the different people in the cards in section 5 of the worksheet. They should **suggest** what they could do to make their treatment more successful and the drawbacks they are likely to encounter. [O1, O2, O3]

Answers to Worksheet 5.20

1 IVF versus normal fertilisation

Normal fertilisation	IVF
Mature egg cell is already in the fallopian tube or uterus.	Mature eggs are extracted fom the ovary and grown in the laboratory.
Sperm enter the egg through sexual intercourse.	Selected sperm are injected into the mature egg sample in the laboratory.
Sperm need to swim up to the egg cell in order to fertilise the egg.	After fertilisation, the embryo is grown in the laboratory for a while.
If fertilised the egg will implant into the uterus directly.	When ready the embryo is inserted into the uterus for implantation.

2 How does IVF occur?

Flow diagram to include: mature follicles extracted from the ovary; follicles are placed in nutrient medium and grown in the lab; when mature, sperm is inserted into the dish containing the mature follicles; the eggs are fertilised; the fertilised eggs are grown in new medium and grown for a few days; mature embryos are inserted into the uterus

3 Role of hormones in IVF

1. LH and FSH are used to stimulate the maturation of several eggs.
2. Several eggs may become mature at once.
3. Risk of multiple births, which can be risky for the mother and the children.

4 Advantages and disadvantages of using IVF

Advantages	Disadvantages
Couples who would not normally be able to have children have the opportunity of having children of their own.	Very low success rates.
	Very expensive treatments as fertility hormones are drugs that are needed before and after the process.
	Can be emotionally stressful for the mother.
	Removal of the mature follicles and insertion of the embryo are painful invasive techniques which risk infection.
	Multiple births are possible which put the mother and children at risk.
	Many embryos may be made during each IVF cycle – these are normally discarded, raising ethical issues.

5 What advice could you give?

A. Outline all the risks for someone of her age having a baby – the risks are greatly increased and the success rate is significantly lower. Advice that she loses weight before she starts any treatment.

B. Discourage them from IVF, they are young enough to keep trying and seeking alternative medical advice, until their mid-thirties.

C. Outline all the risks and if they wish to carry on, encourage them.

Lesson 21: IVF evaluation

Lesson overview

Note: the content of this lesson is Higher Tier only.

AQA specification reference

AQA 4.5.3.6

Learning objectives

- Evaluate data regarding *in-vitro* fertilisation and use this to draw conclusions.
- Evaluate the scientific, emotional, social and ethical issues of *in-vitro* fertilisation.

Learning outcomes

- Draw conclusions from data about the effectiveness of *in-vitro* fertilisation. [O1]
- Describe the issues involved in undergoing IVF treatment. [O2]
- Evaluate the scientific, emotional, social and ethical issues of *in-vitro* fertilisation. [O3]

Skills development

- WS 1.3 Appreciate the power and limitations of science and consider any ethical issues which may arise.
- WS 1.4 Explain everyday and technological applications of science. Evaluate associated personal, social, economic and environmental implications, and make decisions based on the evaluation of evidence and arguments.
- WS 1.5 Evaluate risks both in practical science and the wider societal context, including perception of risk in relation to data and consequences.

Maths focus

2c Construct and interpret frequency tables and diagrams, bar charts and histograms

2g Use a scatter diagram to identify a correlation between two variables

Resources needed Article from BBC News Magazine 'I wish IVF had never been invented' (www.bbc.co.uk). Worksheet 5.21

Digital resources

Key vocabulary eugenics, IVF

Teaching and learning

Engage

- Show students the testimonials of people who have had IVF and have been successful and unsuccessful (www.bbc.co.uk). This is a newspaper article from several readers. Ask different students to **read out** different parts of the article, and the rest should **identify** all the disadvantages from the article. [O1]

Challenge and develop

- Students should look at the table given in section 1 of Worksheet 5.21. They should **draw** the necessary graphs and **answer** the questions about the data. [O1]
- Ask students to **read** the section from the Student Book (pages 212–213) and **answer** all the questions. [O1, O2, O3]
- Ask students to work in small groups. Provide students with a card sort with statements about who can be helped by IVF and some of the issues and problems associated with it from section 2 of the worksheet. Ask students to **sort** the cards into those that are scientific, emotional, social or ethical issues. [O2, O3]

Explain

- Ask students to **summarise** the issues under each of the headings – scientific, social, emotional, financial and ethical. They should **write** a short paragraph explaining the issues for and against under each area. [O2, O3]

Consolidate and apply

- Students should **write** a short account of IVF from a mother's point of view. They should **explain** what she might go through and what fears she may have. They should **describe** the tests, procedures and what medication they may have to take on a daily basis. **Discuss** the success criteria for this piece of writing and share this with the students. [O2, O3]
- Ask students to **peer assess** each other's accounts using the derived success criteria. Students should **feed back** two good things about the account and one thing that could be improved. [O2, O3]

Extend

Ask students able to progress further to:

- **find out** more about what happens to the embryos that are not used. They should **find out** how freezing embryos might affect them, and what kinds of research embryos are used for. [O2, O3]

Plenary suggestions

Ask students to **answer** the question of whether they would undergo IVF, if, in the future, they or their partner had infertility. Ask them to **justify** their responses using data from the lesson. They should **prepare** a 1-minute speech on what they would do and why. [O1, O2, O3]

Answers to Worksheet 5.21

1 The effect of age on IVF

1. graph of age on *x* axis and either cycle resulting in live birth or cycle resulting in pregnancy on *y* axis
2. graph of cycles resulting in pregancy on *x* axis and cycles resulting in live birth on *y* axis
3. Conclusions: older women have a much reduced success rate with IVF; the percentage of live births is proportional to the percentage of pregancy, irrespective of the age of the women. So once the woman has become pregant, the age does not significantly impact on the ability to produce a living child.
4. Only older women have more embryos transferred, as the likehood of them becoming pregnant is significantly reduced. Therefore, they are less likely to produce multiple births.
5. Compared with 2010, there is a higher percentage of live births in 2011; 32% versus 40.1% for women under 35.

2 Scientific, emotional, financial and ethical issues

Scientific: 1, 3, 6, 8, 9, 10, 12, 13

Social: 2, 3, 5, 7, 14, 16

Emotional: 5, 6, 8, 9, 13, 14

Financial: 4, 6, 8, 12, 14, 15, 16

Ethical: 4, 6, 8,10, 11, 12, 13, 14

Lesson 22: Systems working together

Lesson overview

Note: the content of this lesson is Higher Tier only.

AQA specification reference

AQA 4.5.3.7

Learning objectives

- Describe the effects of adrenaline.
- Understand that automatic control systems may involve nervous responses and chemical responses.
- Understand that combinations of hormones work to produce a response.

Learning outcomes

- Describe what impact adrenaline has on the body. [O1]
- Explain how hormones are controlled by nervous and chemical processes. [O2]
- Explain how hormones can work together to produce a response. [O3]

Skills development

- WS 1.2 Use a variety of models such as representational, spatial, descriptive, computational and mathematical to solve problems, make predictions and to develop scientific explanations and understanding of familiar and unfamiliar facts.
- WS 3.5 Interpret observations and other data (presented in verbal, diagrammatic, graphical, symbolic or numerical form), including identifying patterns and trends, making inferences and drawing conclusions.
- WS 4.1 Use scientific vocabulary, terminology and definitions.

Resources needed Animation 'More hormones – adrenaline and ADH' from www.abpischools.org.uk. Worksheet 5.22

Digital resources

Key vocabulary adrenaline, adrenal medulla

Teaching and learning

Engage

- Ask students to work in pairs. They should **recall** all they can about the hormone adrenaline from previous lessons, including where it is found and its function in the body. [O1]
- Ask students to **explain why** adrenaline is called the 'fight, flight or fright' hormone. [O1]

Challenge and develop

- Remind students of what adrenaline does to the body using the animation from www.abpischools.org.uk.
- Ask students to **explain** why adrenaline has these effects (on Worksheet 5.22) and how they help to prepare the body for emergency. [O1]
- Students should **identify** the stimulus–response pathway for a situation of fight, flight or fright. They should **describe** the stimulus and how the nervous system and endocrine system work together to produce a response. [O1, O2]
- Ask students to **read** the section on 'Hormones working together' in the Student Book (page 215) and to **answer** questions 5–6. [O3]

Explain

- Ask students to **look at** the list of hormones in section 2 of Worksheet 5.22. Ask students to **identify** those that work together and which systems are involved. Students should **explain** the effect of the hormones on the systems and processes they control. [O3]

Consolidate and apply

- Ask students to **imagine** they going on a rollercoaster ride. Ask students to **describe** the stimuli that will excite their nervous system, which hormones will be released into their system and what effects these will have on their body. [O1, O2, O3]

Extend

Ask students able to progress further to:

- **find out** about other systems, e.g. the digestive system, and **identify** which hormones are involved and how they work together. [O3]

Plenary suggestions

Ask students to answer the true or false questions in section 3 of the worksheet. [O1, O2, O3]

Answers to Worksheet 5.22

1 What does adrenaline do in the body

Effect on the body	**Reason for these effects**
Heart – heart rate increases	More blood carrying glucose and oxygen flowing around the body so cellular aerobic respiration can increase.
Lungs – breathing rate increases	To provide more oxygen in the blood and remove more carbon dioxide arising from increased respiration.
Liver – more glycogen converted to glucose	So there is more glucose in the blood for aerobic respiration.
Muscles – increases blood flow, supplying more glucose and oxygen	Muscle cells can respire more actively in response to fight, flight or fright and be prepared to move.
Pancreas – more glucagon released and insulin production reduced	Glucagon increases the breakdown of glycogen to glucose in the liver and muscles.
Brain – increased blood supply	Increases glucose and oxygen to the brain so decisions can be made for survival.

2 Hormones working together

ADH – affects water balance in the body

insulin, glucagon, adrenaline, thyroxine, cortisol – affect heart rate, breathing rate, rate of metabolism, glucose levels in the blood

FSH, LH, oestrogen, progesterone – affect the menstrual cycle and pregnancy

growth hormone, thyroxine – affect metabolic rate and rate of cell division

testerosterone, thyroxine – affect secondary sexual characteristics and rate of metabolism in males

3 True or false?

T, T, T, F, T, T, F

Lesson 23: Contraception

Lesson overview

AQA specification reference

AQA 4.5.3.5

Learning objectives

- Define the purpose of contraception.
- Describe hormonal methods and non-hormonal methods of contraception.
- Explain how these methods are effective at preventing pregnancy.

Learning outcomes

- Name and describe the different non-hormonal methods of contraception. [O1]
- Explain how non-hormonal methods of contraception work. [O2]
- Explain how hormonal methods of contraception work. [O3]

Skills development

- WS 1.2 Use a variety of models such as representational, spatial, descriptive, computational and mathematical to solve problems, make predictions and to develop scientific explanations and understanding of familiar and unfamiliar facts.
- WS 3.5 Interpret observations and other data (presented in verbal, diagrammatic, graphical, symbolic or numerical form), including identifying patterns and trends, making inferences and drawing conclusions.
- WS 4.1 Use scientific vocabulary, terminology and definitions.

Resources needed Worksheet 5.23.1 (low demand); Worksheet 5.23.2 (standard and high demand)

Digital resources

Key vocabulary cervix, combined contraceptive pill, progestogen, progestogen-only pill

Teaching and learning

Engage

- Ask students to work in pairs. Ask them to **devise** as many reasons as possible for using contraception. Take suggestions from the pairs and make a class list on the board. [O1]
- Show students the graph of world population from page 216 of the Student Book and the data on unintended pregnancies from Worksheet 5.23.1 (low demand) or Worksheet 5.23.2 (standard and high demand). Ask students to **draw conclusions** from the data and **summarise** what the impact would be if there were no contraceptives. [O1]
- Ask students to **devise** a list of questions they would like to be answered from this lesson and **summarise** some ideas they already have about methods of birth control. [O1, O2]

Challenge and develop

- **Discuss** all the methods of contraception that do not involve hormones. **Discuss** how each one works. If possible, show students real examples of these so they can see how they work. Ask students to **complete** section 2 of the worksheet, where they **match** the methods to the way they work. [O1, O2]
- Students should **read** the sections on 'Hormonal methods' of contraception on page 217 of the Student Book. Students should **answer** questions 5–6 in the Student Book. [O1, O2]
- Discuss with students the various myths there are surrounding birth control methods (you could show a video). **Discuss** some of the ideas in the video and ask students to **compare** these to the questions they devised at the start of the lesson. [O1, O2]

Explain

- Students should complete section 3 of the worksheet. They should **annotate** the diagram of the female reproductive system to **explain how** each method of contraceptive using hormones works .[O2, O3]

Consolidate and apply

- Ask students to work in groups of three. They should **plan** and **produce** a role play of a doctor's surgery where the doctor is advising a young couple about different contraceptive methods. The group should jointly **plan** the questions and answers for the role play. Provide different groups with three different methods of contraceptive to discuss, including at least one hormonal method. [O1, O2, O3]

Extend

Ask students able to progress further to:

- **find out** about the history of contraceptive development and how more effective methods have been produced over time. [O1, O2]

Plenary suggestions

Students should return to the questions they devised at the start of the lesson and the summaries they made. They should try and attempt to **answer** their own or each other's questions and **add** to their summary of contraceptive methods with ideas from the lesson. [O1, O2]

Answers to Worksheet 5.23

1 Why is contraception used?

Low demand: the population in the world is increasing. There is still a significant number of unwanted pregnancies in the UK alone. 40% of unwanted pregnacies end in abortion.

Standard and high demand: the population of the world is continuing to increase, and unless checked, could reach 15 billion. There are not enough resources in the world to sustain this population. It is therefore imperative that unplanned pregancies are reduced to a minimum. Also, many young, single women experience unwanted pregnacies, many of which end in abortion, but for the children that are born, they are not always born into stable family situations.

2 How do contraceptives work?

1E, 2C, 3A, 4B, 5D

3 Methods of contraception

a. Combined pill affects the release of FSH, so follicles do not mature in the ovary. There is no ovulation. Also the vagina becomes thick with mucus, preventing the sperm from reaching.
b. Progestogen-only pill – thickens the mucus at the neck of the womb so the sperm cannot reach the uterus. In some cases can prevent ovulation.
c. IUD – a device inserted into the vagina. It releases copper which changes the fluids in the womb which stops sperm from surviving.

Lesson 24: Which contraceptive?

Lesson overview

AQA specification reference

AQA 4.5.3.5

Learning objectives

- Describe the advantages and disadvantages of different contraceptive methods.
- Use data to evaluate the effectiveness of different contraceptive methods.

Learning outcomes

- Compare the effectiveness of different contraceptive methods. [O1]
- Plot and analyse data to evaluate the effectiveness of different contraceptive methods. [O2]
- Evaluate the quality of the data collected. [O3]

Skills development

- WS 1.4 Explain everyday and technological applications of science. Evaluate associated personal, social, economic and environmental implications, and make decisions based on the evaluation of evidence and arguments.
- WS 3.5 Interpret observations and other data (presented in verbal, diagrammatic, graphical, symbolic or numerical form), including identifying patterns and trends, making inferences and drawing conclusions.
- WS 3.7 Be objective, evaluate data in terms of accuracy, precision, repeatability and reproducibility and identify potential sources of random and systematic error.

Maths focus

- 4a Translate information between graphical and numeric form
- 4c Plot two variables from experimental or other data

Resources needed Worksheet 5.24

Digital resources

Key vocabulary condom, IUD

Teaching and learning

Engage

- Ask students to work in pairs. They should **recall** five different methods of contraception and to rank them in order of how effective the methods might be at preventing pregnancy. Discuss their ideas and produce a class list. [O1]

Challenge and develop

- Show students the data from section 1 of Worksheet 4.24. Ask students to **plot** the data and to **draw conclusions** about the effectiveness of different contraceptives. Ask students to **compare** this data with the ideas they had at the start. [O2]
- Students should work in small groups and **consider** how this data could have been collected. They should **consider how** to ensure that representative data is collected and how the results might be peer-reviewed. [O3]
- Students should **consider** the advantages and disadvantages of different contraceptives.
- Low demand students should **state** the advantages and disadvantages for the given contraceptive methods.
- Standard demand students should **explain** the advantages and disadvantages of given contraceptive methods.

- High demand students should **explain** and **evaluate** the advantages and disadvantages of different contraceptive methods, **ranking** them in order of most effective to least effective, and **explain why**. [O1]

Explain

- Show students the pie chart in section 2 of Worksheet 5.24. Ask students to **explain** the data and **suggest why** some forms of contraceptive are used more than others. Students should **explain** whether the data is likely to be peer-reviewed and how representative the data is likely to be. Students should **identify** reasons why the data could be unrepresentative. [O1, O2, O3]

Consolidate and apply

- Ask students to create an advertisement for the contraception method they decide is the most effective. They should explain how the methods works and why it is the best. They should use data where possible to justify their ideas. [O1, O3]

Extend

Ask students able to progress further to:

- **find out** if contraceptive methods are different in different parts of the world, or if the data is the same as suggested in the pie chart. They should **plot** any data they find and **evaluate** the methods used to collect this data. [O1, O2, O3]

Plenary suggestions

Return to the class list produced at the start of the lesson and ask students to work in pairs to **decide** if there are any changes to be made. [O1]

Ask students to work in small groups and **debate** whether everyone should be given free contraceptives or not, and which contraceptives should be given. Students should **identify** arguments for and against this question. [O1]

Answers to Worksheet 5.24

1 How effective are different contraceptives?

Plot of type of contraception on the *x* axis and percentage of pregnancies prevented on the *y* axis.

The diaphragm is the least effective method of contraception, and oral contraceptives and IUD are the most effective.

2 Explaining the evidence

1. The male condom followed by the contraceptive pill and male sterilisation are the most common forms of contraception. Vaginal barrier is the least used method of contraception.
2. The male condom may be most common because it also protects against STIs. It is also cheap and easy to apply and readily avaiable. Male sterilisation is very common as a permanent contraceptive solution. The contraceptive pill is also common because it is very effective and easy to take. There are some suggestions that it also helps prevent some forms of ovarian cancer. The vaginal barrier may not be used because it needs to be correctly inserted and may be uncomfortable during intercourse. Traditional methods are quite common, may be due to religious reasons, as Catholics believe that using contraception is against their faith.
3. It is unlikley that the data has been peer-reviewed as it represents a study done over a long period of time.
4. It may not have accounted for all people across different age groups. Some ethnic groups may not want to disucss such matters or share their information openly. People in general may find this topic difficult to discuss and be open about.

Lesson 25: Auxins

Lesson overview

AQA specification reference

AQA 4.5.4.1; AQA 4.5.4.2

Learning objectives

- Recall that plants produce hormones to coordinate and control growth, and responses to light and gravity.
- Describe how unequal distributions of auxins cause unequal growth rates in plant shoots and roots.

Learning outcomes

- Describe the term 'tropism'. [O1]
- Explain how tropisms can affect plants. [O2]
- Explain how unequal distributions of auxins cause unequal growths rate in roots and shoots. [O3]

Skills development

- WS 1.2 Use a variety of models such as representational, spatial, descriptive, computational and mathematical to solve problems, make predictions and to develop scientific explanations and understanding of familiar and unfamiliar facts.
- WS 2.4 Carry out experiments appropriately having due regard for the correct manipulation of apparatus, the accuracy of measurements, and health and safety considerations.
- WS 2.6 Make and record observations and measurements using a range of apparatus and methods.

Resources needed Plant which has been growing on its side for at least a week; normal plant; Venus fly trap; mimosa plant. Worksheet 5.25.1 (low demand); Worksheet 5.25.2 (standard demand); Worksheet 5.25.3 (high demand); Practical sheet 5.25; Technician's notes 5.25

Digital resources

Key vocabulary auxin, gravitropism, phototropism, tropism

Teaching and learning

Engage

- Show students a real example of a plant. Ask them to **discuss** in pairs the directions that the plant is growing in and why. [O1, O2]
- Show students an example of a plant that was grown on its side. Ask students to **describe** the differences between the two plants. Ask them to **suggest** reasons why the plants have grown in this way. [O1, O2]
- Show students a Venus fly trap and if possible a mimosa plant, or show a video of these plants in action. **Discuss** what the plant is doing and why. [O1]

Challenge and develop

- **Describe** the term 'tropism', the different types of tropisms that exist and how they enable plants to act. [O1, O2]
- Ask low demand students to **match** the statements to the correct tropisms on Worksheet 5.25.1.
- Standard demand students should **name** the tropisms shown in the images on Worksheet 5.25.2.
- In addition to the standard demand task, high demand students should **explain** what survival benefits the tropisms bring to each situation (Worksheet 5.25.3).
- Ask students to **carry out** the practical activity outlined in Practical sheet 5.25. [O1, O2]
- Show students how auxins stimulate plant growth and how their unequal distributions affect the growth of roots and shoots. **Discuss** what happens to different parts of the shoot and root and how this causes unequal growth. [O3]

Explain

- Ask students to **annotate** the diagrams of the shoots to **explain** the responses to light in section 2 of their worksheet. They should **show** the distribution of auxins in the shoots and draw cells to show how the auxin alters growth. [O3]
- Ask students to **annotate** the diagrams of the roots to **explain** the responses to gravity in section 3 of their worksheet. They should **show** the distribution of auxins in the shoots and draw cells to show how the auxin alters growth. [O3]

Consolidate and apply

- Ask students to work in small groups. Students should **produce** a short animation, if possible, using models they make themselves, or **use role play** and models to **explain** what plant tropisms are and how they work. [O1, O2, O3]

Extend

Ask students able to progress further to:

- **find out** more about other types of plant tropisms such as chemotropisms, hydrotropisms and other types of stigmatropisms. They should **explain** how they work and the survival implications behind them. [O1, O2]

Plenary suggestions

Ask students to compare and contrast plant tropisms to hormonal effects in humans.

Answers to Worksheet 5.25

1 What is a tropism?

Low demand: 1. positively phototropic; 2. positvely geotropic; 3. negatively geotropic; 4. negatively phototropic

Standard and high demand: A. positive phototropism; B, negative geotropism; C. postive geotropism; D. thigmatropism

High demand:

A. By having a positive phototropism, plant shoots can grow towards the light which is needed for photosynthesis.
B. By having a negative geotropism effect, plant shoots grow upwards towards the light and not downwards away from it.
C. Roots are positviely geotropic so that they grow downwards to where there is likely to be more water, which they can absorb for photosynthesis.
D. Thigmotropism responses can provide either protection or food for the plant.

2 Explaining phototropism

Higher distribution of auxin shown on the dark side of the shoot with longer cells on this side and shorter cells on the side with the light, which causes the tip to bend.

3 Explaining geotropism

Greater distribution of auxin on the lower side of the root. This inhibits cell growth and so there are shorter cells on this side compared with the upper side, which causes the root to bend downwards.

Lesson 26: Applications of auxins

Lesson overview

AQA specification reference

AQA 4.5.4.2

Learning objectives

- Explain how auxins coordinate and control responses to light and gravity.
- Explain that auxins act on 'stem cells' in plants called meristems.
- Describe some applications of auxins.

Learning outcomes

- Describe how light and gravity affect auxins. [O1]
- Explain how auxins act on stems cells. [O2]
- Explain some applications of auxins. [O3]

Skills development

- WS 2.6 Make and record observations and measurements using a range of apparatus and methods.
- WS 3.1 Present observations and other data using appropriate methods.
- WS 3.6 Present reasoned explanations including relating data to hypotheses.

Resources needed Auxins animation from www.kscience.co.uk. Worksheet 5.26.1 (low demand); Worksheet 5.26.2 (standard demand); Worksheet 5.26.3 (high demand); Practical sheet 5.26; Technician's notes 5.26

Digital resources

Key vocabulary meristem, plant cuttings, selective weedkiller, stem cells, tissue culture

Teaching and learning

Engage

- Remind students about how auxins work. Show students the animation about auxins from www.kscience.co.uk to remind them of ideas from the previous lesson.
- Ask students to work in pairs and **talk about** the ideas they see in the animation. Ask them to **summarise how** auxins work in the shoots. [O1]

Challenge and develop

- Ask students to **carry out** the practical on Practical sheet 5.26. They will **investigate** auxins in plant shoot tips, **make observations** and **draw conclusions**. They should **review** the results in a week or so. [O1]
- Low demand: show students some results from a simple investigation (section 1 of Worksheet 5.26.1). Ask students to **describe** what the results are showing. [O1]
- Standard and high demand: show students some results from shoots with gelatin placed on the tips (Worksheet 5.26.2, standard demand; Worksheet 5.26.3, high demand). Ask them to **explain** these results. [O1]
- **Discuss** the role of plant growth using auxins. Introduce the idea of meristems and how they are employed in providing stem cells, which are used in cloning plants. [O2]
- Show images of plants being sprayed with auxins as weedkillers. Ask high demand students to **explain** how this might be possible. [O3]

Explain

- Show high demand students the data of how auxins affect cells in plant shoots and roots on Worksheet 5.26.3. Ask them to **draw conclusions** from the data and **explain how** it accounts for the use of auxins as weedkillers. [O2, O3]
- Students should **write** an explanation of how auxins and stem cells are used in growing new plants. [O1, O2]

Consolidate and apply

- Students should **compare** and **contrast** the applications of auxins. [O1, O2, O3]
- Ask low and standard demand students to **read** page 222 in the Student Book and **answer** questions 1 and 2. [O1, O2]
- High demand students should read pages 222–223 and answer questions 3–6 on page 223 of the Student Book. [O1, O2, O3]

Extend

Ask students able to progress further to:

- **find out** more about the experiments performed by Charles Darwin and others on shoot tips, proving the existence of plant hormones and their actions. [O1]]

Plenary suggestions

Students should **summarise** everything they know about auxins in a paragraph. They should reduce this to five bullet points and then to one sentence. [O1, O2, O3]

Answers to Worksheet 5.26

1 Explaining the actions of auxins

Low demand: the shoot with the aluminium foil did not bend to the light but grew straight. The normal tip bent to the light; the aluminium blocked the light from the tip, proving that the tip is responsible for growing to the light.

Standard and high demand: the stem with the gelatin gel between it and the replaced tip continued to grow towards the light. The tip with the metal foil between it and the shoot grew straight up. This shows that the auxin was able to diffuse from the tip to the rest of the shoot. The foil, however, did not let the auxin diffuse through, so the shoot grows upright.

2 Explaining the uses of auxins

1. As the concentration of auxin increases to 10^{-10} mol dm^{-3}, the growth of root cells increases to a maximum. After this, their concentration decreases at a steady rate. Growth is inhibited after 5×10^{-8} mol dm^{-3}.

   As the concentration of auxin increases, the growth of shoots increases rapidly to a maximum at 5×10^{-6} mol dm^{-3}. After this maximum, concnetrations rapidly decrease. Auxin concentrations become inhibitory after 4×10^{-4} mol dm^{-3}.
2.
   a. Concentrations of auxin between 10^{-12} and 10^{-10} mol dm^{-3} will promote root growth increase and can be used in cuttings to promote the growth of roots.

   b. Concentrations of 10^{-9} to 5×10^{-6} will increase stem cell growth and can be used to promote the growth of stem cells.

   c. Concentrations of 5×10^{-6} mol dm^{-3} will greatly increase stem cell growth and can be used to rapidly enhance the growth of weeds, making them weak and killing them.

Lesson 27: Required practical: The effect of light and gravity on the growth of germinating seeds

Lesson overview

AQA Specification reference

AQA 4.5.4.1

Learning objectives

- Describe how an experiment is planned for a specific purpose.
- Make and record observations and translate data from one form to another.
- Interpret observations and other data, identifying patterns and trends, make inferences and draw conclusions.

Learning outcomes

- Plan and evaluate an investigation to identify the effect of light on the growth of seedlings. [O1].
- Record and analyse observations and produce labelled scientific drawings. [O2]
- Describe and explain observations using ideas about tropisms. [O3]

Skills development

- WS 2.2 Plan experiments or devise procedures to make observations, produce or characterise a substance, test hypotheses, check data or explore phenomena.
- WS 2.3 Apply a knowledge of a range of techniques, instruments, apparatus, and materials to select those appropriate to the experiment.
- WS 2.4 Carry out experiments appropriately having due regard for the correct manipulation of apparatus, the accuracy of measurements and health and safety considerations.
- WS 3.7 Be objective, evaluate data in terms of accuracy, precision, repeatability and reproducibility and identify potential sources of random and systematic error.

Maths focus

- 4c Plot two variables from experimental or other data.

Resources needed Recently germinating cress or radish or white mustard plants; one group of seedlings already prepared where the light intensity has been applied at three different distances the week before the lesson. Worksheet 5.27; Practical sheet 5.27; Technician's notes 5.27

Digital sources

Key vocabulary phototropism

Teaching and learning:

Engage

- Show students the experiment 'Responding to gravity' described on page 224 of the Student Book, relating to the effect of gravity on maize seedlings. Ask students to work in pairs and **answer** questions 1–4. [O1]
- Show students some recently germinating radish, cress or white mustard plants. Tell students that they need to **investigate** the effect of light on the plants. Ask students to work in pairs and **identify** some questions that could be asked about how light intensity affects plant growth on their worksheet. They should **discuss** all the variables involved with this investigation and list them, and **identify** how they could be controlled. [O1]
- Pairs should join as fours and **share** ideas. [O1]

Challenge and develop

- Low and standard demand students should work in small groups and **plan** an investigation around the question they selected on the worksheet. Students should **select** appropriate equipment, **identify** the independent, dependent and control variables, and **decide** how to carry out the investigation so that reliable and reproducible evidence can be collected. [O1]

- High demand students should **devise** an investigation to test the effect of blue and red light on gravitropism, given sheets of coloured filters of red, green and blue. [O1]
- Introduce the idea of systematic and random error. Ask students to **consider** these in their planning. Ask students to **devise** their own success criteria for a good plan. **Discuss** the ideas as a class and **devise** an agreed set of success criteria. Students should **identify** any random or systematic errors in their investigation and **explain** ways these may be limited on their worksheet. [O1, O2]

Consolidate and apply

- Ask groups to **peer assess** each other's plans, **identifying** three things that are good with the plan and one thing that needs to be improved. Ask groups to **make** the necessary improvements. [O1]
- Students should **carry out** their procedure given in Practical sheet 5.27 and set up the investigation to investigate the effect of phototropism on shoots. They may need to carry this out over a few days to collect the data required. [O1, O2]
- Show students the experiment and data from page 225 of the Student Book where the angle of the stems was measured. Ask students to **complete** questions 7–10. Ask students to **identify** the strengths and weaknesses of this investigation in their worksheet and to **analyse** the data. [O3]

Explain

- Ask students to **make** the appropriate scale drawings and observations from the images in the Phototropism section of the Student Book (questions 5–6 on page 224). [O2]
- Students should **draw conclusions** from the data and **suggest explanations**. [O3]

Extend

Ask students able to progress further to:

- **find out** about how to investigate the effect of different wavelengths of light and **devise** their own plan to investigate this question. They should **identify** the equipment needed and **explain** how accurate and reliable observations could be made. [O1, O2]

Plenary suggestions

Ask students to **consider** alternative ways they could have carried out the investigation. They should work in pairs to **devise** a different way to carry out the investigation they have planned. [O1, O2]

Ask students to **summarise** the important factors in planning and carrying out this investigation. [O1, O2]

Answers to Worksheet 5.27

Sections 1 and 2: an example question:

Does the intensty of light affect the degree of curvature – independent variable – intensity of light;dependent variable – degree of curvature; control variables – amount of water, amount of nutrients in the soil, same type and age of plant

Suggested method: place a few plants, e.g. at least three, at different distances from a light source and leave them for a few days. Measure the angle of curvature every 8 hours,or take photos of the shoots. Add the same amount of water to all the plants and ensure they are planted in the same soil. Use plants of the same species which were germinated at the same time.

3 Evaluation

1. Strengths: Dependent variable is measured accurately using a protractor; There are no subjective ways of collecting data; There is a small hole so the light falls exactly on the selected place on the seedling; The temperature, nutrients and amount of water are carefully controlled

   Weaknesses: Light has to be shone on the seedlings in order to measure the angle of curvature; Seedling has to be removed and measured. This could cause damage to the seedling
1. Graph drawn with the time on the *x*-axis and mean angle on the *y*-axis
2. Conclusions – as the time increases, the mean angle of curvatue increases. There is a shallow increase, followed by a sharp increase after 120 minutes.

Lesson 28: Other plant hormones

Lesson overview

Note: the content of this lesson is Higher Tier only.

AQA specification reference

AQA 4.5.4.2

Learning objectives

- Recall that gibberellins are important in seed germination, and ethene in cell division and ripening of fruit.
- Explain the application of the plant hormones ethane and gibberellins.

Learning outcomes

- Describe the functions of plant hormones in plants. [O1]
- Describe some commercial applications of plant hormones. [O2]
- Evaluate the artificial use of plant hormones. [O3]

Skills development

- WS 1.2 Use a variety of models such as representational, spatial, descriptive, computational and mathematical to solve problems, make predictions and to develop scientific explanations and understanding of familiar and unfamiliar facts.
- WS 1.4 Explain everyday and technological applications of science. Evaluate associated personal, social, economic and environmental implications, and make decisions based on the evaluation of evidence and arguments.
- WS 3.5 Interpret observations and other data (presented in verbal, diagrammatic, graphical, symbolic or numerical form), including identifying patterns and trends, making inferences and drawing conclusions.

Resources needed Some real plants, fruits, germinating seeds and larger fruits. Worksheet 5.28

Digital resources

Key vocabulary dormant, ethene, gibberellins, hydrocarbon

Teaching and learning

Engage

- Ask students to work in pairs and **recap** the applications of auxins from the previous lessons. Show students some real plants, fruits, germinating seeds and larger fruits to introduce other plant hormones and their uses in the plant. [O1]

Challenge and develop

- Show students the data on how much ethene is produced by a ripening tomato from Worksheet 5.28. Students should explain the shape of the curve. [O1]
- Discuss possible applications of this data. [O1]
- Ask students to work in small groups and use the Internet and Student Book to **research** more completely the role of gibberellins and ethene in plants. They should **annotate** the diagram in the worksheet to **explain** the function of each hormone. They should **identify** whether the hormones have different roles in different types of plants. [O1]
- Ask students to work in groups and **find out** about the commercial applications of plant hormones in seed germination, fruit ripening, and promotion of flowering and controlling fruit size. Students should **explain how** plant hormones are harnessed and what benefits they have. [O1, O2, O3]

Explain

- Students should **use** the Student Book and other sources to **explain** the advantages and disadvantages of using plant hormones. [O3]

Consolidate and apply

- Show students the images of different applications of plant hormones in the worksheet. They should **identify** the hormones in use and **write** a summary of what the function of the hormone is and how it is applied. [O1, O2, O3]

Extend

Ask students able to progress further to:

- **find out** how plant hormones were first isolated and discovered and what experiments were done to **identify** their commercial applications. [O3]

Plenary suggestions

Ask students to answer the true or false questions in section 4 of the worksheet. [O1, O2, O3]

Answers to Worksheet 5.28

1 Interpreting data on ethene levels

As the fruit ripening process increases, the levels of ethene increase. When the fruit is completely ripe, the amount of ethene is at its maximum. The amount of ethene is related to the ripening of the fruit. Ripe fruit give off the most ethene.

2 How do plant hormones act in plants?

Giberrelins are involved in initiating seed germination.

Ethene controls cell division and ripening of fruits.

Giberrelins are also involved in the growth of shoots and promoting flowering.

3 Commercial applications of hormones

bigger grapes – ethene for fruit ripening; giberrelins and auxins for growing larger fruit

ripening tomates – ethene

ripening bananas – ethene

growing sugar cane – auxins and giberrelins

seedless melon and organges – auxins

flowering – giberrelins

germinating seeds – giberrelins

4 True or false?

1. F
2. F
3. F
4. T
5. F
6. T
7. T

Lesson 29: Maths skills: The spread of scientific data

Lesson overview

Learning objectives

- Be able to use range bars on graphs.
- Understand how box and whisker plots can be used to show the spread of data.
- Understand how to use percentiles.

Learning outcomes

- Determine the interquartile range and plot range bars. [O1]
- Compare the spread of data using range bars, and box and whisker plots. [O2]
- Use percentiles to draw conclusions. [O3]

Skills development

- WS 3.1 Present observations and other data using appropriate methods.
- WS 3.2 Translate data from one form to another.
- WS 3.3 Carry out and represent mathematical and statistical analysis.

Maths focus

- 4a Translate information between graphical and numeric form

Resources needed Graph paper, calculators, pencils and rulers. Worksheet 5.29.1 (low demand); Worksheet 5.29.2 (standard demand); Worksheet 5.29.3 (high demand)

Digital resources

Key vocabulary percentile, quartile, range bar, repeatability

Teaching and learning

Engage

- Show students the graph from page 228 of the Student Book. Ask students to work in pairs and **identify** what the bars and circles on the graph might represent. Ask students to **discuss** why the bars are of different heights. [O1]
- Model how to draw a range bar graph. Ask students to then **complete** section 1 of Worksheet 4.29 and draw their own range graph. [O1]

Challenge and develop

- Show students a box and whisker plot. Ask students to **identify** what further information can be obtained from these graphs compared with the range graph. [O2]
- Explain and **discuss** how to construct a box and whisker plot and how to determine the lower and upper quartile ranges.
- Low demand students then **plot** a box and whisker plot of the data in section 2 of Worksheet 5.29.1. [O1, O2]
- Standard demand students then **plot** a box and whisker plot of the data in section 2 of Worksheet 5.29.2 and **explain** the differences between the data sets. [O1, O2]
- High demand students then **plot** the data as a box and whisker plot and as a range graph. **Explain** the differences between the two different ways of representing the data and **explain** the strengths and weakness of each. [O1, O2]
- Discuss percentile graphs. Ask students to **identify** other applications of percentile graphs.

Explain

- Students should **explain** how to construct a percentile graph from a box and whisker plot. [O3]

Consolidate and apply

- Students should construct a percentile graph from the data in section 3 of their worksheet. [O1, O2, O3]
- Students should **read** pages 228–229 of the Student Book and **answer** all the questions. [O1, O2, O3]

Extend

Ask students able to progress further to:

- **find out** about situations where it is preferable to use a percentile graph instead of a range plot. [O3]

Plenary suggestions

Students should **write** a short summary to **explain** what range, box and whisker and percentile plots are and where they can be used. [O1, O2]

Answers to Worksheet 5.29

1 Range graphs

1. Mean values:1.58, 1.94, 2.67, 3,00
2. Graph plotted of concentration on *x* axis and mass of grape on *y* axis. First plot should have bar from 1.45 to 1.7 with a cross at 1.58; second plot should have a bar from 1.85 to 2.05, with a cross at 1.94; third plot from 2.65 to 2.70, with a cross at 2.67; fourth plot from 3.00 to 3.02 with a cross at 3.00.

2 Box and whisker plots

1.

	Concentration 0	**Concentration 5**	**Concentration 20**	**Concentration 50**
Mass of grapes (g)	1.40	1.85	2.65	3.00
	1.43	1.90	2.65	3.00
	1.45	1.90	2.65	3.00
	1.48	1.95	2.68	3.00
	1.50	1.95	2.68	3.01
	1.65	2.00	2.70	3.02
	1.70	2.05	2.70	3.02

Concentration of giberellin sprayed	**Minumum mass of grape (g)**	**Maximum mass of grape (g)**	**Median mass of grape(g)**	**Lower quartile**	**Upper quartile**
0	1.40	1.70	1.48	1.45	1.65
5	1.85	2.05	1.95	1.90	2.00
20	2.65	2.70	2.68	2..65	2.70
50	3.00	3.02	3.00	3.00	3.02

Standard demand:

3. The graph shows that as the concentration of giberrellin increases, the mass of grapes increases sharply and then less steeply. The higher the concentration of giberrellin, the more repeatable the data, as the bars as closer together. The greatest spread of data is with no giberrellin.
4. Interquartile range 0 = 0.2; 5 = 0.1; 20 = 0.05; 50 = 0.02

High demand:

The whisker and box plot tells us how evenly spread the data is and that there is a pattern within the range. This can also be seen in the range plot, but the spread of the data is not clear from this plot.

When and how to use these pages: Check your progress, Worked example and End of chapter test

Check your progress

Check your progress is a summary of what students should know and be able to do when they have completed the chapter. Check your progress is organised in three columns to show how ideas and skills progress in sophistication. Students aiming for top grades need to have mastered all the skills and ideas articulated in the final column (shaded pink in the Student book).

Check your progress can be used for individual or class revision using any combination of the suggestions below:

- Ask students to construct a mind map linking the points in Check your progress
- Work through Check your progress as a class and note the points that need further discussion
- Ask the students to tick the boxes on the Check your progress worksheet (Teacher Pack CD). Any points they have not been confident to tick they should revisit in the Student Book.
- Ask students to do further research on the different points listed in Check your progress
- Students work in pairs and ask each other what points they think they can do and why they think they can do those, and not others

Worked example

The worked example talks students through a series of exam-style questions. Sample student answers are provided, which are annotated to show how they could be improved.

- Give students the Worked example worksheet (Teacher Pack CD). The annotation boxes on this are blank. Ask students to discuss and write their own improvements before reviewing the annotated Worked example in the Student Book. This can be done as an individual, group or class activity.

End of chapter test

The End of chapter test gives students the opportunity to practice answering the different types of questions that they will encounter in their final exams. You can use the Marking grid provided in this Teacher Pack or on the CD Rom to analyse results. This shows the Assessment Objective for each question, so you can review trends and see individual student and class performance in answering questions for the different Assessment Objectives and to highlight areas for improvement.

- Questions could be used as a test once you have completed the chapter
- Questions could be worked through as part of a revision lesson
- Ask Students to mark each other's work and then talk through the mark scheme provided
- As a class, make a list of questions that most students did not get right. Work through these as a class.

Marking Grid for End of Chapter 5 Test

	Student Name																			
Getting Started [Foundation Tier]	Q. 1 (A01) 1 mark																			
	Q. 2 (A01) 2 marks																			
	Q. 3 (A02) 2 marks																			
	Q. 4 (A02) 3 marks																			
	Q. 5 (A03) 2 marks																			
	Q. 6 (A01) 2 mark																			
	Q. 7 (A01) 1 mark																			
Going Further [Foundation and Higher Tiers]	Q. 8 (A01, A02) 4 marks																			
	Q. 9 (A03) 3 marks																			
	Q. 10 (A01) 2 marks																			
	Q. 11 (A01) 2 marks																			
	Q. 12 (A02) 6 marks																			
More Challenging [Higher Tier]	Q. 13 (A01) 2 marks																			
	Q. 14 (A02) 4 marks																			
	Q. 15 (A03) 4 marks																			
	Q. 16 (AO1) 2 marks																			
	Q. 17 (AO2) 1 mark																			
	Q. 18 (AO2) 1 Mark																			
	Q. 19 (AO2) 2 Marks																			
Most demanding [Higher Tier]	Q. 20 (AO3) 4 Marks																			
	Q. 21 (AO1) 2 Marks																			
	Q. 22 (AO1) 2 Marks																			
	Q. 23 (AO1) 6 Marks																			
	Q. 24 (AO2) 2 Marks																			
	Q. 25 (AO2) 2 Marks																			
	Q. 26 (AO2) 6 Marks																			
	Q. 27 (AO3) 4 Marks																			
Total marks																				
Percentage																				

Check your progress

You should be able to:

recognise the need for homeostasis →	recall that the nervous and endocrine systems are responsible for homeostasis →	describe the components of body control systems and compare nervous and hormonal control
describe the structure of the nervous system and location of the endocrine glands →	describe how neurones are adapted to their role and the transmission of an impulse in reflex arc →	explain how nerve impulses are transmitted across synapses
recall the three regions of the brain and their location →	describe the function of the brain regions →	explain the processes involved and why investigating the brain and treating disorders is difficult
identify the structure and function of parts of the eye →	describe how images are focused on the retina and what causes eye defects →	explain how eye defects can be corrected
identify temperature, water and glucose concentrations as conditions requiring control →	describe how these are controlled by the nervous and/or endocrine system →	explain how negative feedback is involved in homeostasis
identify the reproductive hormones and their role in the development of secondary sexual characteristics →	describe the role of hormones in the menstrual cycle →	explain the roles and interactions of hormones in the menstrual cycle
identify fertility drugs and *in-vitro* fertilisation as possible solutions to infertility →	describe the techniques involved in fertility treatments →	evaluate fertility treatments
identify different methods of contraception →	describe how different methods of contraception work →	evaluate the use of different methods of contraception
identify types of plant hormone and their role in plant growth →	describe the role of redistribution of auxins in tropisms →	describe applications of plant hormones

Worked example

1 Describe how the kidneys produce urine.

The kidneys produce urine by filtering waste substances from the blood. If there is more water than we need in the body, the urine will be dilute. If there is not enough, it will be concentrated.

This is a part answer to the question. The filtration and reabsorption processes should be described in more detail.

During filtration of the blood, small molecules, such as glucose and urea, and ions, pass into the kidney tubule. Useful substances, such as glucose, are reabsorbed back into the blood, along with water and some ions. As the student answer says, the composition of urine as it leaves the kidney will depend on amount of water and ions that the body needs to reabsorb.

As a general point, while a lead-in to the answer is important, the student has simply repeated part of the question in the answer, which reduces space for writing the answer.

2 Explain how the body uses the hormone ADH to regulate its water level.

A hormone called ADH controls the amount of water in the body. If the body is short of water, the pituitary gland produces more ADH. It acts on the kidney to reduce water lost.

This answer could be improved. The student could have described the condition that brings about ADH release as being the blood becoming *more concentrated*, rather than the body being 'short of water'. The student could have been more specific in terms of ADH promoting reabsorption of water by the kidney *tubules*, and the reverse of this, when water levels in the body become too high. The student has omitted a key term – negative feedback.

A scientific point is that ADH is only released from the pituitary gland. It is *produced* elsewhere.

Genetics: Introduction

When and how to use these pages

This chapter builds on ideas about DNA and chromosomes from KS3 – the idea that the nucleus contains chromosomes made from DNA. It links to Chapter 1, especially lesson 1.6 Cell division. There are also links to several Chapter 7 lessons: 7.1 Variation, 7.3 The origin of species by natural selection, 7.6 Darwin and Wallace, and 7.18 Cloning.

Overview of the chapter

In this chapter, students will learn about the human genome and how the genome is used to learn more about disease and human migration patterns. Students will investigate the structure of DNA and how the DNA base code is used to synthesise proteins. They will then find out about mutations and how they can affect protein function. Students will also learn about how cells divide by meiosis to form gametes. Finally, students will investigate the work of Gregor Mendel and learn how to interpret genetic cross diagrams in the form of Punnett squares.

This chapter offers a number of opportunities for the students to use mathematics to calculate percentages and probability. They will investigate how to test a hypothesis and interpret results.

Obstacles to learning

Students may need extra guidance with the following terms and concepts:

- Some students will struggle with the high quantity of scientific terms used in genetics. It is important that they are made aware of the terms used in the specification and are given plenty of opportunity to use them.
- Many students struggle with the concept of having two copies of a gene and that these copies may be different.
- Understanding exactly how proteins are synthesised using the DNA code can become extremely complex. This topic should be avoided with students that will definitely only be entered for the foundation tier.
- Many students think that all mutations are harmful; they should be shown many examples where they are not. Only the high demand students are likely to understand the concept of amino acids having more than one triplet code.
- Many students will struggle with the process of meiosis, the stages are not required and for many students it may just be a case of recalling the idea that four genetically different gametes are produced with half the number chromosomes and not how they are formed.
- Some students are put off by the maths in a genetic cross. Practice and the knowledge that the ratio is likely to be either 1 to 1 or 3 to 1 will help.

Practicals in this chapter

In this chapter students will do the following practical work.

- Extracting DNA from plant material
- Modelling DNA structure
- Investigating probability

	Lesson title	Overarching objectives
1	DNA and genes	Describe a gene as a small section of DNA that codes for a protein.
2	The human genome	Explain the importance of understanding the human genome.
3	Tracing human migration	Discuss the use of the human genome in understanding human migration patterns.
4	The structure of DNA	Describe the structure of DNA as repeating nucleotide units, identify the four bases in DNA and explain that the bases A and T, and C and G, are complementary.
5	Proteins	Describe how proteins are synthesised according to the DNA template of a gene and explain that the genetic code of a gene specifies the protein to be made.
6	Mutations	Describe mutations as changes to the base sequence of DNA, explain that their effects can be positive or negative and describe how mutations can affect protein function.
7	Meiosis	Explain how meiosis halves the number of chromosomes to produce four gametes that are genetically different and explain how fertilisation restores the chromosome number.
8	Asexual and sexual reproduction	Understand the advantages of asexual and sexual reproduction.
9	Genetics	Understand and be able to use genetics terms, such as dominant, recessive, genotype, phenotype, homozygous and heterozygous.
10	Genetic crosses	Know that some human conditions, such as cystic fibrosis, are caused by a recessive allele and be able to complete or construct a Punnett square to predict the outcome of a genetic cross.
11	Tracking gene disorders	Know that some human conditions, such as polydactyly, are caused by a dominant allele and understand the use of a family tree to show the inheritance of a characteristic. Explain economic, social and ethical issues concerned with embryo screening.
12	Gregor Mendel	Research how Mendel worked scientifically to make observations, draw conclusions and then test his hypotheses.
13	Key concept: Genetics is simple – or is it?	Understand that certain characteristics are controlled by a single gene but many characteristics are the result of multiple genes which interact and describe the search for genes that are linked to diseases.
14	Maths skills: Fractions, ratios, proportion and probability	Understand and use fractions, percentages, ratio and proportion, and probability when predicting the outcomes of genetic crosses.

Lesson 1: DNA and genes

Lesson overview

AQA Specification reference

AQA 4.6.1.4; AQA 4.6.1.5

Learning objectives

- Describe the structure of DNA.
- Describe a gene as a small section of DNA that codes for a protein.

Learning outcomes

- Understand the meaning of the term genome. [O1]
- Understand that DNA is found in the nucleus of the cell as chromosomes. [O2]
- Understand that small sections of DNA are called genes and that genes are codes for proteins. [O3]

Skills development

- WS 2.2 Plan experiments or devise procedures to make observations, produce or characterise a substance, test hypotheses, check data or explore phenomena.
- WS 2.3 Apply a knowledge of a range of techniques, instruments, apparatus, and materials to select those appropriate to the experiment.
- WS 2.4 Carry out experiments appropriately having due regard for the correct manipulation of apparatus, the accuracy of measurements, and health and safety considerations.

Resources needed Worksheet 6.1; Practical sheet 6.1; Technician's notes 6.1

Digital resources

Key vocabulary gene, genome

Teaching and learning

Engage

- **Write** the word 'genome' on the whiteboard. [O1]
- Students should **discuss** in groups what they believe the word means. Many should be familiar with this word or at least the word 'gene'. They may have also studied chromosomes in Chapter 1. [O1]
- Groups then **report back** their ideas in a class discussion. [O1]

Challenge and develop

- Develop the class discussion to **explain** that the genome is the entire genetic material of an organism. [O1]
- Ensure that students **understand** that the genetic material is made up of DNA and is found in the nucleus of the cell on structures called chromosomes. If they have already studied mitosis in Chapter 1 they should be familiar with this concept. [O2]
- Low demand students then **complete** the questions about cells and DNA on Worksheet 6.1. [O2]
- Standard and high demand students should **write** a summary explaining the terms genome, gene, DNA and chromosome.

Explain

- Ask students to **look** at the image of chromosomes on page 239 of the Student Book. [O3]
- **Discuss** the idea of pairs of chromosomes having the same types of genes. Can they **recall** how many chromosomes are in the nucleus of a human cell. [O3]
- **Explain** that the genes are codes for proteins and that each cell has two versions of the code. In many cases they are the same; however, for some characteristics the genes may be different. Eye colour is a classic example. Students will revisit this concept when they study genetics. [O3]

Consolidate and apply

- Students then **extract** DNA from plant material by following the instructions on Practical sheet 6.1. [O2, O3]
- Question 2 on the practical sheet could be completed in groups or discussed as a class. [O2, O3]

Extend

Ask students able to progress further to:

- (high demand) **complete** questions 5–7 on page 239 of the Student Book. [O3]

Plenary suggestions

Inform students that when you look at the human genome and that of a chimpanzee there is 96% similarity; however, there are some 40 million differences among the three billion DNA molecules.

Ask the students to **suggest** what conclusions they can make from that information.

Answers to Worksheet 6.1

1. membrane, nucleus, cytoplasm, chromosomes, DNA, gene
2. The genome of an organism is the entire genetic material of that organism.

Answers to Practical sheet 6.1

1. protease
2. independent variable = different plant materials
3. dependent variable = mass of DNA made

   controlled variables = mass of plant material used, the volumes of all liquids used, and the volume of alcohol and mixture added to the test tube

   Apparatus list and method are the same as that used in the lesson except the sample should be dried and weighed at the end, so a balance will be required.

Lesson 2: The human genome

Lesson overview

AQA Specification reference

AQA 4.6.1.4; AQA 4.6.1.5

Learning objectives

- Describe a gene as a small section of DNA that codes for a protein.
- Explain the importance of understanding the human genome.

Learning outcomes

- Understand that knowledge of the human genome has enabled us to search for genes linked to different diseases. [O1]
- Understand that knowledge of the human genome can increase our understanding and treatment of inherited diseases. [O2]
- Explain ethical issues associated with knowledge of the human genome. [O3]

Skills development

- WS 1.1 Understand how scientific methods and theories develop over time.
- WS 1.4 Explain everyday and technological applications of science. Evaluate associated personal, social, economic and environmental implications, and make decisions based on the evaluation of evidence and arguments.
- WS 3.8 Communicating the scientific rationale for investigations, methods used, findings and reasoned conclusions through paper-based and electronic reports and presentations using verbal, diagrammatic, graphical, numerical and symbolic forms.

Resources needed Video about the Human Genome Project; video on the ethics of the Human Genome Project; poster-making equipment; Internet access. Worksheet 6.2

Digital resources

Key vocabulary gene expression, gene therapy, genome, genomics, genome editing

Teaching and learning

Engage

- **Show** students an image of our karyotype. [O1, O2]
- Ask the students to **describe** what they think the image is showing. [O1, O2]
- Once the class has determined that the image shows human chromosomes, **discuss** with the students how they think their set of chromosomes would compare with those in the image. The idea is to bring out the fact that we all have 46 and that they are all made of DNA, most of the genes will be the same but there will be a slight difference that makes us each individual. [O1, O2]

Challenge and develop

- **Remind** students about the human genome from the previous lesson. **Explain** that in 1990 a project started to map all the genetic information on human chromosomes. Ask them to **suggest** some reasons why mapping the human genome is so important. [O1, O2]
- **Show** the students a video about the Human Genome Project. [O1, O2]
- **Discuss** with the students the idea that the Human Genome Project has helped scientists search for genes linked to different types of disease, and increased their understanding and treatment of inherited disorders. [O1, O2]

Explain

- Students should then **produce** a presentation that shows the importance of the Human Genome Project. They should use Worksheet 6.2 to help them and work in small groups. [O1, O2]
- Low demand students may need some **support** in their research. All the information is in the Student Book so the use of the Internet could be avoided if they are provided with samples of images they could add to their poster. Suitable images can be found by searching images for 'Human Genome Project'. [O1, O2]

Consolidate and apply

- Ask students to **suggest** some reasons why the Human Genome Project could cause ethical issues. Students could work in groups to **discuss** their ideas and then feed back to the class. A short video clip about the ethics could be shown as a stimulus to start the conversations. [O3]

Extend

Ask students able to progress further to:

- (high demand) **complete** questions 4–6 on page 241 of the Student Book. [O3]

Plenary suggestions

Ask students to take a **vote** on who would want to know their genetic makeup and who would rather not know. Ask them to **provide** reasons for their choice.

Answers to Worksheet 6.2

Students may find out more information if they use the Internet; the following answers cover the basics and are all found in the Student Book.

1. The genome of an organism is the entire genetic material of that organism.
2. The Human Genome Project (HGP) was a study to map all the genetic information on the chromosomes of a human being.
3. The work began in 1990 and the research was published in 2003; however, the study is still continuing.
4. diabetes, cancer and heart disease
5. It has enabled them to develop effective treatments that target and correct defective genes.
   Gene therapy and genome editing are techniques being developed to do this.

Lesson 3: Tracing human migration

Lesson overview

AQA Specification reference

AQA 4.6.1.4; AQA 4.6.1.5

Learning objectives

- Explain the importance of understanding the human genome.
- Discuss the use of the human genome in understanding human migration patterns.

Learning outcomes

- Understand the meaning of the term migration. [O1]
- Explain how the human genome has been used to trace human migration patterns. [O2]
- Explain some of the ethical issues linked to the Genographic Project. [O3]

Skills development

- WS 1.1 Understand how scientific methods and theories develop over time.
- WS 1.4 Explain everyday and technological applications of science. Evaluate associated personal, social, economic and environmental implications, and make decisions based on the evaluation of evidence and arguments.
- WS 3.8 Communicating the scientific rationale for investigations, methods used, findings and reasoned conclusions through paper-based and electronic reports and presentations using verbal, diagrammatic, graphical, numerical and symbolic forms.

Maths focus

- 2c Construct and interpret frequency tables and diagrams, bar charts and histograms

Resources needed Videos about the Genographic Project and 'The journey of your past' from https://genographic.nationalgeographic.com/science-behind/ (click on 'resources', then 'the science behind'); Internet access; poster-making equipment. Worksheet 6.3.1 (low and standard demand); Worksheet 6.3.2 (high demand)

Digital resources

Key vocabulary genome, migration pattern, Y-chromosome

Teaching and learning

Engage

- **Show** the students the video 'about the Genographic Project' from https://genographic.nationalgeographic.com. [O1]
- **Discuss** what they think the project is all about from the clip. Would they like to take part? [O1]

Challenge and develop

- Ask students what they think the word 'migration' means. [O1]
- Ask students to look at the map on Worksheet 6.3, which shows current migration patterns. [O1]
- Ask them to **suggest** reasons why people migrate. [O1]
- Ensure that students **understand** that migration is not always about immigrants and that people can migrate within their own country. In fact, some of them may have migrated themselves from one part of the country to another. [O1]

Explain

- Ask students to **look** at the map on page 243 on the Student Book, which shows migration of ancient humans. [O1]
- **Discuss** the differences between the two maps with all the students and reasons why the patterns may be different. [O1]
- Ask the students to **suggest** how we have evidence of this ancient movement. **Explain** that there is evidence in the fossil record but by looking at the human genome, scientists have identified marker genes. Different populations will have different marker genes. By finding these marker genes we can trace our ancestors. [O1, O2]
- High demand students could look at the science of this in more detail by watching the 'The journey of your past' video on https://genographic.nationalgeographic.com. [O1, O2]
- Low and standard demand students then **complete** the questions on Worksheet 6.3.1. [O2, O3]
- High demand students then **complete** the questions on Worksheet 6.3.2. They will need information from the
- 'behind the science' section of the website https://genographic.nationalgeographic.com; if there is no Internet access, this could be printed off before the lesson. [O2, O3]

Consolidate and apply

- **Discuss** with students some of the ethical issues linked to the Genographic Project. [O3]
- Students then **complete** questions 5–7 on page 243 of the Student Book. [O3]
- Students could then **produce** a 'wanted' poster advertising for volunteers to join the Genographic Project. They may need more information about the benefits of being part of the study, which can be found on https://genographic.nationalgeographic.com. [O1, O2, O3]

Extend

Ask students able to progress further to:

- **write** a short summary to **explain** why they would or would not like to be part of the Genographic Project. [O3]

Plenary suggestions

Ask the students to suggest which country they would like to migrate to and explain why.

Answers to Worksheet 6.3

1. The study and comparison of current human genomes to help trace human migration patterns across history.
2. The HGP, focused on the whole human genome; the Genographic Project is only looking at part of it.
3. The **Y-chromosomes** –inherited down the male line – and from DNA in mitochondria.
4.
   a. Parts of the DNA that have been changed due to mutations; they are the parts that will be different and can therefore be used to identify differences in our ancestors.
   b. Idea that different populations carry distinct genetic markers. The genetic markers are followed back through the generations to reveal a genetic tree. The branches of the tree can be followed backward to their common African root. The markers present in our genes are used to chart ancient human migrations from Africa across the continents.

Lesson 4: The structure of DNA

Lesson overview

AQA Specification reference

AQA 4.6.1.5

Learning objectives

- Describe the structure of DNA as repeating nucleotide units.
- Identify the four bases in DNA.
- Explain that the bases A and T, and C and G, are complementary.

Learning outcomes

- Recall that DNA is polymer made up of repeating nucleotide units. The nucleotides form two strands joined together in the shape of a double helix. [O1]
- Recall that a nucleotide is made up of a sugar molecule and a phosphate molecule attached to one of four different bases. [O2]
- Identify the four bases in DNA as A,T, C and G, and that in the complementary strands, a C is always linked to a G on the opposite strand and a T to an A. [O3]

Skills development

- WS 1.2 Use a variety of models such as representational, spatial, descriptive, computational and mathematical to solve problems, make predictions and to develop scientific explanations and understanding of familiar and unfamiliar facts.
- WS 2.7 Evaluate methods and suggest possible improvements and further investigations.
- WS 4.1 Use scientific vocabulary, terminology and definitions.

Resources needed Video of DNA structure (search 'DNA structure GCSE biology'). Worksheet 6.4.1 (low demand); Worksheet 6.4.2 (standard and high demand); Practical sheet 6.4; Technician's notes 6.4

Digital resources

Key vocabulary base, complementary, double helix, genetic code, nucleotide, polymer

Teaching and learning

Engage

- Ask students to **crack** the code on section 1 of Worksheet 6.4.1 (low demand) or Worksheet 6.4.2 (standard and high demand). [O1]
- Link this to the idea that DNA is a code for making proteins in the body, which is revision from Lesson 1 in this chapter. [O1]

Challenge and develop

- **Show** students an image or model of DNA; for example, on page 245 of the Student Book. [O1]
- **Ask** students what they notice about the letters running down the centre of the DNA. [O2, O3]
- **Explain** that these letters are bases and they are the part of the DNA that codes for proteins. [O2, O3]
- **Explain** that a sequence of three bases is the code for a particular amino acid. The order of bases controls the order in which amino acids are assembled to produce a particular protein. [O2, O3] (*This concept is part of lesson 5, but it is all that students entered for a foundation paper will require from that lesson. It is suggested that this concept is included here so that foundation tier students may miss out lesson 5 at the discretion of the teacher.*)

Explain

- Use the diagram or model or a video (search 'DNA structure GCSE biology') to **discuss** the complete structure of DNA including ideas about nucleotides, bases and double helix. Students do not need to know the names of the bases, only the letters; however, high demand students may find it useful to be familiar with the names. Low demand students do not need to know the base pairs; however, without knowing this they cannot make their model later. [O1, O2, O3]
- Low demand students should **complete** section 2 'DNA structure' on Worksheet 6.4.1. [O1, O2, O3]
- Standard and high demand students should complete section 2 'DNA structure' on Worksheet 6.4.2. [O1, O2, O3]

Consolidate and apply

- **Introduce** the model-building activity described on Practical sheet 6.4. Divide the class into small groups and ask them to **complete** the activity. It is easiest to do this in groups of about four. Two students concentrate on a strand between them. Each pair of students takes a length of wire – it can be 50 cm, 1 m or longer depending on how much space, time and patience you have. You may wish to give less-able students a base sequence consisting of 20 base pairs. [O1, O2, O3]
- Students then **evaluate** their DNA model by completing the questions on Practical sheet 6.4. [O1, O2, O3]

Extend

Ask students able to progress further to:

- (high demand) **complete** questions 5 and 6 on page 245 of the Student Book. [O3]

Plenary suggestions

Ask the students to **judge** which group has made the best DNA model.

Answers to Worksheet 6.4

1 DNA code

1. Genes are made of DNA. Your DNA contains a genetic code. The code is used to put amino acids together to make proteins.

2 DNA structure

1. nucleotide, base, phosphate, sugar
2. double helix, nucleotides, sugar, phosphate, C, T and G (any order), polymer, three, protein

(High demand): The description should be very similar to the following:

DNA has a structure like a twisted ladder, the structure is called a double helix.
DNA has two strands; each strand is made up of units called nucleotides, joined together.
Each unit contains three parts: bases, a sugar molecule and phosphate group.
There are four different bases each represented by the letters A, T, C and G.
The two strands are complementary with A linked to T and C to G.
As there are many units the whole structure is called a polymer.
A sequence of three bases is the genetic code for a particular amino acid.
The order of bases controls the order in which amino acids are assembled to produce a particular protein.

Lesson 5: Proteins

Lesson overview

(The concept that 'A sequence of three bases is the code for a particular amino acid. The order of bases controls the order in which amino acids are assembled to produce a particular protein' is part of this lesson but it is all that students entered for a foundation paper will require from this lesson. The concept was also included in lesson 4 so that foundation tier students may miss out lesson 5 at the discretion of the teacher.)

AQA Specification reference

AQA 4.6.1.5

Learning objectives

- Describe how proteins are synthesised according to the DNA template of a gene.
- Explain that the genetic code of a gene specifies the protein to be made.

Learning outcomes

- Recall that DNA contains genes and each gene consists of a different sequence of bases. [O1]
- Understand that a sequence of three bases is a code for a particular amino acid. [O2]
- Explain that the base sequence is used to determine the order of amino acids in a protein. [O3]

Skills development

- WS 1.2 Use a variety of models such as representational, spatial, descriptive, computational and mathematical to solve problems, make predictions and to develop scientific explanations and understanding of familiar and unfamiliar facts.
- WS 3.8 Communicating the scientific rationale for investigations, methods used, findings and reasoned conclusions through paper-based and electronic reports and presentations using verbal, diagrammatic, graphical, numerical and symbolic forms.
- WS 4.1 Use scientific vocabulary, terminology and definitions.

Maths focus

- 1c Use ratios, fractions and percentages

Resources needed Video of protein synthesis (search 'protein synthesis GCSE'); poster material. Worksheet 6.5.1 (standard demand); Worksheet 6.5.2 (high demand)

Digital resources

Key vocabulary amino acid, ribosome

Teaching and learning

Engage

- **Write** the term 'amino acid' on the whiteboard. [O1, O2]
- Students should **discuss** in groups what they know about amino acids. If they have completed Chapter 2 they will know that amino acids are used to make proteins. [O1, O2]
- Groups then **report back** their ideas in a class discussion. [O1, O2]

Challenge and develop

- **Explain** to students that amino acids are the basic units of proteins and that we obtain amino acids by eating proteins. (They will learn more about this in Chapter 3.) [O1, O2]
- Ask the students to **suggest** what must happen to these amino acids so they become proteins again and how the structure of one protein may be different from another. [O2, O3]

- **Remind** students that a sequence of three bases is the code for a particular amino acid. The order of bases controls the order in which amino acids are assembled to produce a particular protein. [O2, O3] (*This concept was covered in Lesson 4; see note at the top of the page.)*
- Show the students a sequence of about 21 bases and ask them to **calculate** the number of amino acids it would put together. [O2, O3]

Explain

- Use a diagram or a video (search 'protein synthesis GCSE') to **explain** how the DNA code is used to make proteins. Explain that proteins are synthesised on ribosomes, according to a template. Carrier molecules bring specific amino acids to add to the growing protein chain in the correct order. They do not need to know the structure of mRNA or tRNA, but it is easier to explain the process if the terms are used. The Student Book refers to them as messenger molecules and carrier molecules; however, other sources of information used may have the correct names. [O1, O2, O3]
- **Explain** that, when the chain is complete, it folds up to form a unique shape. This unique shape enables the proteins to carry out their function. If they have studied enzymes, this can be related to the idea of the lock and key, from Chapter 3. [O1, O2, O3]

Consolidate and apply

- Standard demand students then **complete** the questions on Worksheet 6.5.1, which has images to help them draw their diagram. [O1, O2, O3]
- High demand students then **complete** the questions on Worksheet 6.5.2, which has no images to help them draw their diagram. [O1, O2, O3]

Extend

Ask students able to progress further to:

- (high demand) **complete** Questions 5 and 6 on page 247 of the Student Book. [O3]

Plenary suggestions

Ask the students to pair up; one member of each pair then describes the process of protein synthesis to their partner without the aid of the book.

The second student should then evaluate their description, explain what they got right and what they missed out.

Answers to Worksheet 6.5

1. A pairs with T, and C pairs with G.
2. So that it can leave the nucleus; DNA molecules are too large to leave the nucleus.
3. three
4. Diagram should be similar to that in Student Book (pages 246 and 247).

Lesson 6: Mutations

Lesson overview

(*Students entered for a foundation paper are not required to complete this lesson.*)

AQA Specification reference

AQA 4.6.1.5

Learning objectives

- Model changes to the base sequences of DNA to illustrate mutations.
- Describe the negative and, sometimes, positive effects of mutations.
- Describe how mutations can affect protein function.

Learning outcomes

- Understand that mutations are changes in the base sequence of DNA and describe examples of positive and negative effects of mutations. [O1]
- Understand that a mutation in the base sequence of DNA may alter the shape of a protein. [O2]
- Explain that non-coding parts of DNA can switch genes on or off and that mutations in these area of DNA may affect how a gene is expressed. [O3]

Skills development

- WS 1.2 Use a variety of models such as representational, spatial, descriptive, computational and mathematical to solve problems, make predictions and to develop scientific explanations and understanding of familiar and unfamiliar facts.
- WS 3.5 Interpreting observations and other data (presented in verbal, diagrammatic, graphical, symbolic or numerical form), including identifying patterns and trends, making inferences and drawing conclusions.
- WS 4.1 Use scientific vocabulary, terminology and definitions.

Resources needed Images of mutations, such as sickle cells; images of sickle-cell haemoglobin. Worksheet 6.6

Digital resources

Key vocabulary amino acids, gene mutation

Teaching and learning

Engage

- **Show** students some images of mutations – there are plenty on the Internet, but choose carefully (safe examples could be sickle cells, albino animals or dark peppered moths). [O1]
- Ask the students to **discuss** what might have caused the mutation. [O1]

Challenge and develop

- **Explain** that mutations are changes to the DNA code. Explain that these mutations are not always harmful and some can in fact be beneficial. [O1]
- Ask students to **complete** the questions in section 1 of Worksheet 6.6. [O1]

Explain

- **Remind** the students how genes code for proteins and if the code is wrong the protein may be different. [O2]
- Use images of sickle cell haemoglobin to **illustrate** how a different gene can change the shape of a protein molecule. Ask the students to **identify** differences between the shapes of the normal haemoglobin and the sickle cell haemoglobin. **Explain** that this change in shape changes how well the protein transports oxygen. [O2]
- Ask students to **suggest** how the change in code would have caused the change in shape. (If they need a hint, tell them to think about protein synthesis.) [O2]

- **Remind** students that three bases in the base sequence will correspond to one amino acid in the protein; **explain** that if that code is changed then the amino acid may also change; this could change the shape of the protein, as in the case of haemoglobin. Use page 249 of the Student Book to illustrate this with different examples. [O2]
- Ask students to **complete** the questions in section 2 of Worksheet 6.6. [O2]

Consolidate and apply

- **Explain** that there are different codes for one amino acid. For example, leucine has six different codes, TTA, TTG, CTT, CTC, CTA and CTG. So in some cases a mutation will have no effect as the same amino acid is coded for. Also make sure the students understand that, as we have two copies of a gene, in most cases we will have the correct copy and be able to make the protein. They will come back to this idea later when they look at inheritance. [O2]
- **Remind** students that large sections of DNA do not code for proteins but have other functions. Some sections switch genes on or off. If there is a mutation in one of these genes, then it means the gene may be switched off and the protein is not made. [O3]
- Standard demand students should **complete** Questions 1, 2, 5 and 6 in the Student Book (pages 248-249). [O1, O2, O3]
- High demand students then **complete** the questions in section 3 of Worksheet 6.6. [O2, O3]

Extend

Ask students able to progress further to:

- **explain** the difference between insertion, substitution and deletion mutations using page 249 of the Student Book. [O2]

Plenary suggestions

Ask a volunteer student to stand in front of the whiteboard facing the class. Write five key terms associated with the lesson onto a piece of paper.

The student is then given 30 seconds to talk about mutations. If they use any of the words, then cross them off. After 30 seconds if there are still words on the list, ask for a second volunteer to continue and so on until all the words are used.

Answers to Worksheet 6.6

1 Mutations

1. Radiation and chemicals such as tar in cigarette smoke.
2. Harm – as will show up in the wild and be seen by predators; harm – individual has breathing problems; benefit – camouflaged; benefit – to bacteria but not to humans; benefit – stops them getting malaria

2 Modelling mutations

1. seven
2. Addition of base A – ACATTAG; substitution of G for T – CCTCC; deletion of C – CGA TTA; repetition of GCC

3 Mutations consequences

1.
   a. The code is now for valine not leucine so there will be a different amino acid in the protein; this may change the shape of the protein and the active site so that the substrate will no longer fit.
   b. TCC has changed to TCT; both are codes for serine so no change to the protein.
2. Gene that switches the gene on may have a mutation.

Lesson 7: Meiosis

Lesson overview

AQA Specification reference

AQA 4.6.1.1; AQA 4.6.1.2; AQA 4.6.1.8

Learning objectives

- Explain how meiosis halves the number of chromosomes for gamete production.
- Explain how fertilisation restores the chromosome number.
- Understand that the four gametes produced by meiosis are genetically different.

Learning outcomes

- Recall that meiosis halves the number of chromosomes for gamete production. [O1]
- Understand that fertilisation restores the chromosome number. [O2]
- Explain how in meiosis the cell divides twice to form four gametes, each with a single set of chromosomes, and that all gametes are genetically different. [O3]

Skills development

- WS 1.2 Use a variety of models such as representational, spatial, descriptive, computational and mathematical to solve problems, make predictions and to develop scientific explanations and understanding of familiar and unfamiliar facts.
- WS 3.5 Interpreting observations and other data (presented in verbal, diagrammatic, graphical, symbolic or numerical form), including identifying patterns and trends, making inferences and drawing conclusions.
- WS 4.1 Use scientific vocabulary, terminology and definitions.

Maths focus

- 1d Make estimates of the results of simple calculations
- 2e Understand simple probability

Resources needed Video clip of sperm swimming to an egg; video or images of meiosis; string or wool. Worksheet 6.7.1 (low demand); Worksheet 6.7.2 (high demand)

Digital resources

Key vocabulary gamete, genetic variation, meiosis, sex determination, X-chromosome, Y-chromosome

Teaching and learning

Engage

- Show students a video clip of sperm swimming to an egg. There are several examples on the Internet. [O1]
- **Discuss** with the students what they understand by the terms ‘egg’, ‘sperm’, ‘gametes’, ‘zygote’, ‘fertilisation’ and ‘sexual reproduction’. [O1]

Challenge and develop

- Ask students to **recall** how many chromosomes there are in a human cell. Then suggest to them if eggs and sperm each have 46 chromosomes how many would be in the zygote? Why would that be a problem? How many chromosomes do they think eggs and sperm have so that the zygote will end up with 46? [O1, O2]
- **Ask** students if they know the difference between males’ and females’ chromosomes, **introduce** the idea of the X- and Y-chromosomes and explain why it is the sperm that determines the sex of the individual. [O1, O2]
- Ask students to **look** at Figure 6.25 on page 251 of the Student Book and predict the probability of having a boy or a girl. [O1, O2]
- Ask low demand students to **complete** the questions in section 1 of Worksheet 6.7.1. [O1, O2]
- Ask standard and high demand students to **complete** the questions in section 1 of Worksheet 6.7.2. [O1, O2]

Explain

- **Remind** the students about mitosis; **discuss** with them what they can remember about the process. Ensure they understand that mitosis produces two genetically identical cells with the same number of chromosomes as the original cell that divided. **Explain** that gametes cannot be produced by mitosis, instead they are produced by meiosis. [O1]
- Use images or videos clips to **illustrate** meiosis (knowledge of the stages is not required, so if using videos turn off any commentary as it will use terms not needed by the students). It is important that students **understand** that copies of the chromosomes are made at the start, the cell divides twice to form four gametes each with a single set of chromosomes, and each gamete is genetically different. [O3]
- **Remind** students that fertilisation will result in restoring the chromosomes number and that the zygote will have 46 chromosomes – 23 from the egg and 23 from the sperm. [O2]
- Draw four sperm cells and four egg cells on the board. Number the sperm 1–4 and the eggs A–D. Ask the high and standard demand students to **calculate** how many different combinations there are from fertilisation of any one of the four eggs with any one of the four sperm (for low demand students, tell them there are 16). Then **remind** them that males produce millions of sperm. Use these ideas to **illustrate** the idea that the combinations are endless, which is why siblings can show a lot of variation. [O2]

Consolidate and apply

- Ask low demand students to **complete** the questions in section 2 of Worksheet 6.7.1. [O3]
- Ask standard and high demand students to **complete** the questions in section 2 of Worksheet 6.7.2. [O3]

Extend

Ask students able to progress further to:

- **explain** the difference between mitosis and meiosis. [O3]

Plenary suggestions

Divide the class into pairs and ask them to use four pieces of string to show their partner what happens to chromosomes when a cell divides by meiosis.

Answers to Worksheet 6.7

1 Gametes

1. Fertilisation – the point when two gametes fuse.
   Gametes – cells with a single set of chromosomes.
   Sexual reproduction – reproduction that involves making unique individuals.
   Zygote – cell formed by fusion of two gametes.
2.
   a. 23
   b. 46
   c. Divides by mitosis and will develop into the embryo or baby.
3. Diagram should be similar to Figure 6.22 on page 250 of the Student Book.

2 Meiosis (low demand)

1. Diagram should be similar to Figure 6.25 on page 251 of the Student Book.
2. copied, twice, halved, gametes, different

2 Meiosis (standard and high demand)

1. Diagram should be similar to Figure 6.25 on page 251 of the Student Book.
2. First division –pairs separate; half go to one end and the other half to the other end, which reduces the number of chromosomes. Second division – chromosomes split as in mitosis.
3. All the gametes are genetically different, there are millions of sperm fertilising the egg so the possible combinations are endless. Some students may mention the exchange of material when the chromosomes pair up.

Lesson 8: Asexual and sexual reproduction

Lesson overview

AQA Specification reference

AQA 4.6.1.1; AQA 4.6.1.3

Learning objectives

- Understand that asexual reproduction involves just one parent and produces genetically identical offspring.
- Understand that sexual reproduction leads to variety in the offspring.

Learning outcomes

- Recall that there are two types of reproduction – asexual and sexual. [O1]
- Explain how asexual reproduction involves just one parent and produces genetically identical offspring. [O2]
- Explain how some organisms reproduce by both methods depending on the circumstances. [O3]

Skills development

- WS 1.1 Understand how scientific methods and theories develop over time.
- WS 3.8 Communicating the scientific rationale for investigations, methods used, findings and reasoned conclusions through paper-based and electronic reports and presentations using verbal, diagrammatic, graphical, numerical andsymbolic forms.
- WS 4.1 Use scientific vocabulary, terminology and definitions.

Resources needed Video clip of aphids reproducing asexually and sexually. Worksheet 6.8.1 (low demand); Worksheet 6.8.2 (standard and high demand)

Digital resources

Key vocabulary asexual reproduction, clone, natural selection, sexual reproduction, variation

Teaching and learning

Engage

- Show students a video clip of aphids reproducing asexually and sexually. [O1]

Challenge and develop

- Ask students to **suggest** the advantages to the aphids of reproducing asexually or sexually. Students should work in groups and then each group should **report** back their suggestions to the class. [O1]
- Develop the class discussion to draw up a **list** of advantages of the two types of reproduction. [O1, O2]
- Students then **check** the class list against those listed on page 252 of the Student Book. **Discuss** the list in the Student Book with the students to ensure they understand terms such as 'selective breeding' and 'natural selection', which they might not have yet covered. [O1, O2]
- Ask low demand students to **complete** section 1 of Worksheet 6.8.1. [O1, O2]
- Ask standard and high demand students to **complete** a summary of the advantages of asexual and sexual reproduction. [O1, O2]

Explain

- **Explain** that aphids use both types of reproduction depending on the circumstances. In early summer they reproduce quickly as there is plenty of food so asexual reproduction is better. In late summer they start sexual reproduction to introduce variation into the population. [O1]
- Use images to **illustrate** other examples of organisms that use both types of reproduction. Students need to be familiar with malarial parasites, fungi, strawberries and onions. Explain that strawberries and onions reproduce sexually using flowers but strawberries can use runners and onions can use bulbs for asexual reproduction. The malarial parasite reproduces asexually in humans and sexually in mosquitoes. We have known for some time that fungi such as moulds reproduce using spores; recent discovery has shown that they use sexual reproduction as well. This may be important in producing new strains of antibiotic. [O3]

Consolidate and apply

- Ask low demand students to **complete** section 2 of Worksheet 6.7.1. [O1, O2, O3]
- Ask standard and high demand students to **produce** a presentation about both methods of reproduction. The students work in groups using the instructions on Worksheet 6.7.2. [O1, O2, O3]

Extend

Ask students able to progress further to:

- **evaluate** another group's poster. [O3]

Plenary suggestions

Read out some statements about reproduction. Students should then indicate in some way if the statement is about asexual or sexual reproduction. The statements can be lifted from the Student Book.

Answers to Worksheet 6.8

1 Types of reproduction (low demand)

Asexual – faster than sexual reproduction; only one parent needed; many identical offspring can be produced when conditions are favourable; more time and energy efficient as do not need to find a mate.

Sexual – produces variation in the offspring; natural selection can be speeded up by humans in selective breeding to increase food production; if the environment changes, variation gives a survival advantage by natural selection.

2 Both methods (low demand)

Fungi – if using only the book then they will simply say grow asexually most of the time, but sometimes reproduce sexually if different strains meet. However, if more information is used then: asexual – produce spores or grow hyphae; sexual – spores produced by meiosis.

malarial parasite – asexual – in humans; sexual – in mosquito

strawberry – asexual – produce runners; sexual – flowers produce seeds

Types of reproduction (standard and high demand)

Ideas may include:

Fungi – if using only the book then they will simply say grow asexually most of the time, but sometimes reproduce sexually if different strains meet. However, if more information is used then: asexual – produce spores or grow hyphae; sexual – spores produced by meiosis.

malarial parasite: asexual – in humans as conditions favourable; sexual – in mosquito, this introduces variation

strawberry: asexual – produces runners that quickly result in new plants; sexual – flowers produce seeds, which introduces variation and can survive the winter

onion: asexual – produces bulbs to store energy over the winter; sexual – flowers produce seeds for variation

Lesson 9: Genetics

Lesson overview

AQA Specification reference

AQA 4.6.1.6; AQA 4.6.1.7

Learning objectives

- Understand and be able to use genetics terms, such as dominant, recessive, genotype, phenotype, homozygous and heterozygous.
- Know that some human conditions are caused by a recessive allele.

Learning outcomes

- Recall the difference between an allele and a gene. [O1]
- Explain the terms, dominant, recessive, genotype, phenotype, homozygous and heterozygous. [O2]
- Explain how cystic fibrosis is caused by a recessive allele. [O3]

Skills development

- WS 3.5 Interpreting observations and other data (presented in verbal, diagrammatic, graphical, symbolic or numerical form), including identifying patterns and trends, making inferences and drawing conclusions.
- WS 3.8 Communicating the scientific rationale for investigations, methods used, findings and reasoned conclusions through paper-based and electronic reports and presentations using verbal, diagrammatic, graphical, numerical andsymbolic forms.
- WS 4.1 Use scientific vocabulary, terminology and definitions.

Resources needed Image of mice of different colours; Internet access. Worksheet 6.9.1 (low demand); Worksheet 6.9.2 (standard demand); Worksheet 6.9.3 (high demand)

Digital resources

Key vocabulary allele, carrier, dominant, heterozygous, homozygous, genotype, phenotype, recessive

Teaching and learning

Engage

- **Show** students an image of mice showing all different colours (search 'mice different colours'). [O1]
- Ask the students to **suggest** what causes the different colours. [O1]

Challenge and develop

- Use the introduction to **explain** the difference between a gene and an allele. Explain that the fur colour is controlled by a single gene but that there are different forms of that gene. These different forms are called alleles. [O1]
- Ask students to **recall** how many copies of each gene they have. Use Figure 6.3 on page 239 of the Student Book to remind them of what they learnt in lesson 1 of Chapter 6 about genes and chromosomes. [O1]
- **Remind** the students that the genes on the chromosome inherited from each parent may not be the same.
- **Introduce** the terms 'dominant', 'recessive', 'genotype', 'phenotype', 'homozygous' and 'heterozygous'; use the example of black and white mice. The black fur colour is dominant over the white. [O1, O2]
- Ask low demand students to **complete** section 1 of Worksheet 6.9.1. [O1, O2]
- Ask standard demand students to **complete** section 1 of Worksheet 6.9.2. [O1, O2]
- Ask high demand students to **complete** section 1 of Worksheet 6.9.3. [O1, O2]

Explain

- **Explain** how cystic fibrosis is an example of a genetic disorder controlled by a single gene. Describe the symptoms and explain that it is the result of a mutation. Explain the term 'carrier' as somebody who is heterozygous for the condition. [O3]

Consolidate and apply

- Students then research more about cystic fibrosis and produce a presentation. [O3]
- Low demand students **use** Worksheet 6.9.1 to help them. [O1, O2, O3]
- Standard demand students **use** Worksheet 6.9.2 to help them. [O1, O2, O3]
- High demand students **use** Worksheet 6.9.3 to help them. [O1, O2, O3]

Extend

Ask students able to progress further to:

- **explain** why the Human Genome Project may help with treatment for cystic fibrosis. [O3]

Plenary suggestions

Write a list of key terms on the whiteboard. Use the ones for the lesson and others used in the topic such as gene and chromosome. Students **choose** six terms and enter them into a 3 × 2 'Bingo' grid. Read out the definitions and students **cross off** terms as they **identify** them from the definitions. When a student has a grid all correct, they shout 'Bingo!', then **read** the terms back, giving the definitions as well.

Answers to Worksheet 6.9

1 Genetic terms

(low demand):

gene, allele, dominant, recessive, phenotype, genotype, homozygous, heterozygous

(standard and high demand):

1. Different forms of the same gene.
2. Two, one inherited from each parent.
3. Black as all the offspring were black.
4. bb
5. (Standard demand answer): The original black guinea pigs and the brown guinea pigs are homozygous as their two alleles are the same.

   (High demand answer): The black guinea pig's offspring from the first mating will have the genotype Bb so will be heterozygous. When mated, some of their offspring will inherit the brown gene from both parents and be brown guinea pigs, which are homozygous bb.

2 Cystic fibrosis

1. Sufferers produce mucus that is thicker and stickier than normal, making it difficult to breathe.
2. CFTR
3. chromosome 7
4. two
5. Recessive, as only shows up when homozygous.
6. They have the normal allele and cystic fibrosis allele, as they have the normal allele they can still make the correct protein.
7. (High demand only): There is a mutation in both their alleles for the protein, the base code is incorrect so the protein formed contains the wrong order of amino acids and is therefore the wrong shape, being the wrong shape means it cannot perform its function.

Lesson 10: Genetic crosses

Lesson overview

AQA Specification reference

AQA 4.6.1.6; AQA 4.6.1.7

Learning objectives

- Use the terms dominant, recessive, genotype, phenotype, homozygous and heterozygous.
- Know that some human conditions, such as cystic fibrosis, are caused by a recessive allele.
- Complete or construct a Punnett square to predict the outcome of a genetic cross.

Learning outcomes

- Explain how cystic fibrosis is inherited by completing a Punnett square. [O1]
- Understand the concept of probability in predicting the results of a single gene cross. [O2]
- Construct a Punnett square to make predictions using the theory of probability. [O3]

Skills development

- WS 1.2 Use a variety of models such as representational, spatial, descriptive, computational and mathematical to solve problems, make predictions and to develop scientific explanations and understanding of familiar and unfamiliar facts.
- WS 3.5 Interpreting observations and other data (presented in verbal, diagrammatic, graphical, symbolic or numerical form), including identifying patterns and trends, making inferences and drawing conclusions.
- WS 4.1 Use scientific vocabulary, terminology and definitions.

Maths focus

- 1c Use ratios, fractions and percentages
- 2e Understand simple probability
- 3a Understand and use the symbols: =, <>, >, ∝, ~

Resources needed Worksheet 6.10.1 (low demand); Worksheet 6.10.2 (standard demand); Worksheet 6.10.3 (high demand)

Digital resources

Key vocabulary Punnett square

Teaching and learning

Engage

- **Write** the three possible genotypes for cystic fibrosis on the whiteboard. These were introduced in lesson 9. [O1]
- Ask students to **match** the genotypes to three phenotypes: non-sufferer, carrier and person with cystic fibrosis. [O1]

Challenge and develop

- **Explain** that if we know the genotype of the parents we can predict the probability of the offspring having cystic fibrosis. [O1]
- **Introduce** the Punnett square by asking students to look at Figure 6.33 on page 256 of the Student Book. [O1]
- **Explain** that it shows that if parents have the genotype Cc and cc, there is a 50% chance they will have a child with cystic fibrosis. [O1, O2]

Explain

- Ask low demand students to **complete** the Punnett squares in Question 1 of Worksheet 6.10.1 to predict the outcome of two other combinations of genotypes, without looking at page 103. [O1, O2]
- Ask standard and high demand students to **predict** the outcome of these two crosses: Cc × Cc and CC × cc by drawing their own Punnett squares, without looking at page 103. (*Note that only students entered for a higher tier paper are expected to be able to construct a Punnett square; foundation candidates just need to be able to complete them.*) [O1, O2, O3]

Consolidate and apply

- Low demand students to **complete** Worksheet 6.10.1. [O1, O2]
- Standard demand students to **complete** Worksheet 6.10.2. [O1, O2, O3]
- High demand students to **complete** Worksheet 6.10.2. [O1, O2, O3]

Extend

Ask students able to progress further to:

- **complete** Questions 3 and 4 on page 257 of the Student Book. [O3]

Plenary suggestions

Ask students to discuss the following statement: 'A couple have three boys and are expecting another child. What is the probability of the child being a daughter?'.

Answers to Worksheet 6.10.1

Genetic crosses

1. a. CC, non- sufferer; Cc, non- sufferer; Cc, non- sufferer; cc, cystic fibrosis
   b. 3 to1
   c. 25%
2. a. CC, non- sufferer; Cc, non- sufferer; CC, non- sufferer; Cc, non-sufferer
   b. 4 to 0
   c. 0%
3. Diagram should show a 3:1 ratio of brown to blue eyes.
4. a. short hair,
   b. long hair,
   c. Hh, short hair; hh, long hair; Hh, short hair; hh, long hair
   d. 1 to 1
   e. 50%
5. a. red
   b. RR, rr, Rr
   c. all boxes are Rr red flowers

Answers to Worksheet 6.10.2

1. Diagram should show a 3:1 ratio of brown to blue eyes.
2. Answers as per Question 4 above.
3. (a–c, as per question 5 above)
   d. 3 to 1, Punnett square drawn showing correct ratio.

Answers to Worksheet 6.10.3

1. Answers as per Question 4 (Worksheet 6.10.1).
2. Answers as per Question 3 above (Worksheet 6.10.2).
3. Appropriate letters chosen, e.g. W and w. The daughter will be Ww as the father must be ww. The daughter's partner will also be ww. So the Punnett square should show the correct cross between Ww and ww to give a 1:1 ratio. So the answer is 50%.

Lesson 11: Tracking gene disorders

Lesson overview

AQA Specification reference

AQA 4.6.1.6; AQA 4.6.1.7

Learning objectives

- Understand the use of a family tree to show the inheritance of a characteristic.
- Explain economic, social and ethical issues concerned with embryo screening.

Learning outcomes

- Recall that a family tree shows how conditions are passed down through a family. [O1]
- Use a family tree to identify genotypes of individuals. [O2]
- Debate economic, social and ethical issues concerned with embryo screening. [O3]

Skills development

- WS 1.3 Appreciate the power and limitations of science and consider any ethical issues which may arise.
- WS 1.2 Use a variety of models such as representational, spatial, descriptive, computational and mathematical to solve problems, make predictions and to develop scientific explanations and understanding of familiar and unfamiliar facts.
- WS 3.5 Interpreting observations and other data (presented in verbal, diagrammatic, graphical, symbolic or numerical form), including identifying patterns and trends, making inferences and drawing conclusions.
- WS 4.1 Use scientific vocabulary, terminology and definitions.

Maths focus

- 1c Use ratios, fractions and percentages
- 2e Understand simple probability
- 3a Understand and use the symbols: =, <>, >, ∝, ~

Resources needed Worksheet 6.11.1 (low demand); Worksheet 6.11.2 (standard and high demand)

Digital resources

Key vocabulary embryo screening, family tree, *in-vitro* fertilisation

Teaching and learning

Engage

- **Ask** the students to look at the family tree on page 258 of the Student Book. [O1]
- Ask the students to **discuss** any conclusions they can make from the family tree. [O1, O2]

Challenge and develop

- **Explain** how the tree shows which individuals have cystic fibrosis and how each family member is connected. [O1]
- **Remind** students that cystic fibrosis is a recessive condition. Can they identify the genotype of Harriett? [O1, O2]
- **Ask** high demand students to explain why Charles must have the genotype of Cc not CC. [O1, O2]
- **Introduce** the idea of polydactyly. Ask students to complete Question 1 on Worksheet 6.11.1 (low demand) or 6.11.2 (standard and high demand). [O1, O2]

Explain

- **Introduce** the idea of embryo screening. Explain that it is used in families where there is a known history of a genetic disorder; family trees are used to first predict if there may be an issue. [O3]
- **Explain** that there are many economic, social and ethical issues concerning embryo screening. [O3]
- Ask students to **read** the section on embryo screening on page 259 of the Student Book. They should then work in groups to list some of the economic, social and ethical issues concerning embryo screening. [O3]
- Each group **reports** back what they have found out so far and a list is drawn up on the board putting their ideas into the three areas: economic, social and ethical issues. [O3]

Consolidate and apply

- Low demand students **complete** Worksheet 6.11.1 and then debate the issue as a class. [O3]
- Standard and high demand students then **use** Worksheet 6.11.2 to research the topic further and then debate the issue as a class. [O3]

Extend

For students able to progress further:

- Genetic disorders can also be checked for later in a pregnancy. **Suggest** why this might cause different ethical issues. [O3]

Plenary suggestions

Students **write down** which way they voted in the debate and their reason for voting that way.

Answers to Worksheet 6.11

Family tree

Low demand:

1. Ian = polydactyly, male, Lucy = normal female
2. 50% Punnett square drawn showing correct ratio

Standard and high demand:

1. Ian = Pp – polydactyly, male; Lucy = pp – normal, female; Kelly = Pp – polydactyly, female; Luke = pp – normal, male
2. 50% Punnett square drawn showing correct ratio

Lesson 12: Gregor Mendel

Lesson overview

Learning objectives

- Plan experiments to explore phenomena and test hypotheses.
- Draw conclusions from given observations.
- Evaluate data in terms of reproducibility.

Learning outcomes

- Describe how Mendel tested his hypotheses. [O1]
- Interpret the results of Mendel's investigations. [O2]
- Evaluate the reproducibility of Mendel's results. [O3]

Skills development

- WS 1.1 Understand how scientific methods and theories develop over time.
- WS 1.2 Use a variety of models such as representational, spatial, descriptive, computational and mathematical to solve problems, make predictions and to develop scientific explanations and understanding of familiar and unfamiliar facts.
- WS 1.6 Recognise the importance of peer review of results and of communicating results to a range of audiences.
- WS 2.2 Plan experiments or devise procedures to make observations, produce or characterise a substance, test hypotheses, check data or explore phenomena.
- WS 3.5 Interpreting observations and other data (presented in verbal, diagrammatic, graphical, symbolic or numerical form), including identifying patterns and trends, making inferences and drawing conclusions.
- WS 3.7 Being objective, evaluating data in terms of accuracy, precision, repeatability and reproducibility and identifying potential sources of random and systematic error.
- WS 3.8 Communicating the scientific rationale for investigations, methods used, findings and reasoned conclusions through paper-based and electronic reports and presentations using verbal, diagrammatic, graphical, numerical and symbolic forms.
- WS 4.1 Use scientific vocabulary, terminology and definitions.

Maths focus

- 1c Use ratios, fractions and percentages
- 2e Understand simple probability

Resources needed Video about Mendel (search 'Mendel genetics GCSE'). Worksheet 6.12.1 (low demand); Worksheet 6.12.2 (standard demand); Worksheet 6.12.3 (high demand)

Digital resources

Key vocabulary hypothesis, pure line, reproducibility valid

Teaching and learning

Engage

- **Show** students a short video about Mendel. [O1]
- Ask the students to **suggest** some facts they know about genetics that Mendel would not have known at the time of his investigations. [O1]

Challenge and develop

- **Describe** Mendel's investigation with purple and white flowers. [O1]
- Ask students to **suggest** the dominant characteristic and to calculate the ratio when he got the results of 705 purple flowers and 224 white flowers, without looking in the Student Book. [O2]
- **Explain** that from these results he came up with a hypothesis. Ask them to **suggest** how he could test his hypothesis out on another characteristic, such as seed colour. [O1]

Explain

- **Introduce** the idea of peer review. Explain that Mendel's results were ignored for many years, and then in the 1900s they were rediscovered and peer review took place, in which the investigations were reproduced. [O3]
- **Ask** students to discuss the answers to question 5 and 6 on page 261 in the Student Book. Low demand students could be given the ratios. [O3]

Consolidate and apply

- Low demand students should **complete** Worksheet 6.12.1 and then produce a poster about Mendel's investigations. [O1, O2]
- Standard demand students **use** Worksheet 6.12.2 to produce a poster about Mendel's investigations. [O1, O2]
- High demand students **use** Worksheet 6.12.3 to produce a poster about Mendel's investigations. [O1, O2, O3]

Extend

Ask students able to progress further to:

- **draw** a Punnett square to explain the 1:1 ratio Mendel got when he crossed one of the yellow-seed-producing plants with a green-seed-producing plant. [O2]

Plenary suggestions

Ask students to discuss why Mendel's experiments are an excellent example of working scientifically.

Answers to Worksheet 6.12.1

Mendel

1. Gregor Mendel was a monk in at a monastery in Brünn, now Brno, in the Czech Republic, in the nineteenth century.
2. A wide range of varieties was available.
   For each trait chosen, differences are sharply defined, with no intermediate forms.
   Fertilisation is easily controlled. A pea plant will fertilise itself, or can be cross-fertilised.
   Peas are easy to cultivate.
   Peas grow and flower, and seed can be collected for sowing the following growing season.
3. Plants with no other characteristics in the offspring other than the one you want.
4.
   a. purple
   b. purple
   c. Idea that he got 705 purple and 224 white.
5. The inheritance of each characteristic was in 'units', which were passed on to descendants.
   Each plant had a pair of 'factors' governing a particular characteristic – in a dominant and recessive form.
   One of these factors, at random, would go into a gamete.
6. He crossed yellow and green seeds; in the first generation, all the plants produced yellow seeds.
   He crossed one of these yellow-seed-producing plants with a green-seed-producing plant. He predicted a 1:1 ratio of plants in the offspring. He obtained 58 yellow seeds and 52 green seeds.

Lesson 13: Key concept: Genetics is simple – or is it?

Lesson overview

AQA Specification reference

AQA 4.6.1.6; AQA 4.6.1.7

Learning objectives

- Explain how certain characteristics are controlled by a single gene.
- Understand that many characteristics are the result of multiple genes which interact.
- Describe the search for genes that are linked to disease.

Learning outcomes

- Understand that some characteristics are controlled by a single gene, such as: fur colour in mice; and red-green colour blindness in humans. [O1]
- Understand that most characteristics are a result of multiple genes interacting, rather than a single gene. [O2]
- Understand that the study of the genomes of humans and pathogens could help with treating disease. [O3]

Skills development

- WS 1.1 Understand how scientific methods and theories develop over time.
- WS 3.5 Interpreting observations and other data (presented in verbal, diagrammatic, graphical, symbolic or numerical form), including identifying patterns and trends, making inferences and drawing conclusions.
- WS 3.8 Communicating the scientific rationale for investigations, methods used, findings and reasoned conclusions through paper-based and electronic reports and presentations using verbal, diagrammatic, graphical, numerical and symbolic forms.
- WS 4.1 Use scientific vocabulary, terminology and definitions.

Resources needed Ishihara cards; sheets of A3 paper, coloured pens. Worksheet 6.13

Digital resources

Key vocabulary complex diseases, Mendelian inheritance

Teaching and learning

Engage

- **Show** students some colour-blind cards (Ishihara cards). The images can be projected onto the whiteboard and the students asked to identify the number or pattern or use the online test (www.color-blindness.com – search 'ishihara-38-plates-cvd-test/#prettyPhoto'). [O1]
- Ask the students to **suggest** why some people are colour blind and others are not. [O1]

Challenge and develop

- **Explain** that red–green colour blindness is controlled by a single gene. (High demand students could discuss the idea that it is sex-linked, although this is not in the specification.) [O1]
- **Remind** the students about the fur colour in mice they saw in lesson 9. **Explain** that this is also controlled by a single gene. However, many characteristics are controlled by a number of genes, which makes predicting outcomes of crosses difficult. [O1, O2]
- Ask students to **suggest** if the characteristics Mendel investigated are controlled by single genes or multiple genes. Explain that he would not have got the clear ratios if they were controlled by multiple genes. [O1, O2]

Explain

- **Explain** that when multiple genes are involved you get continuous variation of a characteristic. For example, height in humans – we are not just tall or short. Students may know that the environment also plays a part in this – something they will look at in more detail in Chapter 7. [O2]
- **Ask** students to **recall** a genetic disorder that is controlled by a single gene. **Explain** that there are also inherited diseases that are controlled by the interaction of many genes and the environment, such as heart disease and diabetes. These are called complex diseases. [O2]
- **Explain** that identifying the genes involved in these diseases could help in the treatment, something they looked at in lesson 2. Explain that studying the genome of the pathogens that cause diseases would also help in the treatment of diseases. [O3]
- Low demand students **complete** Questions 1–4 on pages 262 and 263 of the Student Book. [O1, O2, O3]
- Standard demand and high demand students **complete** Questions 1–7 on pages 262 and 263 of the Student Book. [O1, O2, O3]

Consolidate and apply

- Ask students to **produce** a summary of what they know about genetics in the form of a spider diagram. They should include ideas from all the lessons in the chapter. They should use Worksheet 6.13 to help them and work in groups. [O1, O2, O3]

Extend

Ask students able to progress further to:

- **discuss** why the discovery of DNA has been referred to as a 'turning point in our understanding of life'. [O3]

Plenary suggestions

Share their spider diagrams with the rest of the class.

Lesson 14: Maths skills: Fractions, ratio, proportion and probability

Lesson overview

Learning objectives

- Understand and use fractions and percentages.
- Understand and use ratio and proportion.
- Understand and use probability when predicting the outcomes of genetic crosses.

Learning outcomes

- Convert fractions into percentages when linked to genetic diagrams. [O1]
- Understand that in the study of genetics the most common ratios are 1 to 1 and 3 to 1. [O2]
- Use probability to predict the outcomes of genetic crosses. [O3]

Skills development

- WS 1.2 Use a variety of models such as representational, spatial, descriptive, computational and mathematical to solve problems, make predictions and to develop scientific explanations and understanding of familiar and unfamiliar facts.
- WS 3.1 Presenting observations and other data using appropriate methods.
- WS 3.3 Carrying out and represent mathematical and statistical analysis.
- WS 3.5 Interpreting observations and other data (presented in verbal, diagrammatic, graphical, symbolic or numerical form), including identifying patterns and trends, making inferences and drawing conclusions.
- WS 3.7 Being objective, evaluating data in terms of accuracy, precision, repeatability and reproducibility and identifying potential sources of random and systematic error.
- WS 4.1 Use scientific vocabulary, terminology and definitions.

Maths focus

- 1a Recognise and use expressions in decimal form
- 1c Use ratios, fractions and percentages
- 1d Make estimates of the results of simple calculations
- 2c Construct and interpret frequency tables and diagrams, bar charts and histograms
- 2d Understand the principles of sampling as applied to scientific data
- 3a Understand and use the symbols: =, <>, >, ∝, ~

Resources needed Practical sheet 6.14; Technician's notes 6.14

Digital resources

Key vocabulary fraction, percentage, probability, proportion, ratio

Teaching and learning

Engage

- **Ask** students to think about the roll of a dice. What is the probability of rolling a 6? [O3]
- Ask the students to **suggest** the probability of two people who are cystic fibrosis carriers having a child who has cystic fibrosis. [O1]

Challenge and develop

- **Use** the two answers to explain that the probability can be expressed in different ways: 1/6 or 16.7%, 1/4 or 25%; students should understand that these are fractions and percentages. [O1]
- **Ask** students how to convert fractions to percentages – most students should know, but if not, ensure they do using the example of 1/4 and 25%. [O1]
- Use a Punnett square diagram to **show** the difference between ratio and fractions. A common error in exams is not to state the answer clearly when asked about the outcome of a genetic cross: 3 to 1, 1 in 4 and 25% can all be used but with care. [O1, O2]

Explain

- **Explain** that, although we can predict the outcome of a genetic cross, there are many other factors involved so that the exact probability is not known. [O3]
- Low demand students complete Questions 1–3 on pages 264 and 265 of the Student Book. [O1, O2, O3]
- Standard demand students complete Questions 1–5 on pages 264 and 265 of the Student Book. [O1, O2, O3]
- High demand students complete Questions 1–6 on pages 264 and 265 of the Student Book. [O1, O2, O3]

Consolidate and apply

- **Ask** students to **complete** the investigation on Practical sheet 6.14. They should work in small groups or pairs. Ensure they understand what a frequency table (tally chart) is before you start. [O3]

Extend

Ask students able to progress further to:

- **explain** the difference between proportion and probability. [O2, O3]

Plenary suggestions

Ask a volunteer student to stand in front of the whiteboard facing the class. Write the five key terms associated with the lesson onto a piece of paper.

The student is then given 30 seconds to talk about ideas from the lesson. If they use any of the words, then cross them off. After 30 seconds, if there are still words on the list, ask for a second volunteer to continue, and so on until all the words are used.

Answers to Practical sheet 6.14

Answers to questions 1–7 will depend on the students' results.

8. The more times you toss the coin, the more accurate the results are likely to be; so if there were more than five groups in the class, they could use all the results or each group could toss the coins more times.

When and how to use these pages: Check your progress, Worked example and End of chapter test

Check your progress

Check your progress is a summary of what students should know and be able to do when they have completed the chapter. Check your progress is organised in three columns to show how ideas and skills progress in sophistication. Students aiming for top grades need to have mastered all the skills and ideas articulated in the final column (shaded pink in the Student book).

Check your progress can be used for individual or class revision using any combination of the suggestions below:

- Ask students to construct a mind map linking the points in Check your progress
- Work through Check your progress as a class and note the points that need further discussion
- Ask the students to tick the boxes on the Check your progress worksheet (Teacher Pack CD). Any points they have not been confident to tick they should revisit in the Student Book.
- Ask students to do further research on the different points listed in Check your progress
- Students work in pairs and ask each other what points they think they can do and why they think they can do those, and not others

Worked example

The worked example talks students through a series of exam-style questions. Sample student answers are provided, which are annotated to show how they could be improved.

- Give students the Worked example worksheet (Teacher Pack CD). The annotation boxes on this are blank. Ask students to discuss and write their own improvements before reviewing the annotated Worked example in the Student Book. This can be done as an individual, group or class activity.

End of chapter test

The End of chapter test gives students the opportunity to practice answering the different types of questions that they will encounter in their final exams. You can use the Marking grid provided in this Teacher Pack or on the CD Rom to analyse results. This shows the Assessment Objective for each question, so you can review trends and see individual student and class performance in answering questions for the different Assessment Objectives and to highlight areas for improvement.

- Questions could be used as a test once you have completed the chapter
- Questions could be worked through as part of a revision lesson
- Ask Students to mark each other's work and then talk through the mark scheme provided
- As a class, make a list of questions that most students did not get right. Work through these as a class.

Marking Grid for End of Chapter 6 Test

	Getting Started [Foundation Tier]							Going Further [Foundation and Higher Tiers]					More Challenging [Higher Tier]			Most demanding [Higher Tier]			Total marks	Percentage
Student Name	Q. 1 (AO1) 1 mark	Q. 2 (AO1) 1 mark	Q. 3 (AO1) 1 mark	Q. 4 (AO2) 1 mark	Q. 5 (AO2) 1 mark	Q. 6 (AO2) 2 marks	Q. 7 (AO3) 2 marks	Q. 8 (AO1) 1 mark	Q. 9 (AO1) 1 mark	Q. 10 (AO1) 2 marks	Q. 11 (AO2) 4 marks	Q. 12 (AO3) 2 marks	Q. 13 (AO1) 2 marks	Q. 14 (AO2) 4 marks	Q. 15 (AO3) 4 marks	Q. 16 (AO1) 2 marks	Q. 17 (AO2) 4 marks	Q. 18 (AO3) 4 marks		

Check your progress

You should be able to:

recall the definition of the genome	→	describe a gene as a section of DNA that controls a particular characteristic	→	understand that genes work by coding for the production of a particular protein, and non-coding genes switch genes on and off
recall that the entire human genome has been mapped	→	describe the importance of the mapping of the human genome	→	explain the mapping of the human genome in understanding historical human migration
describe the structure of DNA as repeating units called nucleotides arranged in a double helix	→	describe the structure of nucleotides as a sugar molecule, phosphate groups and bases	→	explain that strands of DNA are complementary
identify how a gene works in coding for the production of a protein	→	describe how a gene controls the order of amino acids in a protein	→	describe in outline how a protein is synthesised
know that there are many different types of protein	→	explain that the structure of a protein is related to its molecular shape	→	explain that a protein folds into a unique shape that enables it to function
identify that organisms reproduce by asexual and sexual reproduction	→	describe advantages and disadvantages of each type	→	explain that the method of reproduction may depend on circumstances
identify meiosis as the cell division used to produce gametes	→	explain the need for meiosis in producing gametes	→	explain that the gametes produced by meiosis are genetically unique
identify a mutation as a change in the DNA	→	describe how mutations can, on rare occasions, affect how an organism functions	→	explain how mutations can affect protein synthesis and structure
recall that genes exist in different forms called alleles, and know key genetic terms	→	complete Punnett squares to show the inheritance of characteristics controlled by single genes	→	construct Punnett squares to predict the outcome of genetic crosses
recall that some disorders are inherited	→	describe the inheritance of polydactyly and cystic fibrosis	→	make judgements about embryo screening

Worked example

A couple's second child was born with a condition called phenylketonuria (PKU). The condition is caused by a recessive gene.

The family tree shows that there is a history of PKU in the mother's (8) family. The father (7) was unaware of the condition in his family.

The couple would like another child. A genetic counsellor draws a family tree.

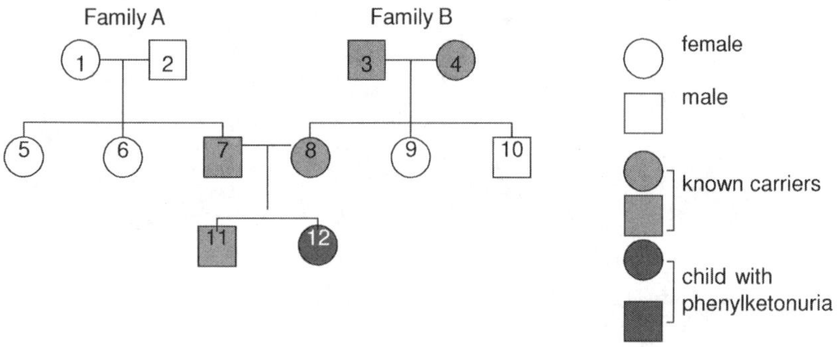

1 The genetic counsellor talks about the father's family's medical history.

a What is the genotype of the father (7)?

Pp

b What can be deduced about the genotype of the father's parents, 1 and 2? Explain your answer.

One is a carrier.

c Draw a genetic diagram to show how 12 came to have PKU.

		Mother (Pp) gametes	
		P	p
	P	PP normal	Pp normal (a carrier)
	p	Pp normal (a carrier)	pp has PKU

The combination of alleles that produced 12

The student has identified the genotype correctly.

The answer would have been better with more explanation. One parent (or both) must be carriers. For 7 to be a carrier, he must have inherited one recessive allele. So, one parent must have passed on a recessive allele. The other parent would have passed on a dominant allele (otherwise 7 would have PKU). Their other allele *could* have been the recessive passed down to any of the three children.

The student has included the parental genotypes, the gametes produced, the possible genotypes and phenotypes produced, and has highlighted the combination of alleles that must have produced 12.

Variation and evolution: Introduction

When and how to use these pages

This chapter builds on ideas that variation occurs between individual organisms as a result of genetic and environmental factors. It uses ideas of genetic inheritance, relating to mitosis and meiosis to explain how variation occurs and can be manipulated to suit our purposes. The development of ideas around inherited characteristics is also explored.

Overview of the chapter

In this chapter, students will learn about the scientific discoveries leading to our present ideas of genetics, including the work of Mendel, Lamarck, Wallace and Darwin. They will learn how scientists performed their investigations and reached their conclusions, and how their work supported and refuted ideas in the future. They will consider the changes humans can cause in the variation within populations through selective breeding, cloning and genetic modification techniques. Students will debate the ethics of human involvement and be made aware of all the issues at hand. They will learn to question whether the advantages of particular techniques justify the disadvantages. Students examine the different processes of plant and animal cloning, and the ethical considerations behind these. Finally, students will study the causes of extinction and explore our role in facilitating these processes.

This unit offers a few opportunities for the students to carry out their own investigation. They will identify variation in a given population, and use investigation to model the process of natural selection. They will use and evaluate the methods used by Kettlewell in his investigations with the peppered moths, to develop their own investigations and hypotheses. They will investigate the process of plant cloning by taking cuttings.

Obstacles to learning

The students may need extra guidance with the following terms and concepts:

- Students may think that all variation is genetic, and often confuse environmental and genetic factors.
- Students often confuse the terms 'cloning', 'selective breeding' and 'genetic engineering'. They should be clearly aware of the differences in how these processes are carried out.
- Students often think that evolution is still a theory and don't realise that there is a wealth of scientific evidence to support it.

Practicals in this chapter

In this unit students will do the following practical work:

- Investigate variation within a given population
- Use a model to investigate natural selection
- Develop a hypothesis and plan an investigation based on the work of Kettlewell on the process of evolution
- Make their own plant clones using runners or cuttings

	Lesson title	Overarching objectives
1	Variation	Recognise examples of variation in plants and animals. Explain how variation arises. Explain the importance of variation in survival.
2	The theory of evolution	Recall that all species of living things have evolved from simple life forms. Describe early ideas about evolution, including that of Lamarck. Explain how evolution occurs through natural selection.
3	The origin of species by natural selection	Explain the evidence that led Darwin to propose the theory of evolution by natural selection. Describe the process of natural selection.
4	Fossil evidence	Understand how, and the situations in which, fossils are formed. Understand how fossils are used as evidence for evolution of species from simpler life forms.
5	How much have organisms changed?	Understand why the fossil record is incomplete. Use the fossil record to understand how much, or how little, organisms have changed as life developed on Earth.
6	Darwin and Wallace	Recognise how Darwin and Wallace proposed, independently, the theory of evolution. Describe how Alfred Wallace gathered evidence for evolution, including warning coloration and mimicry.
7	A new species	Describe how a new species is produced under the influences of natural selection. Describe the definition of the term new species.
8	Evidence of natural selection and evolution?	Understand how scientific theories develop over time. Plan experiments to test hypotheses.
9	Evolution: fitting the pieces of the jigsaw	Describe the work of Mendel, Darwin and Wallace. Explain how they contributed to the theory of evolution. Appreciate that many scientists have contributed to the gene theory.
10	Antimicrobial resistance	Recall that bacteria develop that are resistant to antibiotics, which is evidence of evolution. Understand the mechanism by which antibiotic resistance develops.

		Understand the effects of the development of antibiotic resistance on the treatment of disease.
11	Combating antimicrobial resistance	Recall that bacteria develop that are resistant to antibiotics, which is evidence of evolution. Understand the mechanism by which antibiotic resistance develops. Understand the effects of the development of antibiotic resistance on the treatment of disease.
12	Selective breeding	Describe the process of selective breeding. Recall how selective breeding enables humans to choose desirable characteristics in animals and plants. Explain how selective breeding can lead to inbreeding.
13	Producing new plant varieties	Describe the process of selective breeding. Recall how selective breeding enables humans to choose desirable characteristics in plants. Evaluate the benefits and risks of selective breeding in plants.
14	Genetic engineering	Explain what is meant by the term genetic engineering. Introduce applications of the process. Explain how genetic engineering works.
15	Genetically modified crops: the science	Explain the benefits of genetic modification in a range of crops. Explain the concerns about genetic modifications in crops. Outline the arguments for and against genetic engineering.
16	Is genetic modification safe?	Explore the benefits of genetic modification in medicine. Explain the concerns that people have about genetic modification. Explain the possible safety issues of genetic engineering in agriculture and medicine.
17	Ethically wrong, or essential?	Explain the benefits of, and concerns about, genetic modification. Explain the ethical issues of genetic engineering in agriculture and medicine.
18	Cloning	Describe how cuttings and tissue culture are used to produce new plants. Describe the use of embryo transplants and adult cell cloning in animals.
19	The tree of life	Describe how living things have been classified into groups using a system devised by Linnaeus. Describe how new models of classification have developed.

20	Extinction…or survival?	List the causes of extinction. Explain how new predators, competitors and diseases can lead to extinctions.
21	Maths skills: Using charts and graphs to display data	Understand when and how to use bar charts. Understand how to show sub-groups on bar charts. Understand how to plot histograms.

Lesson 1: Variation

Lesson overview

AQA Specification reference

AQA 4.6.2.1

Learning objectives

- Recall that differences in the characteristics of individuals in a population is called variation.
- Understand the genetic and environmental differences leading to variation.

Learning outcomes

- Identify and investigate specific examples of variation in plants and animals. [O1]
- Explain whether features leading to variation are environmental or genetic. [O2]
- Explain how variation can be linked to increased survival, with examples. [O3]

Skills development

- WS 2.4 Carry out experiments appropriately having due regard for the correct manipulation of apparatus, the accuracy of measurements, and health and safety considerations.
- WS 2.6 Make and record observations and measurements using a range of apparatus and methods.
- WS 3.1 Present observations and other data using appropriate methods.

Maths focus

- 2b Find arithmetic means
- 4a Translate information between graphical and numeric form
- 4c Plot two variables from experimental or other data

Resources needed Worksheet 7.1.1 (low demand); Worksheet 7.1.2 (standard demand); Worksheet 7.1.3 (high demand); Practical sheet 7.1; Technician's notes 7.1

Digital resources PowerPoint slides

Key vocabulary environmental variation, evolution, genetic variation

Teaching and learning

Engage

- Show students the images in the PowerPoint slides of organisms from the same family of dogs and plants. Ask them to use the table in the worksheet and, working in pairs, write as many differences between the organisms as they can. [O1]
- Use the PowerPoint slide to introduce the ideas of genetic and environmental variation. [O1]
- Ask students to work in pairs and **write** a list of as many environmental factors and as many inherited factors as they can think of in one minute. [O1]

Challenge and develop

- Ask students to **complete** both investigations from Practical sheet 7.1. In one they will investigate variation in pea plants and in the other they will investigate the relationship between height and shoe size. [O1, O2]
- Students should present graphs of their data using a suitable scale and appropriate type of graph. Provide them with as little guidance as possible; but **discuss** the success criteria for drawing a good graph. [O1, O2]

Explain

- Students should **analyse** and **explain** the findings from the graphs. They should **explain** whether the features they have investigated arise from environmental or genetic variation and **justify** their ideas. Students should **discuss** whether there are any survival benefits to the features they have investigated. [O1, O2, O3]
- Students should **complete** section 2 of Worksheet 7.1.1 (low demand), 7.1.2 (standard demand) or 7.1.3 (high demand), which relates to data similar to that obtained by Mendel. [O1, O2]

Consolidate and apply

- Ask students to work in pairs to **complete** section 3 of their worksheet. They **classify** the features as being either environmental or genetic, and should **justify** their ideas. [O1, O2]

Extend

Ask students able to progress further to:

- **write** an essay titled 'Variation and survival' to include why it is important for a species to show variation and how it can lead to the survival of the species. The Tasmanian devil could be used. [O1, O2, O3]

Plenary suggestions

Students should read the scenarios from the PowerPoint and **explain** in each case what features within our species would be important and how variation could lead to increased success of the species.

Answers to Worksheet 7.1

1 What is variation?

1.

Similarities between the dogs	Differences between the dogs	Similarities between the pea plants	Differences between the pea plants
same size	colour and patches	colour of leaves	colour of petals
eye colour	colour of paws	colour of stem	number of flowers
shape of ears	weight	shape of flowers	height of plants
size of nose			number of pods
paw size			number of petals

2. Environmental factors: height of pea plants (depends on shade and nutrients in the ground); size of dog (depends on food and exercise); weight of dog (depends on diet and exercise); number of flowers (some might have got picked or eaten).
   Genetic factors: pea plant – colour of flowers, number of pods, shape of flowers, colour of leaves and stem; dogs – colour of coat and markings, eye colour, shape of ears and nose, paw size.

2 Investigating variation

Low demand: genetic factors are: round or wrinkled seeds; yellow or green seeds; flowers on the top or on the side
environmental factors could be: length of stem – could depend on shading or wind speed.

Standard demand: colour of flowers, number of peas in pod, size of the pods, size of the flowers

High demand: he could have planted the same number of seeds from the same family in two different locations; one with good sunlight, rich soil and plenty of water and the other in poor environmental conditions. He could have seen how the plants differed in appearance as a result of their environment. He should have a large sample size for the results to be reliable.

3 Environmental versus genetic

In the LHS of Venn diagram – genetic: foot size, leg length, type of leaf, prickles; In the RHS – environmental: glasses, clothing, earrings; In the centre – height (genetic but depends on diet and sunlight); skin colour (tanning or genetic); hair colour (natural or dyed); leaf size and number of leaves (genetic but depends on nutrients and predators).

Answers to Practical sheet 7.1

Investigation 1

1. There should be a wide variation in height and there should be a correlation between height and foot size.
2. Both could be a factors. Although height is determined by genetics, if a person did not have the right nutrients, they would not grow properly.
3. The bigger your feet the easier it is to walk on soft ground. Being tall means that it is easier to stand out and from a survival point of view, it may be easier to find a mate. Also you can reach foods that are out of reach to other people.
4. size of hand, vs height or weight vs height

Investigation 2

1. Students may find variation in a number of factors – size, mass, circumference, colour, height-to-diameter ratio.
2. More likely to be genetic, as commercial foods will be grown in environments with plenty of nutrients.
3. The more mass there is in fruits, the greater the chance of seeds to germinate and grow as there are plenty of nutrients from the fruit.
4. How many seeds there are vs diameter, size, mass, circumference, colour, height to diameter ratio

Lesson 2: The theory of evolution

Lesson overview

AQA Specification reference

AQA 4.6.2.2

Learning objectives

- Recall that all species of living things have evolved from simple life forms.
- Explain how evolution occurs through natural selection.

Learning outcomes

- Explain the term 'evolution'. [O1]
- Describe and apply Lamarck's theory of evolution. [O2]
- Consider and explain the evidence for evolution. [O3]

Skills development

- WS 1.1 Understand how scientific methods and theories develop over time.
- WS 3.5 Interpret observations and other data (presented in verbal, diagrammatic, graphical, symbolic or numerical form), including identifying patterns and trends, making inferences and drawing conclusions.
- WS 4.1 Use scientific vocabulary, terminology and definitions.

Resources needed Worksheet 7.2.1 (low demand); Worksheet 7.2.2 (standard demand); Worksheet 7.2.3 (high demand)

Digital resources PowerPoint slides

Key vocabulary evolution, natural selection

Teaching and learning

Engage

- Ask students to work in pairs and **discuss** how such a great variety of life on Earth came about. Ask them to **share their ideas** with the class. Be sensitive to students' faith and encourage them to share their beliefs.
- Show students the picture depicting the theory of evolution from the PowerPoint slide. Ask students to **discuss** the image in pairs and **explain** what is being shown. [O1]
- Discuss the image as a class and how the theory of evolution suggests that all life evolved from single-celled animals. Discuss some of the evidence for this (common DNA, fossil record). [O1, O3]

Challenge and develop

- Show students the ideas of Lamarck using the PowerPoint slide and the Student Book page 276. Ask students to **summarise** Lamarck's ideas in bullet points. [O2]
- Ask students to **answer** the questions from section 1 of Worksheet 7.2.1 (low demand), Worksheet 7.2.2 (standard demand) or Worksheet 7.2.3 (high demand) and **identify** other characteristics that Lamarck might have considered as possible factors for inheritance. Students should **explain** the flaws in Lamarck's theory. [O2]
- Introduce the role of mutation in variation and evolution using the PowerPoint. Discuss the example of sickle cell anaemia. [O2]

Explain

Low, standard and high demand:

Ask students to **read** the case study on section 2 of their worksheet about the white and black peppered moths. They should answer the differentiated questions from the worksheet to explain how evolution has worked and how it differs from Lamarck's theory.

Consolidate and apply

- Ask students to **look at** the flow diagram from section 3 of the worksheet. They should **use** the diagram and pages 276–277 of the Student Book to **explain how** the polar bear came to evolve. Students should **identify** any features of the flow diagram that may be missing. [O1, O2, O3]

Extend

Ask students able to progress further to:

- **find** three other examples where organisms have evolved in a way that clearly refutes Lamarck's ideas. [O1, O2, O3]

Plenary suggestions

Students could **write a letter** to Lamarck as his scientific peer. They could **describe** circumstances where his theory fits in with the evidence and then **give examples** where it does not. They should **summarise** the ways he might think about changing his theory. [O1, O2, O3]

Answers to Worksheet 7.2

1 Applying Lamark's theory

1.
    a. gymnast – good balance, strong muscles
    b. cyclist – strong leg muscles, good working knees
    c. mountaineer – lots of power in arms and legs, no fear of heights
2. For example, parent may be good at playing a musical instrument. They would pass on the characteristics of being able to read music well and having a good sense of timing and ability to play the instrument onto their children.
3. All the children would be as good, if not better than their parents at playing the instrument.They may be good at playing a variety of other instruments as well. It is unlikely that all the offspring will inherit these characteristics.

2 The role of mutations

1. Low demand: the peppered moth would not have survived and all would have been wiped out in the towns and cities.

   Standard demand: the peppered moth would have passed on the ability to camouflage well into light-coloured backgrounds and fly well. However, these characteristics would not have helped the moth to survive.

   High demand: mutations are also needed to provide variety, so that if circumstances change, organsisms best suited to the new changes are more likely to survive and the species as a whole survives.

3 Explaining evolution

Bears become plentiful and exist in large numbers. However, there is competition among the bears. The most camouflaged and fittest bears are the best hunters. Due to the large numbers, mutations occur and a variety of different coat colours emerge. The environment changes and some of the bears are isolated from the rest by glaciers. Those with the lightest coats are best camouflaged by the environment and pass on this characteristic to their offspring. The bears with the white coats survive best and eventually evolve into polar bears.

Lesson 3: The origin of species by natural selection

Lesson overview

AQA Specification reference

AQA 4.6.3.1

Learning objectives

- Explain the evidence that led Darwin to propose the theory of evolution by natural selection.
- Describe the process of natural selection.

Learning outcomes

- Identify the conditions needed for evolution to take place. [O1]
- Describe examples of evolution. [O2]
- Explain the evidence for evolution. [O3]

Skills development

- WS 1.1 Understand how scientific methods and theories develop over time.
- WS 3.5 Interpret observations and other data (presented in verbal, diagrammatic, graphical, symbolic or numerical form), including identifying patterns and trends, making inferences and drawing conclusions.
- WS 4.1 Use scientific vocabulary, terminology and definitions.

Maths focus

- 2c Construct and interpret frequency tables and diagrams, bar charts and histograms

Resources needed www.evolution-of-life.com (click on 'observe' then view 'Darwin on the evolution trail' animation). Worksheet 7.3.1 (low and standard demand); Worksheet 7.3.2 (high demand); Practical sheet 7.3; Technician's notes 7.3

Digital resources Error! Hyperlink reference not valid.PowerPoint files

Key vocabulary evolution, natural selection

Teaching and learning

Engage

- Ask students to **imagine** what would happen if a group individuals were stranded on an island and could not leave. Ask: what might happen to the humans over tens of thousands of years? Ask students to explain the conditions needed for the humans to survive and evolve; include ideas of more offspring needed than parents, stable death and birth rates, variation and passing of successful characteristics to offspring. Use the PowerPoint files to support the discussion. [O1]
- Introduce students to Charles Darwin. **Explore** what they already know about him. Explain his background and the reasons why he was such a successful scientist. Introduce his famous voyage on HMS *Beagle*. [O1]

Challenge and develop

- Show students the PowerPoint images of the Galapagos islands showing where they are and how they are different from each other. Introduce students to the importance of these islands, explaining that they are too far away for animals to cross from the mainland to live on these islands and the significance of the geographical separation. [O1, O2]
- Show students the video about Charles Darwin and his observations. Discuss the main points from the video, and ask students to **summarise** what made Darwin such a successful scientist. [O1, O2, O3]

- Ask students to **carry out** the activity from Practical sheet 7.3. Here they will **model** how different types of 'beak' can be adapted to different food types. They will **collect** and **record** data to show how some are better adapted to different types of food. [O2]
- Ask students to **explain** the significance of their findings. [O2, O3]

Explain

- Low and standard demand: students should **write** a short summary to explain the evidence they have encountered for evolution (Worksheet 7.3.1). [O3]
- High demand: students should **write** a scientific paper to explain all the evidence they have seen so far on evolution and what factors contribute to it (Worksheet 7.3.3). [O3]

Consolidate and apply

- Ask students to **read and sort** the mystery cards relating to the Asian shore crab and the blue mussel from the worksheet. They should **predict** how the mussels and the crabs are likely to evolve in the future. [O2]

Extend

Ask students able to progress further to:

- **find** three other examples of evolution in action, e.g. Italian wall lizards and cane toads, and create a two-minute presentation to the class about these cases. They should **explain** why evolution is essential for the survival of the species. [O3]

Plenary suggestions

Ask students to **prepare a debate** about the evidence of evolution to a group of religious people who do not accept the idea. They should ensure their arguments are based on real examples and scientific evidence and be prepared to **explore** all explanations for the evidence.

Answers to Worksheet 7.3

1 Evidence for evolution

Low and standard demand: summary should include evidence from white and black peppered moths and finches with different beaks.

High demand: scientific paper should include ideas about the need for more offspring that reach adulthood, examples from finches, peppered moths and e.g. the giraffe of how favourable characteristics are passed on and lead to changes in the species.

2 Evolution in action

All mussels will eventually develop the ability to detect the crab and instaneously develop a thicker shell in their presence. Crabs will either develop thicker pincers, or adapt to use an alternaive food source.

Answers to Practical sheet 7.3

6. The species that have survived will continue to do so, and the other species will become extinct.

Lesson 4: Fossil evidence

Lesson overview

AQA Specification reference

AQA 4.6.3.4; AQA 4.6.3.5

Learning objectives

- Understand how, and the situations in which, fossils are formed.
- Understand how fossils are used as evidence for evolution of species from simpler life forms.

Learning outcomes

- Identify different types of fossils. [O1]
- Describe how fossils are made. [O2]
- Explain the importance of fossil evidence in evolution. [O3]

Skills development

- WS 3.5 Interpret observations and other data (presented in verbal, diagrammatic, graphical, symbolic or numerical form), including identifying patterns and trends, making inferences and drawing conclusions.
- WS 3.6 Present reasoned explanations including relating data to hypotheses.
- WS 4.1 Use scientific vocabulary, terminology and definitions.

Resources needed Video of how fossils are made (www.discovery.com and search for 'How fossils are made'). Worksheet 7.4.1 (low demand); Worksheet 7.4.2 (standard and high demand)

Digital resources PowerPoint slides

Key vocabulary fossil, radiometric dating

Teaching and learning

Engage

- Show students the PowerPoint slide showing different images of fossils and ask them to **explain** what is meant by the term 'fossil'. [O1]
- Show real examples or images of different fossils from the worksheet. Ask students to work in pairs and **list** the different features of fossils. [O1]

Challenge and develop

- Show the video (see Resources) to show how fossils are made. [O2]
- Show and outline how fossils are made from the PowerPoint or Student Book. [O2]
- Low demand students should put the **processes in order** from section 2 of Worksheet 7.4.1 to describe fossilisation. [O2]
- Standard and high demand students **answer** the questions about how all the different types of fossils might have been made from Worksheet 7.4.2. They then **answer** all the questions from pages 280–281 of the Student Book. [O2]

Explain

- Students should use role play or models to **explain** how fossilisation takes place and show this to the class. [O1, O2]
- Students should then **answer** the questions in section 3 of their worksheet. They **explain** why the fossil record is incomplete. [O2, O3]

Consolidate and apply

- Show students the diagram of how fossils are found in rocks. Students should **explain** what the diagram is showing and how this is evidence for evolution. [O2, O3]

Extend

Ask students able to progress further to:

- **find out** more about the evidence from fossils for evolution. They could **make a presentation** on a particular case study. [O2, O3]

Plenary suggestions

Students should **devise a mind map** of how different fossils are made. [O2]

Students should **answer** the true or false questions from the PowerPoint. [O1, O2, O3]

Answers to Worksheet 7.4

1 Features of fossils

1.
   a. Fossils are the dead remains or evidence of the existence (footprints) of a dead organism. Sometimes they are the hard skelelton (shells or bones) of an organism, or the whole organism, which has been trapped in the sediment, or in amber.
   b. They all come from dead organisms. Some are just parts of the organism, some are the whole organism and some are just the evidence that the organism has existed.

2 How are fossils made?

Low demand: 4, 3, 2, 1

Standard and high demand:

Dragonfly – became stuck in the silt. Over time more and more layers buried the remains. There was no oxygen and minerals gradually replaced the skeleton.

3 Fossil record

1. It shows that early fossils are formed in the lower layers of the rock and over time more recent fossils are added to the higher layers of the rock. This shows how different organisms have evolved over time as the species in the top layers are different from some in the lower layers.
2. It is very difficult to find the right conditions for making fossils. Soft-bodied organisms will be decomposed and will not make fossils.
3. They are buried deep underground, or have been eroded by rocks on mountains or hill. They are in difficult places to access.

Lesson 5: How much have organisms changed?

Lesson overview

AQA Specification reference

AQA 4.6.3.5

Learning objectives

- Understand why the fossil record is incomplete.
- Use the fossil record to understand how much, or how little, organisms have changed as life developed on Earth.

Learning outcomes

- Explain why the fossil record is incomplete. [O1]
- Explain the evidence provided by the fossil record for evolution. [O2]
- Interpret diagrams to explain how organisms have changed over time. [O3]

Skills development

- WS 1.1 Understand how scientific methods and theories develop over time.
- WS 2.1 Use scientific theories and explanations to develop hypotheses.
- WS 3.5 Interpret observations and other data (presented in verbal, diagrammatic, graphical, symbolic or numerical form), including identifying patterns and trends, making inferences and drawing conclusions.

Resources needed Worksheet 7.5.1 (low demand); Worksheet 7.5.2 (standard demand); Worksheet 7.5.3 (high demand)

Digital resources PowerPoint slides; useful websites with 'missing links' (e.g. search 'Fossils reveal truth about Darwin's theory' at www.livescience.com)

Key vocabulary fossil record, missing link, palaeontologist

Teaching and learning

Engage

- Ask students to work in pairs and **discuss** how fossils are made. They should **draw** an annotated diagram of the process. Ask pairs to team up with another pair and **share** their ideas. They should **add to** or **amend** their diagrams accordingly. [O1]
- Ask students to **complete** the task in section 1 of the worksheet to **identify** why the fossil record is incomplete. [O1]

Challenge and develop

- Explain the meaning of the term 'missing link' with reference to the archaeopteryx fossil shown on the PowerPoint. **Discuss** why missing links are so important and why they are so hard to find.
- Show the video relating to the idea of missing link in the evolution of humans. Ask students to **summarise** the key points from the video and **explain** how missing links can help provide evidence for evolution (search 'nature scitable missing link fossils' to find 'Episode 10: Missing Link Misunderstood' video). [O2]
- **Discuss** the image of the manatee and its predecessor in section 2 of Worksheet 7.5.1 (low demand), Worksheet 7.5.2 (standard demand) or Worksheet 7.5.3 (high demand) and the PowerPoint. Ask students to **answer** the (differentiated) questions on the worksheet. [O2]
- Show students the evolutionary development of the horse on the PowerPoint slide and **discuss** the features observed. [O3]

Explain

- Ask students to work in pairs and **prepare** a one-minute talk to **explain** missing links and **how** they provide evidence for evolution. [O2]

Consolidate and apply

- Students could **predict** what the features of the missing link are of ancestors given in the worksheet. [O2]
- Ask students to **answer** the questions from section 3 of their worksheet and all the questions from page 283 of the Student Book. [O3]

Extend

Ask students able to progress further to:

- **find out** about the fossil record of another species of organism. They should **explain** how the organisms have changed over time and what the fossil record has revealed. [O3]

Plenary suggestions

Ask students to **look** at the **'**missing links**'** from the images from the PowerPoint and **decide** which two types of organisms have evolved from them. [O2, O3]

Ask students to **write** a paragraph to summarise their learning. They should **use** the key words 'fossil record', 'missing link' and 'evolution' in their answer. [O1, O2, O3]

Answers to Worksheet 7.5

1 Why is the fossil record incomplete?

Fossils are formed under very specific conditions. The remains of the organisms are usually washed into water and buried by the mud or silt. If there is any oxygen at all, the remains would decay and there would be nothing left. The hard parts of the organism are replaced by minerals. If an organism is completely soft bodied like an earthworm, there are no hard parts and a fossil will not be formed. It is a very rare event for all these conditions to be met. Finally, the fossils need to be found. They are usually in rocks beneath the sea or river bed, or brought to the surface by tectonic plate movement. Only weathering and erosion can expose them here, which take a long time.

2 Missing links

They have similar skull features, a turned down snout, similar location of eyes and a long tail for balance.

3 Evidence for evolution

Low demand: started with four toes, then three, then one long and two short toes, then just one strong hoof.

Standard demand: the organisms grew taller, with longer tails, more muscular bodies and more upright. Longer noses and larger ears.

High demand: the ground may have become less swampy, where a wide foot with many toes may have been advantageous, to a harder ground, where a single strong hoof may have been better for running. There may have been fewer predators and hence the need to hide in the swampy marshes was unnecessary, and the ability to run fast was more advantageous.

Lesson 6: Darwin and Wallace

Lesson overview

AQA Specification reference

AQA 4.6.3.2

Learning objectives

- Recognise how Darwin and Wallace proposed, independently, the theory of evolution.
- Describe how Alfred Wallace gathered evidence for evolution, including warning coloration and mimicry.

Learning outcomes

- Identify the independent contributions made by Darwin and Wallace in the theory of evolution. [O1]
- Describe the evidence collected by Wallace on mimicry and warning coloration. [O2]
- Explain how mimicry and warning colorations help organisms improve their chances of survival. [O3]

Skills development

- WS 1.1 Understand how scientific methods and theories develop over time.
- WS 4.1 Use scientific vocabulary, terminology and definitions.
- WS 3.5 Interpret observations and other data (presented in verbal, diagrammatic, graphical, symbolic or numerical form), including identifying patterns and trends, making inferences and drawing conclusions.

Resources needed Internet access. Worksheet 7.6.1 (low demand); Worksheet 7.6.2 (standard demand); Worksheet 7.6.3 (high demand)

Digital resources PowerPoint slides

Key vocabulary mimicry, warning coloration

Teaching and learning

Engage

- Provide students with the learning log of ideas relating to evolution from Worksheet 7.6. Ask them to work in pairs and **recall** the ideas of Lamarck and Darwin from lesson 2 in the chapter. [O1]
- Discuss their ideas and ask them to **add** further ideas they are reminded of from the discussions. [O1]
- Show students the PowerPoint slides of Darwin and Wallace. Explain that they will find out about the contributions of both scientists to the theory of evolution. [O1]

Challenge and develop

- Students should use the Internet to **carry out their own research** on the contributions of Alfred Wallace and Charles Darwin to the theory of evolution. [O1, O2, O3]
- Show students a video of the life of Alfred Wallace. [O1]
- Ask students to **complete** the learning log and include the individual contributions made by Wallace to the theory of evolution. They should use the Student Book also to help them. [O1]
- Introduce students to the work of Wallace on mimicry and how it provided a good example of evolution by natural selection. Show students the PowerPoint slides on warning coloration and use the Student Book.
- Ask students to **explain** how warning coloration works and why this is of benefit to the organism. [O2]
- Now show the PowerPoint slides of mimicry. Ask students to **explain the differences** between the original organisms and the mimics and how the mimics benefit from the feature of mimicry. They should **discuss**, in pairs, how mimicry exists in other forms from the next slide. Students should use the images and features to **derive their own definition** of mimicry. They should **record** their responses on the worksheet. [O3]

Explain

- Students should **summarise** and **explain** the similarities and differences between the theories of Lamarck, Darwin and Wallace using their learning logs and Student Books. [O1, O2]

Consolidate and apply

- Ask students to work in groups of four. They should **prepare** a short role play where one student is a reporter, another is Wallace, another is Darwin and another is Lamarck. Students should **decide together** on suitable questions and answers they could pose **to explain** what each of the scientists thought and what contributions they made to the theory of evolution. They should also **explain how** the scientists collected their respective evidence, how their ideas were received and what impact their ideas had on the world. Students should **present** their role plays. [O1, O2]
- Students should **answer** the differentiated questions on mimicry and warning coloration from Worksheet 7.6.1 (low demand), Worksheet 7.6.2 (standard demand) or Worksheet 7.6.3 (high demand). [O3]

Extend

Ask students able to progress further to:

- **research** mimicry in plants and **explain** further benefits to plants of using mimicry in relation to other species. They should **contemplate** and **explain** the advantages and disadvantages of using mimicry. [O3]

Plenary suggestions

Students should **formulate** a reasoned argument for whether they believe that Charles Darwin should take all the credit for the discovery of evolution, or whether the credit should go to Alfred Wallace, or both. [O1, O2]

Present this thought to students: 'Evolution is thought to be the most important idea humans have come up with'. Ask students to agree or disagree, **explaining** their reasons. [O1, O2]

Answers to Worksheet 7.6

2 Warning coloration

Low demand: blue ringed octopus – white background with brigtly coloured blue eyes, which will show up in deep waters. The eyes are very ditinctive all over the body; blue poison dart frog – bright blue and black patches, which stand out against the green forest canopy.

Standard demand: they need a warning coloration so that predators will immediately recognise them and not eat them in the first place. Once they have been eaten it is too late and it does not benefit the species.

High demand: the number of organisms with warning coloration outweigh the number of organisms without colourations, and are more dominant in the gene pool, as they have survived better. So the mutation that led to the development of a warning colouration was passed on from generation to generation until it has become the dominant gene in the gene pool.

3 Mimicry

Low demand:

1. Because predators will be convinced that the non-venomous is the same as the venomous and will not eat it.
2. No. Some mimics, like the bee orchid, is a completely different species to the bee it is mimicking.

Standard demand: predators might soon learn that the mimic is a fake and then begin to eat more of this species, as there would be fewer numbers of the real organism.

High demand: if the original organism should undergo a mutation and change appearance, then the mimic could soon die out. The popultions of the mimic need to be kept smaller than the original organism. Predators could learn the differences between the mimic and the orginal species. Predators may evolve to eat the orginal species, and so the mimic is at risk too.

Lesson 7: A new species

Lesson overview

AQA Specification reference

AQA 4.6.3.2

Learning objectives

- Understand that when natural selection operates differently on populations, a new species is produced.
- Understand that during evolution, new species are formed when populations become so different that they can no longer interbreed.

Learning outcomes

- Identify the similarities and differences between similar species. [O1]
- Describe examples of different species. [O2]
- Describe the factors affecting the formation of new species. [O3]

Skills development

- WS 1.2 Use a variety of models such as representational, spatial, descriptive, computational and mathematical to solve problems, make predictions and to develop scientific explanations and understanding of familiar and unfamiliar facts.
- WS 3.5 Interpret observations and other data (presented in verbal, diagrammatic, graphical, symbolic or numerical form), including identifying patterns and trends, making inferences and drawing conclusions.
- WS 4.1 Use scientific vocabulary, terminology and definitions.

Resources needed Worksheet 7.7.1 (low demand), Worksheet 7.7.2 (standard demand) or Worksheet 7.7.3 (high demand)

Digital resources PowerPoint slides

Key vocabulary hybrid, interbreeding, speciation

Teaching and learning

Engage

- Ask students to **look at** the image of the different species of finches from the Galápagos Islands. They should discuss all the similarities and differences between them, in pairs, and **recall** why the different species arose. [O1]
- Students should now **look at** the images of different species of foxes. Discuss first as a class, and then students should **answer** the questions on Worksheet 7.7.1 (low demand), Worksheet 7.7.2 (standard demand) or Worksheet 7.7.3 (high demand). [O1]
- Students should **read** the case study of the mule. They should **derive** their own definition of a species as a result. Students should **answer** the questions relating to the definition of a species from their worksheet. [O1]

Challenge and develop

- Show students a video of how new species might arise (go to .www.hhmi.org – click on 'Biointeractive' at the bottom of the page, then search 'anole lizards example speciation').
- Show the PowerPoint slide that summarises speciation and **discuss** this with the class. Ask students to **sort** the card sort from the worksheet into the correct timeline of stages. [O1, O2]

Explain

- Ask students to work in small groups. They should **make their own model** of an organism and its habitat. They should **make a model of a physical barrier** and **explain** how this might give rise to new species. They should **explain** how new species have arisen from the original species. They should **make models** of the new species that have developed over time and **explain** why they might have particular features. If possible they could **make a film** of this; alternatively, they could **present** their model and explanation to the class. [O1, O2, O3]

Consolidate and apply

- Ask students to study the finches from the worksheet. They should **explain how** the finches evolved into different species. [O1, O2, O3]

Extend

Ask students able to progress further to:

- **find out** about the three different types of speciation (allopatric, peripatric and parapatric speciation). They should **find** different examples of each and **present** their ideas to the class in a PowerPoint presentation. [O1, O2, O3]

Plenary suggestions

Students should **use the card sort** they carried out earlier to **make a mind map** of speciation on the worksheet. They should **include** various examples from the lessons and **add** images and drawings to **explain** what is happening at each stage of the process. [O1, O2, O3]

Answers to Worksheet 7.7

1 What is a species?

1. Table
   a. **Similarities:** all have furry coats and bushy tails; all have forward facing eyes and ears on top of their heads; all have four feet and paws

      Differences: Cape fox – lighter brown coloration to blend in with the more desert-like and hot features of South Africa; much longer ears to detect predators and prey and to help lose heat more effectively; thin coat to increase heat loss. Fennec fox – long ears to increase heat loss and improve hearing for predators and prey; much smaller so larger surface area to volume ratio, to increase heat loss; sandy coat to blend into the desert environment. Arctic fox – white coat to blend into the environment; small ears and nose to conserve heat; thicker coat to conserve heat.

   b. You can't tell from just their appearance that they are different species. They could just be large adapatations.
2. Low demand: all the species of horses and donkeys and mules would die out as there is nothing to replace them.

   Standard demand: organisms within the same species that can successfully interbreed to produce fertile offspring.

   High demand: so organisms can be classified accordingly and monitored, so that if species are at risk of extinction, steps could be taken. Also it is important to understand how these phenomena work, so predictions can be made for the future.

2 How does speciation occur?

Card sort: G, B, D, H, E, F, C, A

Lesson 8: Evidence of natural selection and evolution?

Lesson overview

AQA Specification reference

AQA 4.6.3.4

Learning objectives

- Understand how scientific theories develop over time.
- Plan experiments to test hypotheses.

Learning outcomes

- Describe how data can be collected to prove or disprove a hypothesis. [O1]
- Plan an experiment to test a hypothesis. [O2]
- Explain how data can be used to develop scientific theories. [O3]

Skills development

- WS 2.1 Use scientific theories and explanations to develop hypotheses.
- WS 2.2 Plan experiments or devise procedures to make observations, produce or characterise a substance, test hypotheses, check data or explore phenomena.
- WS 3.5 Interpret observations and other data (presented in verbal, diagrammatic, graphical, symbolic or numerical form), including identifying patterns and trends, making inferences and drawing conclusions.

Maths focus

- 4a Translate information between graphical and numeric form

Resources needed Worksheet 7.8; Practical sheet 7.8.1 (low demand); Practical sheet 7.8.2 (standard and high demand)

Digital resources

Key vocabulary lichen, melanism

Teaching and learning

Engage

- Show students the images of the peppered moths from page 289 of the Student Book. Ask students to **discuss** why they were important in confirming the theory of evolution. [O3]
- Ask students to **discuss** in pairs how the scientists might have collected the information on the moths. Ask students to **share** their ideas with the class. [O1]

Challenge and develop

- Show students the information about the black and white peppered moths and Kettlewell's experiment from the website http://peppermoths.weebly.com.
- Students should **select** any relevant information from the website slides and **use** this to **answer** Questions 1–4 on Worksheet 7.8. [O1, O2, O3]
- Ask students to **read** the information from the Student Book relating to how the experiments on the black peppered moth were conducted. They then **answer** questions 1–6 from pages 288 and 289 of the Student Book. [O1]
- Provide students with Practical sheet 7.8.1 (low demand) or Practical sheet 7.8.2 (standard and high demand). Students **develop** their own hypothesis and **plan** how to collect evidence to test their hypothesis. They should **explain** how to collect accurate and reliable data and how the data can confirm or refute their hypothesis. [O1, O2]

Explain

- Students should **write** an explanation of how Kettlewell's experiment provided the evidence needed for Tutt's hypothesis on Worksheet 7.8. They should **explain** what further evidence could have been carried out to ensure the results were reliable (questions 5–10). [O1, O3]

Consolidate and apply

- Ask students to work in small groups. Ask them to **make a model** of a habitat with the features of their test area from the plan of their investigation. They should **explain** the features of the habitat that will enable them to test their hypothesis. They should make models of their test organisms and explain how to carry out their investigation. They should use their model to **simulate** what would happen if their hypothesis were correct and also if it was incorrect. [O1, O2, O3]

Extend

Ask students able to progress further to:

- **find out** about further experiments that scientists have carried out to see evolution in action. They could find out more about the evolution of the Italian wall lizard on the island and what data scientists collected. [O3]

Plenary suggestions

- Ask students to work in pairs. Ask them to **explain** whether Kettlewell's experiment was sufficient to develop Tutt's scientific theory. [O3]
- Students should identify the strengths and weaknesses in Kettlewell's experiments. [O1, O2]
- They should devise one further experiment he could have done to prove his hypothesis completely. [O1, O2]

Answers to Worksheet 7.8

1 Collecting evidence about the peppered moths

1. White peppered moth. It was well camouflaged against the light background of the tree trunks.
2. The industrial revolution meant that a lot of pollution was allowed to escape in the air.
3. The population of the white peppered moth declined.
4. Because they were no longer camouflaged against the trees, as the trees were not covered in soot and were black.The white peppered moth can be easily seen by predators.
5. Tutt believed that pollution was the cause of the change in populations.
6. Kettlewell conducted the investigation to test the hypothesis.
7. He released known numbers of white and black peppered moths into the environment and marked them. After a short while he recaptured them and counted how many of the original balck and white were found. He found a much higher proportion of black compared with white.
8. This proved the theory that fewer white peppered moths were surviving.
9. When the Clean Air Act came into force, Kettlewell predicted that the population of white peppered moth would increase and the population of black peppered moth would decrease. He was right and this proved the theory.
10. Carry out the investigation again over a range of different industiral and non-industrial towns.

Lesson 9: Key concept: Evolution: fitting the pieces of the jigsaw

Lesson overview

AQA Specification reference

AQA 4.6.3.1; AQA 4.6.3.2; AQA 4.6.3.3

Learning objectives

- Describe the work of Mendel, Darwin and Wallace.
- Explain how they contributed to the theory of evolution.
- Appreciate that many scientists have contributed to the gene theory.

Learning outcomes

- Describe the theory of evolution. [O1]
- Describe the contributions made by Mendel, Darwin and Wallace in developing and explaining the theory. [O2]
- Explain how the gene theory supports the theory of evolution. [O3]

Skills development

- WS 1.1 Understand how scientific methods and theories develop over time.
- WS 1.6 Recognise the importance of peer review of results and of communicating results to a range of audiences.
- WS 2.7 Evaluate methods and suggest possible improvements and further investigations.

Resources needed Video on the life of Mendel; video on the work of William Bateson. Worksheet 7.9.1 (low demand); Worksheet 7.9.2 (standard demand); Worksheet 7.9.3 (high demand)

Digital resources

Key vocabulary gene theory, genetic variation, mathematical model

Teaching and learning

Engage

- Briefly recap the work of Darwin and Wallace from previous lessons. [O1]
- Ask students to **look at** the statements about the theory of evolution on section 1 of the worksheet and **cross out** any statements that are not correct. They should **provide** a full description of the theory of evolution. [O1]

Challenge and develop

- Show students a video on the life of Mendel. [O2]
- Ask students to **answer** the questions on Gregor Mendel (section 2) on Worksheet 7.9.1 (low demand), Worksheet 7.9.2 (standard demand) or Worksheet 7.9.3 (high demand). [O2]
- Now show the video on the work of William Bateson, which follows on from the studies of Mendel. [O2]
- Ask students to work in pairs. They should **discuss** how the work of Mendel and Bateson can support the theory of evolution. They should **explain** how the discovery of ‘units’ of inheritance, we now know as genes, can help provide a mechanism to understand how evolution might work. Take feedback from selected pairs. [O3]

Explain

- Ask students to form groups of three. They should **write** a speech for each of the scientists involved to **explain how** each contributed to the theory of evolution. They should **explain** the strengths and shortcomings of their contributions and **explain how** they might have liked to have improved their scientific studies. Ask students to **present** their speeches. [O1, O2, O3]

Consolidate and apply

- Place students in small groups. Ask students to **use** the Student Book and **answer** all the questions from pages 290–291.
- Ask students to now **imagine** a scene between Darwin and Mendel. Ask them to **construct** a role play of a conversation between Darwin and Mendel, where Darwin **explains** to Mendel why the finches from the Galapagos Islands have beaks that suit their habitat, using the hypothesis he has come up with. [O2, O3]

Extend

Ask students able to progress further to:

- **find out** more about Mendel and his investigations. They should **write** a scientific paper on the research that Mendel carried out and comment on the validity and reliability of his data. [O2]

Plenary suggestions

Students could **make up** their own true or false questions based on the learning in the lesson. [O1, O2, O3]

Students could **draw up** a Hall of Fame of evolution; they could **write** a pen portrait of Mendel, Wallace, Darwin and other scientists who have contributed to the theory, and **include** a brief summary of their contributions.

Answers to Worksheet 7.9

1 Theory of evolution

The theory of evolution by natural selection states that:

All life came from simple life forms such as single-celled organisms.

Life first developed more than three billion years ago.

Simple life forms evolved into different and more complex life forms.

Evolution occurs via natural selection.

Individual organisms within a particular species show a wide range of phenotype variation because of differences in their genes.

Mutations in genes cause variation.

Individuals with characteristics most suited to the environment are more likely to survive to breed successfully.

The genes that have enabled these individuals to survive are then passed on to the next generation.

If two populations of one species become so different in phenotype that they can no longer interbreed to produce fertile offspring, they have formed two new species. New species have evolved from an original species.

2 Gregor Mendel

(low demand)

1. large sample size used, recorded all his results and counted the numbers of offspring with particular characterisics, looked for patterns, used maths, developed a hypothesis from his data, published his findings
2. because he recorded all the numbers and was able to find a mathematical link

(standard demand)

1. he was not a prominent scientist, he wrote his paper in German, the subject matter was about peas, which had been widely studied previously, no-one at the time to see how it linked to evolution, which was the greatest topic of the time
2. there would have been much more activity following his findings and further research from different scientists; we may have discovered genes far earlier

(high demand)

1. It provided an explanation for evolution.
2. carried out his findings on animals, asked a prominent scientist to peer-review his findings and write a paper about them in English

Lesson 10: Antimicrobial resistance

Lesson overview

AQA Specification reference

AQA 4.6.3.7

Learning objectives

- Recall that bacteria develop that are resistant to antibiotics, which is evidence of evolution.
- Understand the mechanism by which antibiotic resistance develops.
- Understand the effects of the development of antibiotic resistance on the treatment of disease.

Learning outcomes

- Describe how antibiotic-resistant bacteria develop. [O1]
- Explain the factors which lead to antibiotic resistance. [O2]
- Explain the impact of antibiotic resistance. [O3]

Skills development

- WS 1.2 Use a variety of models such as representational, spatial, descriptive, computational and mathematical to solve problems, make predictions and to develop scientific explanations and understanding of familiar and unfamiliar facts.
- WS 1.4 Explain everyday and technological applications of science. Evaluate associated personal, social, economic and environmental implications, and make decisions based on the evaluation of evidence and arguments.
- WS 1.5 Evaluate risks both in practical science and the wider societal context, including perception of risk in relation to data and consequences.

Maths focus

- 4a Translate information between graphical and numeric form
- 4c Plot two variables from experimental or other data

Resources needed Modelling clay; poster paper; sticky notes for each pair. Video about the discovery of penicillin; two short clips: one from the Chief Medical Adviser and one about the threat from antibiotic-resistant bacteria. Worksheet 7.10.1 (low demand); Worksheet 7.10.2 (standard demand); Worksheet 7.10.3 (high demand)

Digital resources PowerPoint slides

Key vocabulary antimicrobial resistance

Teaching and learning

Engage

- Show students a video about the discovery of penicillin and the headline figures from the PowerPoint about how many lives antibiotics have saved since they were first discovered.
- Now show students a couple of short clips explaining the threat we are under today, one from the Chief Medical Adviser. Also show the further clips about the threat from antibiotic-resistant bacteria. [O1]
- Ask students to work in pairs. They should **discuss** what they already know about 1) antibiotics and what they have learned about 2) antibiotic-resistant bacteria. Have two posters set up for each topic and ask students to **write their ideas** on sticky notes and then place them on the posters. Discuss the class's ideas about antibiotic-resistant bacteria. [O1]

Challenge and develop

- Recap ideas about evolution and mutation. Ask students to **read** the steps of antibiotic resistance from the card sort in section 1 of the worksheet. Students should **sort the cards** into order of sequence. [O1, O2]

- Show students the PowerPoint slide on how to use antibiotics safely.
- Provide students with the data on penicillin resistance from section 2 of the worksheet. Ask them **to answer** the (differentiated) questions on Worksheet 7.10.1 (low demand), 7.10.2 (standard demand) or 7.10.3 (high demand). [O1, O2, O3]
- Introduce and explain what MRSA is using the PowerPoint slide. Show students an animation on how MRSA can spread in a hospital (www.yourgenome.org – search for 'MRSA animation').

Explain

- Ask students to work in pairs and to **make bacterial models** using modelling clay. They should use the models to **explain** the process of how bacteria become resistant to antibiotics. [O1, O2]
- Ask students **to use** their bacterial models **to explain** what happens when antibiotics are used when they are not needed and what happens when the dose is not completed. [O1, O2]

Consolidate and apply

- Students should read pages 292–293 in the Student Book and **answer** all the questions. [O1, O2, O3]
- Ask students to **write a leaflet** for their local GP surgery. They should **explain** what antibiotics are, how they work, how they become resistant and how they should be used. They should **use** data to **persuade** people not to demand the use of antibiotics. [O1, O2, O3]

Extend

Ask students able to progress further to:

- **find** further evidence about the impact of antibiotic resistance and **create** a display to warn people of the seriousness of the threat. [O1, O2, O3]

Plenary suggestions

Ask students to work in pairs and return to the posters with the sticky notes from earlier in the lesson. Ask students to **add** what they have learned from the lesson to the posters, using more sticky notes. [O1, O2, O3]

Ask students to **write** a response of whether they agree or disagree and why, to the statement 'Antibiotic-resistant bacteria are the greatest threat to humanity today'. [O1, O2, O3]

Answers to Worksheet 7.10

1 How bacteria become resistant

Card sort: B, C, H, A, E, I, F, D, G

2 What harm does antibiotic resistance cause?

(low demand)

1. France
2. Netherlands

(standard demand)

The greater the population the higher the dose of penicillin per 1000 people.

(high demand)

France 0.25; Spain 0.22; Hungary 0.84; Portugal 1.14; UK 0.1; Netherlands 0.24; Italy 5.81

Even though France has a relatively low usage of penicillin per head, it has the highest percentage of resistant bacteria. Italy has a relativley high use but a relatively low population of resistant bacteria.The countries with the lowest resistance also have the lowest use.

Lesson 11: Combating antimicrobial resistance

Lesson overview

AQA Specification reference

AQA 4.6.3.7

Learning objectives

- Describe how to reduce the rate of development of antibiotic resistance.
- Understand the requirement for, and the impact of, new antibiotics.
- Recognise the difficulties associated with developing new antibiotics.

Learning outcomes

- Describe some measures which reduce the rate of antibiotic resistance. [O1]
- Describe one new method of producing new strains of antibiotic. [O2]
- Evaluate the advantages and disadvantages of developing new antibiotics. [O3]

Skills development

- WS 1.4 Explain everyday and technological applications of science. Evaluate associated personal, social, economic and environmental implications, and make decisions based on the evaluation of evidence and arguments.
- WS 1.3 Appreciate the power and limitations of science and consider any ethical issues which may arise.
- WS 4.1 Use scientific vocabulary, terminology and definitions.

Resources needed Internet access. Worksheet 7.11.1 (low demand); Worksheet 7.11.2 (standard and high demand)

Key vocabulary AMR, antimicrobial resistance

Teaching and learning

Engage

- Ask students to work in pairs. They should **role play** being in a doctors' surgery where the patient has the flu. They should **devise a discussion** between a patient and a careless doctor and then one between the patient and a good doctor. Ask students to share their discussions. [O1]
- Ask students to **answer** the true or false questions in section 1 of their worksheet. [O1]

Challenge and develop

- Ask students to **watch** a video that shows the development of a new antibiotic (go to www.cbsnews.com and search the news archive for 'advanced antibiotic may fight resistance bacteria').
- Ask students to **read** more about the development of new antibiotics in the section 'Can we solve the problem?' on page 295 of the Student Book. They then **answer** Questions 6 and 7. [O1, O2]
- Ask students to **listen** to the podcast of an interview from the site www.nature.com (find the news pages; enter search terms 'promising antibiotic discovered in microbial dark matter'; the podcast is from January 2015).

Explain

- Low demand students should **complete the gaps** in the sentences (section 2 of Worksheet 7.11.1), after listening to the podcast and reading the Student Book. [O1, O2]
- Standard and high demand students should **annotate** the diagrams presented in section 2 of Worksheet 7.11.2, after listening to the podcast and reading the Student Book. [O1, O2]

Consolidate and apply

- Students should work in small groups. They should **consider** what they have learned about the development of teixobactin and **draw up a list** of advantages and disadvantages on their worksheet. [O2, O3]
- Students should **prepare a case** to present to the drug company to speed up the development of the new antibiotic. [O3]
- High demand students should also **prepare a second case**, this time for the drug company of why they might be reluctant to develop the antibiotic. [O3]
- Select different students to **present** their cases. [O3]

Extend

Ask students able to progress further to:

- **find out** about the development of the iChip and how it works. They should **explain why** it has been successful in leading to the development of new antibiotics. [O2, O3]

Plenary suggestions

Ask students to **write a summary** to explain whether they believe that we should continue to invest in the development of antibiotics, or whether they think it will eventually be a waste of time. Students should **justify** their answers. [O2, O3]

Students should also **consider** the ethical position of companies. Is it right for companies to develop an antibiotic that could save millions of lives and not expect to profit from this? Ask students to work in pairs to come up with an **argument** for and against. [O3]

Answers to Worksheet 7.11

1 How to use antibiotics safely

T, F, F, T, T, F, T, T/F

2 Developing new antibiotics

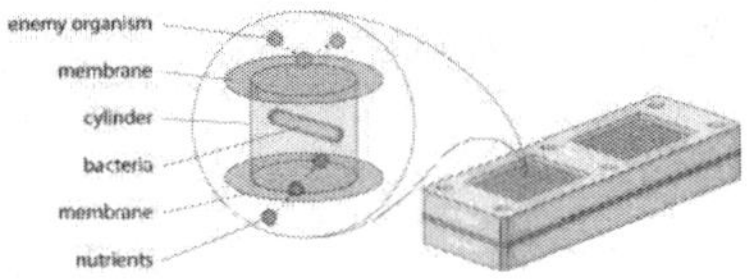

Low demand: antibiotics, resistant, grown, cell, walls, soil, membranes, natural, 50 000, resistance, toxic, mice, ten

Standard and high demand: the first photo shows the iChip. It works by collecting samples of bacteria and antibiotics from their natural environment. It is more effective than growing bacteria in the lab. It separates out the individual bacterial samples, allowing them to grow naturally. Also antibiotics, naturally present in the soil, can be isolated and tested.

The next photo is the teixobactin antibiotic. It is made by some bacteria to kill off competing bacteria. It was isolated and tested on mammalian cells and found to be non-toxic. This has proved to be one of the most promising new types of antibiotic to be produced so far.

Teixobactin works by destroying the cell walls of certain bacteria. This means that even if the bacteria mutate, they will always be vulnerable to an attack on their cell wall. Resistant bacteria are destroyed by this type of antibiotic.

3 Advantages and disadvantages of developing new antibiotics

Advantages	Disadvantages
The world will soon face the threat of bacteria that are resistant to all our current antibiotics. The development of new antibiotics could save millions of lives worldwide.	Testing and development of new antibiotics is slow.
Through the development of new antibiotics we have discovered new and better ways of growing bacterial cultures, which means we can study and learn more about the way they work.	It costs a lot of money to develop and test a new antibiotic.
The more we learn, the more we can find out new ways to tackle the problem of antibiotic-resistant bacteria.	If a new antibiotic is to be produced, it is likely to be used in very small quantities and as a last resort. The drug companies will not earn much profit.
	Once the new antibiotics are used, it may not be long before a resistant strain of bacteria evolves. This would make all that work in producing the new antibiotic wasted.

Lesson 12: Selective breeding

Lesson overview

AQA Specification reference

AQA 4.6.2.3

Learning objectives

- Describe the process of selective breeding.
- Recall how selective breeding enables humans to choose desirable characteristics in animals and plants.
- Explain how selective breeding can lead to inbreeding.

Learning outcomes

- Recognise examples of selective breeding. [O1]
- Describe the steps involved in selective breeding. [O2]
- Explain the advantages and disadvantages of selective breeding. [O3]

Skills development

- WS 1.3 Appreciate the power and limitations of science and consider any ethical issues which may arise.
- WS 1.4 Explain everyday and technological applications of science. Evaluate associated personal, social, economic and environmental implications, and make decisions based on the evaluation of evidence and arguments.
- WS 4.1 Use scientific vocabulary, terminology and definitions.

Resources needed Videos on selective breeding. Worksheet 7.12.1 (low demand); Worksheet 7.12.2 (standard and high demand)

Digital resources PowerPoint files

Key vocabulary breed, inbreeding, selective breeding

Teaching and learning

Engage

- Show students the PowerPoint slide of images of different animals and plants. Ask students to work in pairs and **decide** what they might have in common. [O1]
- Introduce the idea of selective breeding by showing students two videos from www.marthastewart.com (search videos for 'selective breeding') and www.hhmi.org (go to the 'biointeractive' page and search for 'dog breeding'). [O1]

Challenge and develop

- Describe the process of selective breeding using the PowerPoint, and/or the Student Book. Ask students to **answer** questions 1 and 2 from the Student Book (page 297). [O1, O2]
- Show students the video of how humans selectively breed organisms at http://channel.nationalgeographic.com (search for 'super cow creators').
- Show students the PowerPoint slides, highlighting the problems with selective breeding. Students should **answer** the questions from section 1 of their worksheet. [O1, O2]

Explain

- Ask students to **complete** section 2 of the worksheet. They should **explain how** to selectively breed a chick that produces many large eggs with large yolks. [O2]

Consolidate and apply

- Students should work in small groups.
- All students should **sort** the cards from section 3 of the worksheet into advantages and disadvantages of selective breeding. [O3]
- Standard and high demand students (Worksheet 7.12.2) should **evaluate** whether the advantages outweigh the disadvantages and write an argument to support their case. [O3]

Extend

Ask students able to progress further to:

- **find out** about selective breeding in an animal of their choice. They should **explain how** the characteristics are chosen and what impact this will have on the species as a whole over many generations. [O2, O3]

Plenary suggestions

Students should **prepare** and **present** a one-minute talk on 'The world without selective breeding'. [O1, O2, O3]

Answers to Worksheet 7.12

1 What is selective breeding?

1. They are uniquely different from normal lions and are less likely to breed successfully in the wild.
2. They would have died out. This is because white are more easily spotted by prey and it is harder for them to hunt successfully.
3. because they are unusual
4. This can lead to problems related to inbreeding. Harmful recessive genes could also be passed on, leading to a greater susceptibility to disease and disorders. Also more useful genes from animals with less desirable traits, may no longer be passed on through and may become lost from the gene pool.

2 How is selective breeding carried out?

Select hen A and cockerel Y (most number of eggs laid)

↓

Offspring are produced. Some of the offspring will produce greater number of large eggs. These are selected and breed with each other.

↓

Offspring are produced. Some of the offspring will produce greater number of large eggs. These are selected and breed with each other.

↓

Breeding programme continues until all the hens lay lots of large eggs.

3 Advantages and disadvantages of selective breeding

Advantages: B, D, F, H, J

Disadvantages: A, C, E, G, I

Lesson 13: Producing new plant varieties

Lesson overview

AQA Specification reference

AQA 4.6.2.3

Learning objectives

- Describe the process of selective breeding.
- Recall how selective breeding enables humans to choose desirable characteristics in plants.
- Evaluate the benefits and risks of selective breeding in plants.

Learning outcomes

- Recall examples of selective breeding in plants. [O1]
- Describe how plants are selectively bred. [O2]
- Explain the advantages and disadvantages of selective breeding in plants. [O3]

Skills development

- WS 1.3 Appreciate the power and limitations of science and consider any ethical issues which may arise.
- WS 1.4 Explain everyday and technological applications of science. Evaluate associated personal, social, economic and environmental implications, and make decisions based on the evaluation of evidence and arguments.
- WS 4.1 Use scientific vocabulary, terminology and definitions.

Resources needed Selection of flowers, fruits and vegetables. Worksheet 7.13.1 (low demand); Worksheet 7.13.2 (standard and high demand)

Digital resources PowerPoint slides

Key vocabulary environmental change, genetic variation, mixed population

Teaching and learning

Engage

- Show students the PowerPoint images and story of the Irish potato blight. Ask students to work in pairs and **explain** why this problem occurred in the first place and why so many potatoes were wiped out. [O1]
- Show students the PowerPoint slide of plants that are selectively bred. Ask students to work in pairs and **devise** a list of all the different characteristics we might want from plants. They should join with another group and **add to** their list. Take feedback from the class and write up a class list on the board. [O1]

Challenge and develop

- Provide students with examples of a range of flowers, fruits and vegetables. Ask them to **complete** the task from section 1 of their worksheet and **identify** the characteristics that we might want from these examples. [O1, O2]
- Show students the PowerPoint slides and ask them to **read** pages 298–299 of the Student Book on how selective breeding takes place. Ask students to **answer all** the questions in the Student Book on these pages. [O1, O3]
- Show students the video of how wheat has become selectively bred (available at: http://www.saps.org.uk/secondary/teaching-resources/818-norman-borlaug).
- Students should work in pairs and **discuss** the implications of selectively breeding our crops today. [O1, O2, O3]

Explain

- Ask students to work in small groups. Provide them each with a vegetable, fruit or flower. Ask them to **work together** to **produce** a short presentation for 'Gardeners' question time' to **explain** how to selectively breed the plants to produce a particular characteristic. They should **produce** their own flow-diagram poster of the processes involved and **explain** special procedures that need to be carried out on the plants. [O1, O2]

Consolidate and apply

- Show students the case study of the banana in section 2 of their worksheet. Ask students to **recall** the problems of selective breeding in animals and **write** the ones that apply to plants in the worksheet. They should i**dentify** any further problems involved specifically in selective breeding of plants. [O3]
- Ask low demand students to **answer** the questions given on Worksheet 7.13.1. [O3]
- Standard and high demand students should **write a leaflet** to plant and crop growers to **explain** how to avoid the problems they have identified in the worksheet. [O3]

Extend

Ask students able to progress further to:

- **find out** about other crops that have been affected because they have been selectively bred. Students should **weigh up** these disasters against the benefits of selective breeding in plants and **summarise** the arguments for and against. [O1, O2, O3]

Plenary suggestions

Ask students to **imagine** a world without selective breeding in plants. They should work in pairs and compare this to a world without selective breeding in animals and explain how this would be different. [O1, O2, O3]

Answers to Worksheet 7.13

1 Why do we selectively breed plants?

Example table:

Name of plant, fruit or vegetable	Desirable qualities for selective breeding
Grapes	Seedless, large, disease resistant, sweet, quantity made per vine
Tomatoes	Different colours (red, yellow), size (large and small), sweetness, disease resistance, quantitiy made per plant
Potatoes	Roundness, disease resistant, production per plant, taste, ability to cook, mash, bake

2 Selective breeding disaster

1. The use of selective breeding to such an extent has meant there is little variation in the plants, so all will be susceptible to the same disease.
2. used more than one variety, and not just relied on the single variety

Lesson 14: Genetic engineering

Lesson overview

AQA Specification reference

AQA 4.6.2.4

Learning objectives

- Explain what is meant by the term genetic engineering.
- Give examples of how plant crops have been genetically engineered to improve products.
- Describe how fungus cells are engineered to produce human insulin.

Learning outcomes

- Recall examples of genetic engineering applications. [O1]
- Describe the benefits of genetic engineering. [O2]
- Explain how the process of genetic engineering works. [O3]

Skills development

- WS 1.2 Use a variety of models such as representational, spatial, descriptive, computational and mathematical to solve problems, make predictions and to develop scientific explanations and understanding of familiar and unfamiliar facts.
- WS 1.4 Explain everyday and technological applications of science. Evaluate associated personal, social, economic and environmental implications, and make decisions based on the evaluation of evidence and arguments.
- WS 4.1 Use scientific vocabulary, terminology and definitions.

Resources needed Pipe cleaners of different colours, scissors, small boxes and tubs to represent cells. Worksheet 7.14.1 (low demand); Worksheet 7.14.2 (standard and high demand)

Digital resources PowerPoint slides

Key vocabulary genetic engineering, GM crops, vector

Teaching and learning

Engage

- Show students the PowerPoint slide of different GM organisms and ask them to **discuss** in pairs what the images might have in common. [O1]
- Introduce the term 'genetic engineering' or 'GM'. Ask pupils to work in pairs and **write down** everything they already know about the process on sticky notes. Collect all their ideas and place these on a single poster for the end of the lesson. [O1]

Challenge and develop

- Show students a video relating to the need for GM crops and examples of GM (www.saps.org.uk – search video clips for 'The case for genetically modified crop plants').
- Show students the PowerPoint slide with other examples of genetic engineering. Ask students to **complete** the table in section 1 of the worksheet. [O1]
- Show students the process of how genetic engineering works using the PowerPoint slide.
- Low demand students can **put the processes** in order in section 2 of Worksheet 7.14.1.
- Standard and high demand students should **produce** their own annotated diagrams to **explain** how disease resistance can be genetically modified in plants (Worksheet 7.14.2, section 2). [O2]
- Low and standard demand students should **read** the section on Producing human insulin on page 300 of the Student Book. They then **answer** Questions 1 and 2.

- High demand students **read** the section from the Student Book on The genetic engineering technique and **answer** Questions 5 and 6. [O1, O2, O3]

Explain

- Ask students to imagine they are explaining GM to their grandma, who does not have any idea of what it means. Ask them to **write** a letter **explaining** the benefits, including examples and how it works. [O1, O2, O3]

Consolidate and apply

- Ask students to work in small groups. They should **use** the pipe cleaners, boxes, and tubs **as a model to explain** how the blight-resistant Desiree potato might have been produced. [O1, O2, O3]
- Students should modify their models to explain how fungal cells are genetically engineered to produce human insulin.

Extend

Ask students able to progress further to:

- **find out** about the historical development of GM technology and which types of organisms are intended for GM modification. They should **make up** their own timeline of events, **explaining** the advances that have occurred. [O1, O2]

Plenary suggestions

Return to the list drawn up at the start of the lesson. Ask students to **correct** any incorrect ideas they had, and to **add** new ideas as a result of their learning. [O1, O2, O3]

Students should think about all possible applications of GM for the future. Ask them which GM modifications would most benefit 1) humans and 2) the planet, and how might they be achieved. [O1, O2, O3]

Answers to Worksheet 7.14

1 Examples of GM applications

Name of organism	Genetic modifications	Where did the genes come from?	Benefits to the organism	Benefits to humans
Soybean	Resistant to herbicides	Wild plants that are resistant to herbcides	Will not die when sprayed with herbicides	Crops can be sprayed with herbicides, killing the weeds but not the crop.
E. coli or yeast	Can make human insulin	Humans	None	Human insulin is made easily and cheaply in vast supplies and is suitable for humans without side effects.
Rabbit	Can glow under UV radiation	Jellyfish	None	None

2 How does genetic engineering occur?
(low demand): 4, 1, 8, 2, 3, 6, 5, 7

Lesson 15: Genetically modified crops: the science

Lesson overview

AQA Specification reference

AQA 4.6.2.4

Learning objectives

- Explain the benefits of genetic modification in a range of crops.
- Explain the concerns about genetic modification.
- Explain the ethical concerns about genetic engineering.

Learning outcomes

- Recall a range of examples of genetically modified crops and their benefits. [O1]
- Describe some of the disadvantages of genetically modified crops. [O2]
- Explain the debate over genetically modified crops using scientific ideas. [O3]

Skills development

- WS 1.3 Appreciate the power and limitations of science and consider any ethical issues which may arise.
- WS 1.4 Explain everyday and technological applications of science. Evaluate associated personal, social, economic and environmental implications, and make decisions based on the evaluation of evidence and arguments.
- WS 3.5 Interpret observations and other data (presented in verbal, diagrammatic, graphical, symbolic or numerical form), including identifying patterns and trends, making inferences and drawing conclusions.

Maths focus

- 4a Translate information between graphical and numeric form

Resources needed Video on the development of genetically modified papaya plants from www.bbc.co.uk (the BBC Bitesize website). Worksheet 7.15

Digital resources

Key vocabulary food security, GM crops, larva, order

Teaching and learning

Engage

- Ask students to work in pairs. They should **imagine** they are managing their own crop.
- All students should **identify** and **write** down on sticky notes all the potential risks that may attack their crop and livelihood. They will come back to the list later on in the lesson. [O1]
- Low demand students could **use** the cards in section 1 of Worksheet 7.15 to help them. [O1]

Challenge and develop

- Ask students to **watch** the video on the development of genetically modified papaya plants from the BBC Bitesize website. [O1]
- **Discuss** all the genetic modifications plants can undergo to enable them to provide stronger healthier crops. **Write** a class list on the board. [O1]
- Ask students to work in pairs. They should **select** one of these modifications and **explain how** it might be carried out. Take feedback from the pairs. [O1]
- Students should **read** the information on the cards in section 1 of Worksheet 7.15 outlining the possible threats from GM crops, and **analyse** the graphs given in section 2.
- Standard and high demand students should **use** the values from the graph in their analysis. [O2]

- Students should **evaluate the evidence** for each of the potential problems, and **decide** what level of threat they pose to humans and to the wider environment. [O2]

Explain

- Ask students to **imagine** the worst case scenarios caused by the problems of GM crops. Ask them to **explain** what impact this is likely to have on the environment and on the world. [O2, O3]

Consolidate and apply

- Ask students to **prepare** an article for an agricultural newspaper outlining the arguments for and against GM crops. [O3]
- Students should **read** the Student Book pages 302–303 and **answer** all the questions. [O1, O2, O3]

Extend

Ask students able to progress further to:

- **find** further evidence of the migration of genetically modified genes to other species of wild plants. They should **present** their findings in the form of graphical data and share this with the class. [O1, O2, O3]

Plenary suggestions

Return to the list of potential problems devised at the start of the lesson. Ask students to **make a case** for and against using GM to solve each one, based on the learning in the lesson. [O1, O2, O3]

Answers to Worksheet 7.15

1 What are the risks of growing a crop?

Cards 1, 2, 4, 6, 7 ,9

2 Problems caused by GM crops

1. Graph 1 shows that the production of geneticallly modified herbicide-resistant plants has increased – the most from 1996 to 2012 – by 100 million hectares. There is a rising increase in the production of both herbicide and insecticide-resistant genetically engineered forms, to about 40 million hectares. The amount of insecticide-only resistant plants has increased but the rate seems to have levelled off, and has increased much more slowly.

   Graph 2 shows that weed species have become most resistant to acetolactate since 1980, with a steep increase from 1990 to date, and continuing to rise. Weeds increased their resistance to atrazine from 1970, with a very steep increase to 1990, where the increase has slowed down. Weed resistance to urease, amides and dinitroanilines, began to rise steeply, but has levelled off. Resistance to glyphosate increased from 1996, when GM crops began to be widely used and continues to increase.
2. A resistant gene is naturally present, or was formed by a mutation. The resistant plants thrived and passed the gene onto their offspring, so a resistant population has developed.
3.

   Concern 1: human risk (1–3); environmental risk: 0

   Concern 2: human risk (1–3); environmental risk (6–8)

   Concern 3: human risk (7–10); environmental risk (7–10)

Lesson 16: Is genetic modification safe?

Lesson overview

AQA Specification reference

AQA 4.6.2.4

Learning objectives

- Explore the benefits of genetic modification in medicine.
- Explain the concerns that people have about genetic modification.
- Explain the possible safety issues of genetic engineering in agriculture and medicine.

Learning outcomes

- Provide examples of using genetic modification in medicine. [O1]
- Describe problems that can arise as a result of using genetic modification in medicine. [O2]
- Explain how genetic markers are used to improve safety. [O3]

Skills development

- WS 1.3 Appreciate the power and limitations of science and consider any ethical issues which may arise.
- WS 1.4 Explain everyday and technological applications of science. Evaluate associated personal, social, economic and environmental implications, and make decisions based on the evaluation of evidence and arguments.
- WS 1.5 Evaluate risks both in practical science and the wider societal context, including perception of risk in relation to data and consequences.

Resources needed Worksheet 7.16.1 (low demand); Worksheet 7.16.2 (standard and high demand)

Digital resources PowerPoint slides

Key vocabulary genetic marker, gene

Teaching and learning

Engage

- Ask students to work in pairs. See if they can **make a list** of all the genetic engineering applications in medicine they currently know of. Ask them to **think** about all the potential applications of genetic engineering and **describe how** the techniques might work. [O1]

Challenge and develop

- Show students a video on 'How Ebola drug ZMapp is made'. This shows how genetic engineering has been used to produce a drug to treat the Ebola virus outbreak last year.
- Low demand students could **put the steps in order** in section 1 of Worksheet 7.16.1.
- Standard and high demand students could **summarise** the steps in an annotated diagram on Worksheet 7.16.2. [O1]
- Introduce some of the safety concerns associated with using genetic modifications, using the PowerPoint slides and the Student Book (pages 304–305). Ask students to work in small groups and **devise** ways to test whether these concerns are valid. [O1, O2]
- Introduce the idea of a genetic marker, using the Student Book and PowerPoint presentation. Ask students to **evaluate** the advantages and disadvantages of using these. [O3]
- Use the PowerPoint presentation and Student Book to show further possible applications of genetic engineering in medicine. Ask students to **highlight** the benefits and disadvantages of each. [O1, O2]

Explain

- Students should **produce** an annotated poster to **explain** some of the applications of genetic engineering in medicine, and how they are achieved. They should **explain** the role of genetic markers and **evaluate** the overall safety considerations of using genetic engineering in medicine. [O1, O2, O3]

Consolidate and apply

- Ask students to work in small groups. Students should **imagine** they will use genetic engineering to develop a vaccine for the flu virus. Ask them to **consider and plan how** they might do this. They should **focus** on the safety precautions they will undertake, to ensure there are no adverse effects to their technology. [O1, O2, O3]
- Students should **answer** all the questions in the Student Book (pages 304–305). [O1, O2, O3]

Extend

Ask students able to progress further to:

- **find out** about further possible applications of genetic engineering in humans – medical and non-medical applications – and **present** these to the class. [O1, O2]

Plenary suggestions

Ask students to **write an argument** either for or against the use of genetic engineering to treat human conditions. [O1, O2, O3]

Answers to Worksheet 7.16

1 Using genetic engineering against the Ebola virus

1. A genetically engineered virus with the Ebola proteins is injected into a tobacco plant.
2. The plant produces many antibodies in response to the Ebola infection.
3. The infected leaves are ground before they die.
4. The antibodies are extracted from the dying leaves.
5. The antibodies are cloned and made fit for human cells.
6. The cloned antibodies are purified and turned into medicine.
7. The antibodies attach themselves to infected human cells in the body and destroy the virus.
8. The tobacco plant is used because it produces many antibodies. Tests done on mice did not produce enough antibodies.

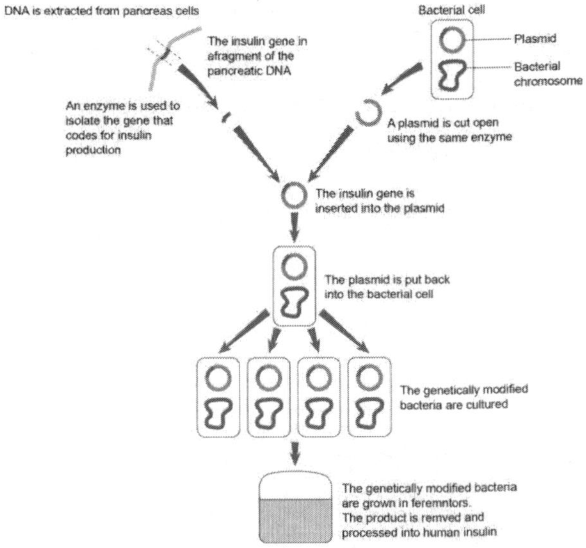

Advantages

Genetic diseases can be treated like sickle cell and cystic fibrosis, which have no cure.

The potential to treat thousands of genetic conditions is high.

A cure for Ebola was found through genetic engineering, saving lives.

Without genetic engineering, there would not be enough human insulin about and millions of people would die from diabetes.

It is a way of providing much larger quantities of medicines than would otherwise be possible.

Disadvantages

Does not have a long-term proven record of safety as it is a developing technology.

Worries over what happens if genetically modified organisms escape or mutate.

Not normal – transferring genes from one organism into another is against nature. This would never happen normally.

Worries over genetic engineering of non-human traits into humans for military purposes.

Lesson 17: Ethically wrong, or essential?

Lesson overview

AQA Specification reference

AQA 4.6.2.4

Learning objectives

- Explain the benefits of, and concerns about, genetic modification.
- Explain the ethical issues of genetic engineering in agriculture and medicine.

Learning outcomes

- Identify the arguments for and against genetic engineering. [O1]
- Present a reasoned argument for and against the ethical considerations of genetic engineering. [O2]
- Explain how gene therapy takes place and the ethical implications involved. [O3]

Skills development

- WS 1.3 Appreciate the power and limitations of science and consider any ethical issues which may arise.
- WS 1.4 Explain everyday and technological applications of science. Evaluate associated personal, social, economic and environmental implications, and make decisions based on the evaluation of evidence and arguments.
- WS 1.5 Evaluate risks both in practical science and the wider societal context, including perception of risk in relation to data and consequences.

Resources needed Infographic from http://us.naturespath.com; video from http://news.discovery.com. Worksheet 7.17.1 (low demand); Worksheet 7.17.2 (standard demand); Worksheet 7.17.3 (high demand)

Digital resources PowerPoint slides

Key vocabulary gene therapy, vector

Teaching and learning

Engage

- Show students the image of the glowing fish from the PowerPoint slide. Ask students to work in pairs and **discuss** the questions on the slide relating to the glowing fish. [O1]
- Show the poster found at the link http://us.naturespath.com (search for 'infographic what are GMOs').
- Ask students to **identify** which information is pro genetic engineering and which is against. Ask them to **decide** what it is about the information that reveals the bias. **Discuss** the problems with bias and the need to examine the source information. [O1, O2]

Challenge and develop

- Use the PowerPoint and the Student Book to introduce the problems with GM crops and the exploitation of poor farmers. Ask students to **discuss** how genetic engineering can be shared fairly between wealthy and poor countries and the disadvantages that may be faced by poorer countries. [O1, O2]
- Show students the PowerPoint slide relating to GM labelling. Ask students to **discuss** whether labelling food can help to justify the disadvantages of genetic engineering. [O1, O2]
- Introduce the concept of gene therapy, using the video at http://news.discovery.com (search for 'how does gene therapy work') and/or ideas from the Student Book, or ask students to **research** at least three possible applications of gene therapy. Ask students to **discuss** the pros and cons of gene therapy and **write** a list of advantages and disadvantages. [O3]

Explain

- Students should **explain** how gene therapy works using a cartoon strip. [O3]

Consolidate and apply

- Ask students to work in groups of four. They should **sort** the cards from section 1 of Worksheet 7.17.1 (low demand), 7.17.2 (standard demand) or 7.17.3 (high demand).
- Low demand students **rank** the applications of genetic engineering according to importance. Standard demand students **identify** the ethical issues behind each application. High demand students **explain** whether the applications justify overlooking the ethical issues at hand. [O1, O2]

Extend

Ask students able to progress further to:

- **find out** about the ethical issues involved in gene therapy techniques. They should **make** a PowerPoint presentation about the advantages and ethical considerations of gene therapy. [O1, O2, O3]

Plenary suggestions

Ask students to work in groups of four. Each person should **represent** one of these four people: a person with religious beliefs, a genetic scientist, a person with a genetic disease and a lover of animal rights. They should **rank** the cards from worksheet activity 2 to those most relevant to their situation and **explain why** the ethical issues are most important to them. [O1, O2, O3]

Ask students to **debate** the questions from the PowerPoint slide in small groups from their own perspective. Discuss their responses. [O1, O2, O3]

Answers to Worksheet 7.17

1 Do the ends justify the means?

Possible answers: low demand – (most important) 4, 5, 10, 1, 3, 6, 7, 9, 11, 12, 8, 2 (least important)

For each numbered application, (a) indicates standard demand response and (b) indicates high demand response.

1. (a) Cloning of pototoes so there is less variety and greater vulnerability to disease. Unnatural. (b) More people will benefit from enhanced protein so their diet is better – justified use.
2. (a) Unnatural – house plant has no choice. (b) Not justified – application is trivial.
3. (a) Less variation in rice varieities so more prone to blight and diseases. Unnatural transfer of genes. (b) Rice is the most staple food for many poor people – so justified use.
4. (a) Unnatural transer of human genes into plants. (b) Protection against many harmful diseases and as edible, means less transfer from infected needles in poor countries – justified use.
5. (a) Unnatural transer of human genes into plants. Plants have no choice. (b) Protection against cancer possible – justified use.
6. (a) Less variation in potato and chestnut varieities so more prone to other diseases. Other forms of blight may emerge. Unnatural transfer of genes. (b) Will improve yields of potato and chestnut – justified use.
7. (a) Unnatural transer of human genes into plants. Plants have no choice. (b) Protection against flu, which kills. However, there are many different types of flu – use not justified.
8. (a) Unnatural transer of spider genes into goats. Goats have no choice. Milk can't be used for the young goats and these goats will not be allowed to reproduce. (b) Use not justified as there are other ways that strong materials can be made without using host organisms.
9. (a) Unnatural transfers of human genes into pigs. Pigs have no choice and will be killed once the heart is grown. (b) Less complications with organ rejection and heart can be made using the patient's own genes. Many pigs can be used, saving thousands of lives – justified use.
10. (a) Unnatural transer of human genes into other organisms, which have no choice. (b) If effective against arthritis and cancer can save many lives and improve the health of millions – justified use.
11. (a) Human DNA should not be altered artifically – unnatural and leads to all kinds of other ethical issues. (b) Genetic diseases could be irradicated –but high risks also involved.
12. (a) Human DNA should not be altered artifically – where is the limit – will it extend to cosmetic features or just life threatening ones? (b)The risk of meddling with the human genome is too serious and should not be tampered with.

Lesson 18: Cloning

Lesson overview

AQA Specification reference

AQA 4.6.2.5

Learning objectives

- Describe how cuttings and tissue culture are used to produce new plants.
- Describe the use of embryo transplants and adult cell cloning in animals.

Learning outcomes

- Explain how plants are cloned using cuttings and tissue culture. [O1]
- Explain how animals are cloned using embryo transplants and adult cell cloning. [O2]
- Evaluate the advantages and disadvantages of each method of cloning. [O3]

Skills development

- WS 1.3 Appreciate the power and limitations of science and consider any ethical issues which may arise.
- WS 1.4 Explain everyday and technological applications of science. Evaluate associated personal, social, economic and environmental implications, and make decisions based on the evaluation of evidence and arguments.
- WS 4.1 Use scientific vocabulary, terminology and definitions.

Resources needed Videos on how to clone a cauliflower and how jellyfish reproduce. Worksheet 7.18.1 (low demand); Worksheet 7.18.2 (standard demand); Worksheet 7.18.3 (high demand); Practical sheet 7.18; Technician's notes 7.18

Digital resources PowerPoint slides

Key vocabulary adult cell cloning, embryo transplant, cuttings, therapeutic cloning, tissue culture

Teaching and learning

Engage

- Show students the images of different organisms from the PowerPoint and ask them to **identify** what they each have in common. [O1]
- Show students a real example of a spider plant, strawberry runner, potato tubers, and/or other suitable plants from which a cloned plant can be obtained. Demonstrate how to take a cutting and turn it into a new plant. Ask students to work in pairs and **explain how** this process works. [O1]
- Introduce or remind students of the word 'cloning'. Ask students to **explain** how it links to mitosis, stem cells and asexual reproduction. [O1]

Challenge and develop

- Ask students to work in pairs and **create** their own plant cuttings from the plants available using the instructions from Practical sheet 7.18. [O1]
- Show students the video that shows how to clone a cauliflower (search 'Cauliflower Cloning – Tissue Culture and Micropropagation' at www.saps.org.uk). [O1]
- Students should **complete** section 1 of Worksheet 7.18.1 (low demand), 7.18.2 (standard demand) or 7.18.3 (high demand). [O1]
- Show students the method to clone GM crops using the Student Book or the PowerPoint slide. Ask students to **spot the differences** between this technique and the one to produce normal plant cells. [O1]
- Show students the two methods used to clone animals either by embryo transplants or animal cell cloning using the Student Book and/or the PowerPoint slide. [O2]
- Students should complete section 2 of their respective worksheets. [O2]

Explain

- Working in small groups, students should **analyse** each cloning technique and **explain** the advantages and disadvantages of each method. Take responses from each group and **discuss** their ideas. [O3]

Consolidate and apply

- Show students the video of how jellyfish reproduce (search 'Moon Jelly Life Cycle' at http://shapeoflife.org). Ask them to **identify** any resemblances to the cloning of plants they have just observed. [O1]
- Ask students to **explain how** Dolly the sheep might have been cloned. [O2]
- Students should use their scientific understanding to **debate** the ethical issues around cloning of plants and animals. [O3]

Extend

Ask students able to progress further to:

- **find out** about which organisms are commonly cloned and the impact this has on our lives. They should **compare** the impact of cloning plants with cloning animals. They could **present** their findings to the class. [O1, O2, O3]

Plenary suggestions

Ask students to **produce** a concept map of all the ideas relating to cloning. They should include the ethical considerations behind cloning. [O1, O2, O3]

Answers to Worksheet 7.18

1 Methods for cloning plants

Low demand: tissue, small, stem, growth medium, hormones, plantlets, cloned, thousands, few, stem cells, leaves, stems, hormone (rooting), roots

Standard and high demand:

Cuttings	Tissue culture
One plant produced from each cutting	Thousands of plants produced from a few cells
Plants need minimal care	Lots of care needed for the small plantlets
Small amounts of rooting powder needed	Large number of specific hormones needed
Plants can be grown in normal conditions	Highly controlled growing situations needed, free of disease
Cheap	Expensive

2 Animal cloning

Low and standard demand: order and labels should be as per the diagram in the Student Book (page 309).

High demand:

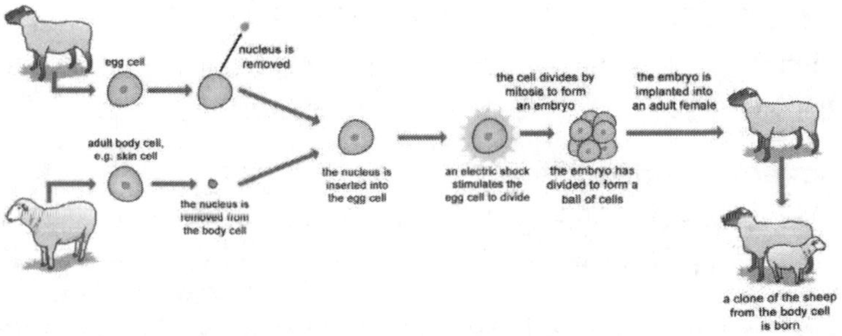

Embryo cloning	Adult cell cloning
Many organisms can be made from a few embryo cells	Only one organism is made
Many host mothers are needed	Only one host mother is needed
Embryo cells need to be grown in special, controlled conditions so expensive	Less successful as it depends on an electric shock to stimulate development of egg cell to form an embryo
Host mothers given hormones to prepare for implantation of the embryo	Host mothers given hormones to prepare for implantation of the embryo

Lesson 19: The tree of life

Lesson overview

AQA Specification reference

AQA 4.6.4

Learning objectives

- Describe how living things have been classified into groups using a system devised by Linnaeus.
- Describe how new models of classification have developed.

Learning outcomes

- Describe and apply the Linnaean system of classification. [O1]
- Explain the development new systems of classification including the three domains. [O2]
- Compare and contrast the Linnaean system to the evolutionary tree. [O3]

Skills development

- WS 1.1 Understand how scientific methods and theories develop over time.
- WS 1.4 Explain everyday and technological applications of science. Evaluate associated personal, social, economic and environmental implications, and make decisions based on the evaluation of evidence and arguments.
- WS 3.5 Interpret observations and other data (presented in verbal, diagrammatic, graphical, symbolic or numerical form), including identifying patterns and trends, making inferences and drawing conclusions.

Resources needed Videos of Linnaean classification (e.g. www.tigtagworld.co.uk) and evolutionary tree (e.g. bbc.co.uk). Worksheet 7.19.1 (low demand); Worksheet 7.19.2 (standard demand); Worksheet 7.19.3 (high demand)

Digital resources

Key vocabulary binomial system, evolutionary tree

Teaching and learning

Engage

- Ask students to work in pairs. Provide them with the card sort of different animals and plants. Ask them to **identify** the similarities and differences between them, and to **group them** accordingly. Ask pairs to join with another group and **compare** their ideas. [O1]
- Ask each group of students to **devise** their own classification key for these organisms to tell them apart. **Discuss** the difficulties with formulating a classification system. [O1]

Challenge and develop

- Introduce the Linnaean classification system using the video (**Error! Hyperlink reference not valid.**www.tigtagworld.co.uk – search for 'film carl linnaeus'). **Discuss** the advantages and limitations of using such a system. [O1]
- Ask students to **classify** the organisms from the card sort using the information on the worksheet. Ask students to **identify** what further information might be helpful in classifying certain organisms. Students could **research** one of the organisms and **find out** its full identification using the Linnaean system. [O1]
- Introduce and **discuss** the limitations of the Linnaean system using the Student Book and how new systems, including the three domains and the binomial system, developed. [O1, O2]
- Introduce the concept of an evolutionary tree and how it works using the video at www.bbc.co.uk (search 'evolutionary tree natural life video'). Emphasise the importance of using genetics and physical characteristics and how difficult it is to exactly pinpoint evolutionary lines. **Discuss** how evolutionary trees developed, using the Student Book, and what tools scientists have used to construct them. [O2]

Explain

- Low demand students should **describe** the features of the Linnaean classification system and the evolutionary tree (section 3 of Worksheet 7.19.1). [O1, O2, O3]
- Standard demand students should **explain** the scientific methods used to construct each classification system (section 3 of Worksheet 7.19.2). [O1, O2, O3]
- Ask high demand students to **compare** and **contrast** the Linnaean system with the current method using an evolutionary tree (section 3 of Worksheet 7.19.3). [O1, O2, O3]

Consolidate and apply

- Ask students to **return** to their card sort from the start of the lesson. They should **explain how** the Linnaean system applies to the organisms and also how an evolutionary tree might be constructed. Students should **identify** any further information they might need in order to fully classify each organism. [O1]
- Students should **read and answer** all the questions from the Student Book (pages 310–311). [O1, O2, O3]

Extend

Ask students able to progress further to:

- **find out** about how evolutionary trees were developed in the first place. They should **present** their findings to the class and **predict how** classification systems might be different in the future. [O2, O3]

Plenary suggestions

Students could **write** a letter to Linnaeus to explain the strengths of his classification systems, but also **explain** why a new one had to be developed and the advantages this has. [O1, O2, O3]

Answers to Worksheet 7.19

1 How does classification begin?

Lion: animal, Chordata, mammal

Turtle: animal, Chordata, reptile

Lobster: animal, invertebrate, athropod

2 Linnean classification system – see above

3 Explaining classification systems

Low demand: Linnaean system consists of kingdom, phylum, class, order, family, genus and species. Organisms are classified mainly by their visual appearances.

Evolutionary tree: organisms are classified according to their evolutionary branches. It is based on the structure of their DNA and other chemicals such as proteins.

Standard demand: Linnanean system largely observation skills used – microscopes, dissections and appearance of external features

Evolutionary tree: complex and expensive techniques involving extracting and analysing DNA bases sequenced with the use of powerful computers.

High demand:

Linnaean	Evolutionary tree
Largely observation skills used – microscopes, dissections and appearance of external features.	Complex and expensive techniques involving extracting and analysing DNA base sequenced with the use of powerful computers.
Complex system based on similarities between features.	Cannot be done without computers.
Room for mistakes e.g. organisms which are mimics.	Room for mistakes as analysis of genetic codes could be misinterpreted.
Difficult to accomodate newly evolved species.	More easy to include newly evolved species.

Lesson 20: Extinction…or survival?

Lesson overview

AQA Specification reference

AQA 4.6.3.6

Learning objectives

- List the causes of extinction.
- Explain how new predators, competitors and diseases can lead to extinctions.

Learning outcomes

- Recognise the causes of extinctions. [O1]
- Explain why these factors may result in extinctions. [O2]
- Explain how extinctions may be prevented. [O3]

Skills development

- WS 2.1 Use scientific theories and explanations to develop hypotheses.
- WS 3.5 Interpret observations and other data (presented in verbal, diagrammatic, graphical, symbolic or numerical form), including identifying patterns and trends, making inferences and drawing conclusions.
- WS 4.1 Use scientific vocabulary, terminology and definitions.

Resources needed Worksheet 7.20.1 (low demand); Worksheet 7.20.2 (standard demand); Worksheet 7.20.3 (high demand)

Digital resources PowerPoint slides

Key vocabulary extinction, invasive species

Teaching and learning

Engage

- Show students the PowerPoint slide of the different organisms. Ask students **to identify** what the images have in common. [O1]
- Introduce the idea that the world has experienced at least five mass extinctions over its history, using the PowerPoint and the Student Book. **Discuss** the possible causes of these extinctions. [O1, O2]

Challenge and develop

- **Discuss** the possible reasons for extinctions today and how they might compare with those of the past. Show students the PowerPoint slide comparing human growth with the number of extinctions. Ask students to work in pairs and **discuss** how extinctions from the past might compare to those in the future. [O1, O2]
- Provide students with the case study cards from the worksheet. Students should answer the question in section 1 (low demand: Worksheet 7.20.1; standard demand: Worksheet 7.20.2; high demand: Worksheet 7.20.3). [O1, O2]

Explain

- Students should **produce** their own concept map of extinction and explain all the factors that are responsible for them (see worksheet). [O1, O2]

Consolidate and apply

- Ask students to work in small groups and **construct a presentation** to the UN to **explain why** extinctions occur and how future extinctions could be prevented. They could use examples from the worksheet to help them. [O1, O2, O3]
- Students should **answer** all the questions from the Student Book (pages 312–313). [O1, O2, O3]

Extend

Ask students able to progress further to:

- **find out** about how conservation programmes are used to fight extinctions. They could **debate** whether these measures are enough and why we are putting organisms at risk, although we are aware of the issues at hand. [O1, O2, O3]

Plenary suggestions

Provide students with the mystery card sort on the extinction of the dinosaurs (section 3 of the worksheets). Students should **sort the cards** and use the information to **decide** what led to their extinction and whether it could have been prevented. [O1, O2, O3]

Answers to Worksheet 7.20

1 Causes of extinctions

Low demand:

Woolly mammoth: climate change and hunting; Ladybirds: new competitor; Red squirrel: new competitor and new disease; Javan tiger: new predator – hunting by man, habitat destruction by man; Dodo – new predator – hunting by man and introduced animals; White clawed crayfish: new, more successful competitor; Coral reef: new predator – human hunting; Mass extinction: single catastrophic events – volcanoes; Galapagos tortoise: introduced new competiors and new predators

Standard demand:

Woolly mammoth: highly adapted to live in ice cold environments. Loss of habitat due to climate change and melting of ice pushed the mammoths closer to human settlements. They were hunted for their fur and meat to extinction.

Ladybirds: new and more aggressive competitor means there is less food for the native ladybirds.

Red squirrel: less food for the red squirrel due to new aggressive competitor, the grey squirrel. They also introduced disease, which is responsible for the decline of red squirrel numbers. New disease is probably more responsible for endangered levels.

Javan Tiger: hunting by humans, new predator, and loss of prey due to the loss of habitat.This is probably the greatest reason for their extinction.

Western black rhino: hunting by humans and loss of habitat and food. Loss of food would have the greatest impact.

Dodo: eggs were eaten by introduced predators – man, rats and dogs. So they could not reproduce faster than they were killed.

White clawed cray fish: introduced more aggressive predator containing a fungal disease, which kills of many of these crayfish.

Coral reef: introduced predator – human taking pieces of coral, and pollutung the seas. Global warming also causes coral bleaching and is killing off the coral reefs.

Mass extinction: single catastrophic events – volcanic action.

Galapagos tortoise: introduced new competitors and new predators eating the eggs of the tortoise. Goats eat the food of the tortoise so they die of starvation.

High demand:

Woolly mammoth: would probably have become extinct eventually due to climate change.

Ladybirds: contain introduced species and do not allow them to mingle in the wild and affect native populations.

Red squirrel: grey squirrel populations should have been contained and not been allowed to enter the wild.

Javan tiger: no habitat destruction. Preserve the forest of the tiger, but keep the populations under control, so they did not increase out of control.

Western black rhino: have strict measures on the control of hunting of these rhinos. Control the destruction of habitats where the rhinos live.

Dodo: restrict the numbers of dodos being eaten and protect the eggs so they will continue to reproduce.

White clawed cray fish: do not let introdcued species into the same habitats as the native species.

Coral reef: global warning will have the greatest effect. Humans could stop producing greenhouse gases. Humans could stop overfishing where the coral reef populations are.

Mass extinction: no way of stopping this.

Galapagos tortoise: eggs of the tortoise could be preserved and allowed to reproduce. Goats should be given alternative food supplies to preserve the orginal supplies of the tortoise.

Lesson 21: Maths skills: Using charts and graphs to display data

Lesson overview

Learning objectives

- Understand when and how to use bar charts.
- Understand how to show sub-groups on bar charts.
- Understand how to plot histograms.

Learning outcomes

- Describe the differences between discrete and continuous data. [O1]
- Draw more complex bar charts. [O2]
- Explain the differences between histograms and bar charts. [O3]

Skills development

- WS 3.1 Present observations and other data using appropriate methods.
- WS 3.3 Carry out and represent mathematical and statistical analysis.

Maths focus

- 2c Construct and interpret frequency tables and diagrams, bar charts and histograms

Resources needed Worksheet 7.21.1 (low demand); Worksheet 7.21.2 (standard and high demand)

Digital resources

Key vocabulary demographic, grouped bar chart, stacked bar chart

Teaching and learning

Engage

- Ask students to **devise** their own definition of discrete and continuous data. Ask students to **discuss** how this data can be represented and what types of graph would be appropriate. [O1]

Challenge and develop

- Show students the success criteria for drawing a bar chart from examples of different types of bar chart. Students should **complete** section 1 of their worksheet.
- Low demand students should **draw** a simple bar graph (Worksheet 7.21.1). [O1, O2]
- Standard and high demand students should **draw** stacked and dual bar graphs to represent the data (Worksheet 7.21.2). [O1, O2]
- Ask students to **read** the section on histograms on page 315 of the Student Book. They then **answer** the questions relating to histograms. [O3]

Explain

- Students should **write** a summary to **explain** the key differences between histograms and bar charts. They should **explain why** different types of bar graphs are useful. [O1, O2, O3]

Consolidate and apply

- Show students the data in section 2 of the worksheet. Ask students to **decide** which is the most appropriate graph to present the data given in the two tables, and **draw** these graphs. [O1, O2, O3]

Extend

Ask students able to progress further to:

- **find out** more about histograms and whether they can be used for continuous and/or discrete data. They should **explain** their advantages and disadvantages. [O3]

Plenary suggestions

Ask students to **spot the mistakes** in the bar graphs and histograms in section 3 of the worksheet.

Answers to Worksheet 7.21

2 Choose the most appropriate graph

Bar graphs, dual or stack bar graphs could be drawn for the data represented in the first table.

A histogram should be drawn for the data in the second table.

3 Spot the mistakes

First graph: bars should be the same width and the scale should be equal on the *y* axis.

Second graph: there should be a range on the *x* axis, not a mean value.

When and how to use these pages: Check your progress, Worked example and End of chapter test

Check your progress

Check your progress is a summary of what students should know and be able to do when they have completed the chapter. Check your progress is organised in three columns to show how ideas and skills progress in sophistication. Students aiming for top grades need to have mastered all the skills and ideas articulated in the final column (shaded pink in the Student book).

Check your progress can be used for individual or class revision using any combination of the suggestions below:

- Ask students to construct a mind map linking the points in Check your progress
- Work through Check your progress as a class and note the points that need further discussion
- Ask the students to tick the boxes on the Check your progress worksheet (Teacher Pack CD). Any points they have not been confident to tick they should revisit in the Student Book.
- Ask students to do further research on the different points listed in Check your progress
- Students work in pairs and ask each other what points they think they can do and why they think they can do those, and not others

Worked example

The worked example talks students through a series of exam-style questions. Sample student answers are provided, which are annotated to show how they could be improved.

- Give students the Worked example worksheet (Teacher Pack CD). The annotation boxes on this are blank. Ask students to discuss and write their own improvements before reviewing the annotated Worked example in the Student Book. This can be done as an individual, group or class activity.

End of chapter test

The End of chapter test gives students the opportunity to practice answering the different types of questions that they will encounter in their final exams. You can use the Marking grid provided in this Teacher Pack or on the CD Rom to analyse results. This shows the Assessment Objective for each question, so you can review trends and see individual student and class performance in answering questions for the different Assessment Objectives and to highlight areas for improvement.

- Questions could be used as a test once you have completed the chapter
- Questions could be worked through as part of a revision lesson
- Ask Students to mark each other's work and then talk through the mark scheme provided
- As a class, make a list of questions that most students did not get right. Work through these as a class.

Marking Grid for End of Chapter 7 Test

	Getting Started [Foundation Tier]							Going Further [Foundation and Higher Tiers]					More Challenging [Higher Tier]			Most demanding [Higher Tier]			Total marks	Percentage
Student Name	Q. 1 (A01) 2 marks	Q. 2 (A01) 1 mark	Q. 3 (A01) 1 mark	Q. 4 (A02) 1mark	Q. 5 (A02) 1 mark	Q. 6 (A02) 2 marks	Q. 7 (A03) 2 marks	Q. 8 (A01) 1 mark	Q. 9 (A01) 1 mark	Q. 10 (A01) 2 marks	Q. 11 (A02) 4 marks	Q. 12 (A03) 2 marks	Q. 13 (A01) 2 marks	Q. 14 (A02) 2 marks	Q. 15 (AO3) 4 Marks	Q. 16 (AO1) 2 Marks	Q. 17 (AO2) 4 Marks	Q. 18 (AO3) 4 Marks		

Check your progress

You should be able to:

distinguish between variation caused by genes and by the environment →	describe how variation contributes to an organism's survival →	understand the mechanism of genetic variation
recall early ideas about evolution →	describe how natural selection leads to a struggle for existence →	explain how environmental change operating with natural selection leads to the evolution of a new species
recall that the theory of evolution was developed independently by Darwin and Wallace →	describe the evidence that led to the development of the theory of evolution by Darwin and Wallace →	explain the evidence for the occurrence of evolution and natural selection
identify reasons why evolution was not, to begin with, accepted →	identify the reasons why genetics and evolution were not linked until the twentieth century →	explain how the scientific work of many scientists contributed to the gene theory
recall and use the classification system developed by Linnaeus →	explain the features used to develop evolutionary trees →	explain how microscopic examination, fossils and biochemistry have led to modern evolutionary trees
identify the causes of extinction →	describe how new species, predators and competitors can lead to extinction →	evaluate circumstances that may lead to another mass extinction

Worked example

Scientists in Germany have investigated the resistance of rats to two poisons – warfarin and bromadiolone.

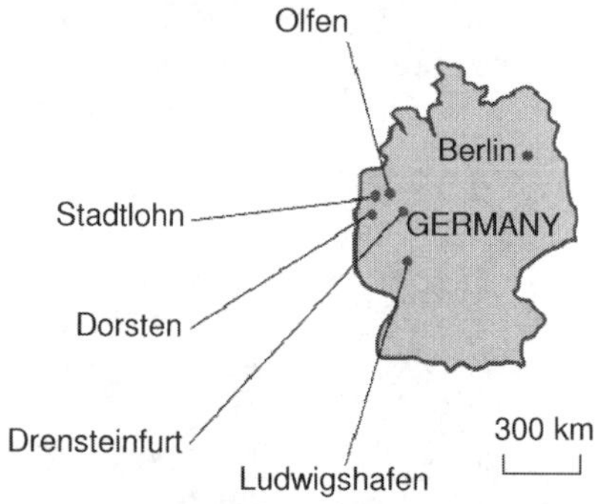

Some of their results are shown below.

Town	Not resistant to either poison (%)	Resistant to warfarin alone (%)	Resistant to both poisons (%)
Dorsten	44	56	0
Drensteinfurt	90	5	5
Ludwigshafen	100	0	0
Olfen	21	21	58
Stadtlohn	5	8	87

1 Explain why the data collected are evidence for natural selection.

In a population, some rats will be resistant to the rat poisons, and survive their use. These will pass on their resistant genes. These will spread through the population as the rats breed, until most of the rats become resistant.

2 Suggest reasons for the differences in the degree of resistance to the poisons.

Resistance is greatest in Stadtlohn and Olfen. There is no resistance to the poisons in Ludwigshafen.

This could be because of the locations of the towns. If the towns are close together, populations can interbreed so that the resistant gene(s) spread. Ludwigshafen is further away.

This is a concise answer, but one that is not fully detailed. The student should have mentioned that some rats in a population will be resistant to the rat poison because of mutations. It is also important to say that these mutations will spread rapidly through the population because of the rats' rapid reproductive rate.

In this answer the student has overlooked the fact that mutations arise spontaneously in populations, as well as being spread by sexual reproduction.

These could also arise at a greater rate in Stadtlohn and Olfen, perhaps because of the numbers of rats in the different towns, which we do not know from the data given.

The data suggest that resistance arises first to warfarin, followed by bromadiolone. You do not need to know the reason for this, but warfarin was a 'first generation' poison, while bromadiolone was developed because rats became resistant to warfarin.

Ecology in action: Introduction

When and how to use these pages

This chapter investigates ecosystems and looks at ideas about the relationships between organisms and how they are adapted to their environment. It links to Chapter 3 lesson 3.11 Learning about plants and minerals, and lesson 3.12 Investigating how plants use minerals. There are also links to Chapter 7 Variation and evolution.

Overview of the chapter

In this chapter, students will learn about ecosystems and biodiversity. They will investigate the effects of abiotic factors on the ecosystem and predator–prey relationships. They will then find out about energy flow through trophic levels and how to construct pyramids of biomass. Students will investigate competition between organisms and how animals and plants are adapted for survival. They will have the opportunity to investigate the population of a species in a selected habitat and will then find out about cycling materials, such as water and carbon. Students will investigate how temperature affects the rate of decay and the impact humans are having on the environment. Finally, students will learn about maintaining a food supply for humans and how biotechnology can be used to produce food.

This chapter offers opportunities for the students to interpret data and look at the principles of sampling as applied to populations in habitats. They also carry out investigations that provide an opportunity to evaluate data and methods.

Obstacles to learning

The students may need extra guidance with the following terms and concepts:

- Students struggle with calculations, such as efficiency of energy transfer. Low demand students can get the numbers the wrong way round to produce answers in excess of 100%. Practice of examples should help.
- Some students will struggle with the correct terminology used in cycles. Using key cards where they have to match the term to the correct definition may help low demand students become more familiar with the terms.
- Some students at low demand will struggle with describing cyclical fluctuations in detail. If they use phrases such as 'it goes up' they need to be encouraged to replace the word 'it' with the animal in question.
- It is important that students are exposed to as many examples of adaptations as possible. They need to be able to identity adaptations of organisms they might not have heard of. Using video clips or pictures of organisms from one environment may help them realise many adaptations are common to different organisms in a habitat. Emphasis on surface area to volume ratios should be made, as this is a common theme throughout the specification.
- Students often mix up global warming and ozone depletion, even though ozone depletion is no longer in the biology specification.

Practicals in this chapter

In this chapter students will do the following practical work:

- Investigating abiotic factors
- Investigating predator–prey relationships
- Investigating why penguins huddle
- Investigating leaf adaptation to water loss
- Demonstrating the water cycle
- Demonstrating the carbon cycle
- Investigating decay
- Investigating biotic indicators
- Required practical: investigate the population size of a common species in a habitat
- Required practical: investigate the effect of a factor on the rate of decay

	Lesson title	Overarching objectives
1	Key concept: learning about ecosystems	Describe what an ecosystem is and explain the importance of high biodiversity.
2	Changing abiotic factors	Explain changes in the distribution of species in an ecosystem.
3	Investigating predator–prey relationships	Explain how changes in one population affect another.
4	Looking at trophic levels	Explain and construct pyramids of biomass.
5	Transferring biomass	Calculate the efficiency of biomass transfers and explain the impact of biomass loss on the numbers of organisms.
6	Competing for resources	Explain how and why animals compete for resources.
7	Required practical: Measure the population size of a common species in a habitat	Estimate the size of a population using sampling techniques.
8	Adapting for survival in animals	Explain how animals are adapted for survival in their habitats.
9	Adapting for survival in plants	Explain how plants are adapted for survival in their habitats.
10	Cycling materials	Explain the stages in the water and decay cycles and why it is important to recycle materials.
11	Cycling carbon	Explain how carbon is recycled.
12	Investigating decay	Describe how different factors affect decay.
13	Required practical: Investigate the effect of temperature on the rate of decay of fresh milk by measuring pH change	Analyse data on the effect of temperature on the rate of decay.
14	Changing the environment	Describe and explain the impact of environmental change.
15	Learning about land use	Evaluate the effects of humans changing the use of land.
16	Changing the landscape	Describe and evaluate the impact of peat bog destruction and deforestation.
17	Thinking about global warming	Describe the causes of global warming and explain how global warming impacts on biodiversity.
18	Looking at waste management	Describe the impact of waste on ecosystems and biodiversity.
19	Investigating pollution	Explain how indicator species measure pollution and compare different methods of measuring pollution.
20	Maintaining biodiversity	Describe the impact of breeding programmes on maintaining biodiversity and explain how habitats are regenerated.
21	Learning about food security	Describe how different factors affect food security.
22	Maintaining food security	Evaluate modern farming techniques.
23	Using biotechnology	Describe some uses of biotechnology and explain the advantages of using biotechnology.
24	Maths skills: Using graphs to show relationships	Calculate reaction or growth rates from the gradient of a graph.

Lesson 1: Key concept: Learning about ecosystems

Lesson overview

AQA Specification reference

AQA 4.7.1.1; AQA 4.7.3.1

Learning objectives

- Describe what an ecosystem is.
- Explain the importance of high biodiversity.
- Explain what is meant by a self-supporting ecosystem.

Learning outcomes

- Understand that an ecosystem is the interaction of a community of living organisms with the non-living parts of their environment. [O1]
- Understand that a high biodiversity will help stabilise an ecosystem. [O2]
- Understand that in a self-supporting ecosystem all the requirements for living organisms to grow and survive are present. [O3]

Skills development

- WS 1.2 Use a variety of models such as representational, spatial, descriptive, computational and mathematical to solve problems, make predictions and to develop scientific explanations and understanding of familiar and unfamiliar facts.
- WS 3.8 Communicating the scientific rationale for investigations, methods used, findings and reasoned conclusions through paper-based and electronic reports and presentations using verbal, diagrammatic, graphical, numerical and symbolic forms.
- WS 4.1 Use scientific vocabulary, terminology and definitions.

Resources needed Video of an ecosystem. Worksheet 8.1.1 (low demand); Worksheet 8.1.2 (standard demand); Worksheet 8.1.3 (high demand)

Digital resources

Key vocabulary biodiversity, ecosystem

Teaching and learning

Engage

- Show students a short video clip of organisms in an ecosystem they are unfamiliar with, such as the African plains. [O1]
- Put the following terms onto the board (ecosystem, habitat, community, population) and ask students to use the video clip to help them explain the difference between these terms. [O1]

Challenge and develop

- Divide the class into small groups to **research** ecosystems. If possible, each group should be given different ecosystems to research. Some suggestions are: ocean, desert, Arctic, garden pond, wheat field, rainforest. Instructions can be found on Worksheet 8.1. [O1, O2]
- Low demand students should **research** just one ecosystem and use the help sheet (Worksheet 8.1.1). [O1, O2]
- Standard demand students should **research** just one ecosystem without the use of the help sheet (Worksheet 8.1.2). [O1, O2]
- High demand students should **research and compare** two different ecosystems (Worksheet 8.1.3). [O1, O2]

Explain

- Low demand students should **present** their research by completing the help sheet. [O1, O2]
- Standard demand students should **present** their research as a poster. [O1, O2]
- High demand students should present their **research** as a written report. [O1, O2]

Consolidate and apply

- Show the students a self-contained artificial biosphere, this could be an image or one that has been made by the school (e.g. http://thescienceclassroom.org – search 'how-to-make-your-own-self-contained-ecosystem-biosphere'). Tell the students that an artificial biosphere is a self-supporting ecosystem within a manufactured container. **Explain** that the plants growing in the biosphere do not require any outside influences apart from light. Ask the students to **suggest** how the plant survives. [O3]
- Use the self-contained biosphere to **explain** the idea of self-supporting ecosystems. [O3]
- Ask students to **suggest** why the self-contained biosphere can exist with such a low biodiversity, yet in the natural ecosystems this would be unusual. They made need help with this so tell them that the biosphere is isolated and left completely untouched – would this happen in a natural ecosystem? [O3]

Extend

Ask students able to progress further to:

- **compare** the artificial biosphere with the natural biosphere of the earth. They should explain why the balance of the materials in the natural biosphere is less likely to be maintained than in the artificial biosphere. [O3]

Plenary suggestions

Name or describe some habitats (desert, rainforest, oak woodland, pine forest, ocean) and ask the students to **list** them in order from relatively high to low biodiversity (rainforest, ocean, oak woodland, pine forest, desert).

Lesson 2: Changing abiotic factors

Lesson overview

AQA Specification reference

AQA 4.7.1.2; AQA 4.7.2.4

Learning objectives

- Identify abiotic factors that affect ecosystems.
- Explain changes in the distribution of species in an ecosystem.
- Describe stable and unstable populations.

Learning outcomes

- List the different abiotic factors that affect an ecosystem. [O1]
- Describe zonation as a gradual change in the distribution of species across a habitat. [O2]
- Understand the difference between stable and unstable populations. [O3]

Skills development

- WS 2.4 Carry out experiments appropriately having due regard for the correct manipulation of apparatus, the accuracy of measurements, and health and safety considerations.
- WS 2.5 Recognise when to apply a knowledge of sampling techniques to ensure any samples collected are representative.
- WS 2.6 Make and record observations and measurements using a range of apparatus and methods.
- WS 3.3 Carrying out and represent mathematical and statistical analysis.
- WS 3.7 Being objective, evaluating data in terms of accuracy, precision, repeatability and reproducibility and identifying potential sources of random and systematic error.

Maths focus

- 1d Make estimates of the results of simple calculations
- 2b Construct and interpret frequency tables and diagrams, bar charts and histograms

Resources needed Video of a rocky shore (e.g. www.arkive.org – search 'habitats rocky shores'); image of a kite diagram from a rocky shore. Worksheet 8.2.1 (low demand); Worksheet 8.2.2 (standard demand); Worksheet 8.2.3 (high demand); Practical sheet 8.2.1 (low and standard demand); Practical sheet 8.2.2 (high demand); Technician's notes 8.2

Digital resources

Key vocabulary abiotic factor, biotic factor, distribution, competition, zonation

Teaching and learning

Engage

- Show students a short video clip of a rocky shore. [O1]
- Ask students to **suggest** how the habitat would change during the day. [O1]

Challenge and develop

- Introduce the terms 'abiotic factor' and 'biotic factor'. Explain that abiotic factors are non-living things that affect a community. Ask the students to discuss the rocky shore in small groups and **list** the abiotic factors and biotic factors affecting it. [O1]
- Groups should then report back to the class to produce a complete list of **abiotic** factors. [O1]
- Students then **investigate** an abiotic factor that affects the growth of plants using Practical sheet 8.2.1 (low and standard demand) or Practical sheet 8.2.2 (high demand). [O1]

Explain

- Show the students an image of a kite diagram that shows the distribution of species on a rocky shore (search Internet for 'kite diagram rocky shore'). Explain that it shows the number of different species in areas (zones) on a rocky shore. [O2]
- Ask low demand students to **describe** patterns in the diagram. [O1, O2]
- Ask standard and high demand students to **explain** patterns in the diagram. [O1, O2]
- Ask students to **imagine** a population of rabbits living on an island where there are no predators. Ask the students to **explain** why the population will reach a maximum number. Explain that this is a stable population as the number of births will be similar to the number of deaths. Ask students to **suggest** how this population may become unstable (e.g. disease, introduction of a predator or increased competition for food). [O3]

Consolidate and apply

- Low demand students **complete** the questions about abiotic factors on Worksheet 8.3.1. [O1, O2]
- Standard demand students **complete** the questions on Worksheet 8.3.2. [O1, O2]
- High demand students **complete** the questions on Worksheet 8.3.3. [O1, O2, O3]

Extend

Ask students able to progress further to:

- **write a summary** that **explains** why the tide plays an important part in zonation on a rocky shore. [O3]

Plenary suggestions

Ask students to compare the abiotic factors affecting a population of penguins in the Arctic with a population of elephants in Africa.

Answers to Worksheet 8.2

Abiotic factors (low and standard demand)

1. any five from – light intensity, temperature, moisture levels, soil mineral content, wind intensity and direction, carbon dioxide levels
2. light intensity – lower in plantation; temperature – will vary but likely to be lower is shade of plantation; moisture levels – lower in plantation; soil mineral content – will depend so difficult to say; wind intensity and direction – intensity likely to be lower in plantation; carbon dioxide levels will be the same

Distribution of organisms (standard and high demand)

1. kite diagram
2. zone 2 moss and grass; zone 3 ferns
3. close to the path it is trampled, away from the path it is outcompeted by ferns
4. eat grasses or mosses, eaten by spiders (in zone 3)

Zonation (high demand)

1. Different plants live at different distances from the pond.
2. The soil is too wet.
3. obtaining enough sunlight and carbon dioxide
4. As they are near the edge they may be covered if levels rise too much or dry out if levels fall too much.

Answers to Practical sheet 8.2

Analysis

3. Leaves growing in the shade have a larger surface area to absorb more light for photosynthesis.

Evaluation

2. They should only be ignored if they were the results of errors in taking measurements, if that is the actual surface area of the leaf then they should be counted – see WS 3.7.

Lesson 3: Investigating predator-prey relationships

Lesson overview

AQA Specification reference

AQA 4.7.2.1

Learning objectives

- Describe how changes in one population affect another.
- Explain interdependent relationships.
- Explain how predator–prey population cycles have cyclical changes.

Learning outcomes

- Understand that a change in a predator population may result in a change in the number of prey and vice versa. [O1]
- Understand that organisms in a community are interdependent. [O2]
- Understand that the populations of some predators and their prey show cyclical fluctuations. [O3]

Skills development

- WS 1.2 Use a variety of models such as representational, spatial, descriptive, computational and mathematical to solve problems, make predictions and to develop scientific explanations and understanding of familiar and unfamiliar facts.
- WS 3.1 Presenting observations and other data using appropriate methods.
- WS 3.2 Translating data from one form to another.
- WS 3.3 Carrying out and represent mathematical and statistical analysis.

Maths focus

- 4c Plot two variables from experimental or other data

Resources needed Image of a food web. Worksheet 8.3.1 (low demand); Worksheet 8.3.2 (standard and high demand); Practical sheet 8.3; Technician's notes 8.3

Digital resources

Key vocabulary cycle, interdependence, out of phase, parasitism

Teaching and learning

Engage

- **Show** students an image of a food web for an ecosystem. [O1]
- Ask the students to **name** the producer in the food web and two consumers. [O1]
- Choose one of the animals in the food web and ask students to **predict** what would happen to the other organisms if that animal was lost from the ecosystem. [O1]

Challenge and develop

- Ask the students to **identify** one predator in the food web and to list its prey. [O1]
- In groups the students should then **discuss** the following ideas: reasons why the prey population would increase or decrease, reasons why the predator population would increase or decrease, how changes in one population would affect the other. [O1]
- Groups should then **report back** to the class to **assess** how much the class already knows about predator–prey relationships. The high demand students may pick up on the fact that if a predator eats more than one type of prey their numbers may not be affected that much if just one source of prey is reduced. [O1]

Explain

- Develop the discussion to introduce the idea that organisms in a community have interdependent relationships. Predator–prey is just one example; see if they can they name any others. [O2]
- Discuss the idea of parasites and cleaner species as other examples of interdependent relationships. Video clips showing cleaner species in action can be found on the Internet. [O2]
- Students then **investigate** predator–prey relationships by following the instructions on Practical sheet 8.3. They **record** their results on Worksheet 8.3. [O1, O2]
- Standard and high demand students should **plot a graph** of their results and then complete the questions (Worksheet 8.3.2). [O1, O2]
- Low demand students should just **complete** the questions (Worksheet 8.3.1). [O1, O2]

Consolidate and apply

- **Discuss** with the class their results. Link the results to cyclic fluctuation of predator–prey. They may not have enough results to show complete cyclic fluctuation, so use Figure 8.8 in the Student Book to help illustrate the concept. [O1, O2]
- Low and standard demand students **complete** Questions 3 and 4 on page 329 of the Student Book. [O2]
- **Explain** to high demand students why the cycles are out of phase. [O3]
- High demand students also **complete** Questions 5 and 6 on page 329 of the Student Book. [O3]

Extend

Ask students able to progress further to:

- **use** ideas about food webs to **explain** why most predator–prey relationships do not follow the perfect patterns seen in the lynx–snowshoe hares graph in the Student Book (Figure 8.8). [O2, O3]

Plenary suggestions

Students often struggle with the difference between 'explain' and 'describe' when answering an exam question.

On the Technician's notes there are statements about a predator–prey graph. Read out or project the statements onto the board. Ask students to identify which are descriptions and which are explanations.

Answers to Worksheet 8.3

Predator–prey questions

Answers will depend on results but the lemming numbers should go down while the owl numbers go up or the owl numbers will go down so the lemming numbers go up. They are unlikely to see cyclic fluctuation.

There should be a drop in both populations after round 8 due to half the lemming population starving.

Lesson 4: Looking at trophic levels

Lesson overview

AQA Specification reference

AQA 4.7.2.1; AQA 4.7.4.1; AQA 4.7.4.2

Learning objectives

- Explain trophic levels.
- Explain and construct pyramids of biomass.
- Explain the difficulties in constructing pyramids.

Learning outcomes

- Identify the different trophic levels in a food chain. [O1]
- Construct pyramids of biomass from given information. [O2]
- Explain the difficulties in constructing pyramids of biomass in terms of measuring the biomass of living organisms. [O3]

Skills development

- WS 1.2 Use a variety of models such as representational, spatial, descriptive, computational and mathematical to solve problems, make predictions and to develop scientific explanations and understanding of familiar and unfamiliar facts.
- WS 3.3 Carrying out and represent mathematical and statistical analysis.
- WS 3.8 Communicating the scientific rationale for investigations, methods used, findings and reasoned conclusions through paper-based and electronic reports and presentations using verbal, diagrammatic, graphical, numerical and symbolic forms.
- WS 4.1 Use scientific vocabulary, terminology and definitions.

Maths focus

- 1c Use ratios, fractions and percentages
- 2c Construct and interpret frequency tables and diagrams, bar charts and histograms

Resources needed Video clip showing food chain in action. Worksheet 8.4.1 (low demand); Worksheet 8.4.2 (standard demand); Worksheet 8.4.3 (high demand)

Digital resources

Key vocabulary apex predator, biomass, decomposer, secrete, trophic level

Teaching and learning

Engage

- **Show** the students a wildlife clip (e.g. the first few minutes of *Blue Planet* episode 1, where the fish are being eaten – www.youtube.com). [O1]
- Then as a class **construct** a food web seen in the clip. [O1]

Challenge and develop

- Choose a food chain from the constructed food web. Introduce the term 'trophic levels' and ask the students to **identify** the different trophic levels (numbers only) seen in the food chain. [O1]
- Introduce the terms 'producer', 'herbivores', 'primary consumers', 'carnivores', 'secondary consumers', 'tertiary consumers', 'apex predators'. [O1]
- Low demand students should **write** the food chain down and **write each term** below the correct organisms in the food chain. [O1]

- Standard and high demand students should **write** the food chain down, **write each term** below the correct organisms in the food chain and **explain** their choices. [O1]
- Ask students to **discuss** what would happen to the apex predators when they die and where the energy in their bodies goes. Use the discussion to **explain** the role of decomposers in food chains. [O1]

Explain

- Introduce the idea of pyramids of biomass and numbers. **Explain** what each shows and why they are different. Use the Student Book to illustrate the concepts. [O2]
- Show students how to construct a pyramid of numbers from data. Students then construct a pyramid of biomass using the data on Worksheet 8.4 (section 1). [O1, O2]

Consolidate and apply

- Students **produce** a poster about feeding relationships. Worksheet 8.4 is used as a guide. To save time on research the students can use the Student Book and the food web constructed at the start of the lesson. [O1, O2, O3]
- Low demand students **research** one habitat to **produce** a poster using Worksheet 8.4.1. [O1, O2]
- Standard demand students **research** one habitat to **produce** a poster using Worksheet 8.4.2. [O1, O2]
- High demand students **research** one habitat to **produce** a poster using Worksheet 8.4.3. [O1, O2, O3]

Extend

Ask students able to progress further to:

- **think** back to the food web they constructed at the start of the lesson and to **suggest** how humans may influence the food web. [O1]

Plenary suggestions

Show the image of a pyramid of numbers containing fleas and ask students to describe and explain the shape of the pyramid of biomass.

Answers to Worksheet 8.4

Pyramid of biomass

Biomass of grass = 40 000 kg 40 000/400 = 200 small squares
Biomass of zebra = 2400 kg 2400/400 = 6 small squares
Biomass of hyena = 1200 kg 1200/400 = 3 small squares
Biomass of lion = 200 kg 200/400 = 1/2 small square (drawn as a single line)

Lesson 5: Transferring biomass

Lesson overview

AQA Specification reference

AQA 4.7.4.3

Learning objectives

- Identify how biomass is lost.
- Calculate the efficiency of biomass transfers.
- Explain the impact of biomass loss on the numbers of organisms.

Learning outcomes

- List ways biomass is lost between trophic levels. [O1]
- Calculate the efficiency of biomass transfer between trophic levels. [O2]
- Understand that biomass loss will limit the number of trophic levels in a food chain. [O3]

Skills development

- WS 1.2 Use a variety of models such as representational, spatial, descriptive, computational and mathematical to solve problems, make predictions and to develop scientific explanations and understanding of familiar and unfamiliar facts.
- WS 3.3 Carrying out and represent mathematical and statistical analysis.
- WS 4.1 Use scientific vocabulary, terminology and definitions.

Maths focus

- 1c Use ratios, fractions and percentages
- 3d Solve simple algebraic equations

Resources needed Video of an ecosystem. Worksheet 8.5.1 (low demand); Worksheet 8.5.2 (standard demand); Worksheet 8.5.3 (high demand)

Digital resources

Key vocabulary biomass, efficiency, trophic level

Teaching and learning

Engage

- Show students a short video clip of a lion hunting its prey. [O1]
- Ask the students to **list** the parts of the prey the lion is unlikely to eat and to **suggest** what happens to these parts. [O1]

Challenge and develop

- Show students an image of an owl pellet (search owl pellets). **Explain** that owls eat their whole prey but only digest part of it; the rest is regurgitated (coughed up) as an owl pellet. [O1]
- **Explain** that these parts of the prey are lost to the food chain; the owl cannot use the energy in the biomass of the food pellet to increase its own biomass. Ask students to **suggest** other reasons why energy in the biomass of an organism is not used to form biomass in the animal that eats it. As a hint ask what happens to the food they eat – do they use it all to put on biomass (or, using everyday language, 'weight')? [O1]

Explain

- Ask students to **draw** a food chain in the centre of a piece of A4 plain paper. They can use Figure 8.6 on page 328 of the Student Book or make up their own. [O1]

- Students should then **annotate** the diagram to show how biomass is lost and why only a small amount of the energy taken in is turned into biomass. They can use Figure 8.13 (page 333) to help them.
- Low demand students can use Worksheet 8.5.1 to help them. [O1]

Consolidate and apply

- Introduce the idea of **calculating** the efficiency of biomass transfer between trophic levels. Use the example in the book to show students how to complete the calculations. They may get confused between energy and biomass. Explain that the organism takes in food, the mass of that food is the biomass. The energy in the food is then used. Only a small percentage is used to increase the biomass of the consumer. This percentage is the efficiency. [O1, O2]
- Students then **complete** the questions about calculating the efficiency of biomass transfer on Worksheet 8.5.1 (low demand), Worksheet 8.5.2 (standard demand) or Worksheet 8.5.3 (high demand). [O1, O2]
- High demand students should **complete** the extension questions on Worksheet 8.5.3. [O1, O2, O3]

Extend

Ask students able to progress further to:

- **use** ideas about efficiency of biomass transfer to **explain** why people should eat more vegetable protein sources than animal protein sources. [O3]

Plenary suggestions

Show students an image of a pyramid of biomass and ask them to explain, using ideas from the lesson, why it is pyramid-shaped.

Answers to Worksheet 8.5

Calculating efficiency of biomass transfer

1. 125
2. efficiency = 125/3056 × 100 = 4%
3. There is less heat loss and less energy used for movement.

Calculating efficiency of biomass transfer extension

1. 8 kJ
2. efficiency = 12/350 × 100 = 3.4%
3. As so much energy is lost, and only about 4% passed on, then there is not enough to support more trophic levels; also, the higher up the trophic levels, the more vulnerable an animal is as it depends on earlier food links not to be interrupted.

Lesson 6: Competing for resources

Lesson overview

AQA Specification reference

AQA 4.7.1.1

Learning objectives

- Describe how competition impacts on populations.
- Explain why animals in the same habitat are in competition.
- Explain interspecific and intraspecific competition.

Learning outcomes

- Describe how competition for food, water, shelter, light and minerals may influence the distribution and population size of animals or plants. [O1]
- Understand that organisms with similar ecological niches are in greater competition. [O2]
- Understand that interspecific competition is between species and intraspecific is within a species. [O3]

Skills development

- WS 1.2 Use a variety of models such as representational, spatial, descriptive, computational and mathematical to solve problems, make predictions and to develop scientific explanations and understanding of familiar and unfamiliar facts.
- WS 3.5 Interpreting observations and other data (presented in verbal, diagrammatic, graphical, symbolic or numerical form), including identifying patterns and trends, making inferences and drawing conclusions.
- WS 3.8 Communicating the scientific rationale for investigations, methods used, findings and reasoned conclusions through paper-based and electronic reports and presentations using verbal, diagrammatic, graphical, numerical and symbolic forms.
- WS 4.1 Use scientific vocabulary, terminology and definitions.

Maths focus

- 4c Plot two variables from experimental or other data

Resources needed Animals competing for a mate video. Worksheet 8.6.1 (low demand); Worksheet 8.6.2 (standard demand); Worksheet 8.6.3 (high demand)

Digital resources

Key vocabulary interspecific competition, intraspecific competition, niche

Teaching and learning

Engage

- Show a video clip of animals competing for a mate (there are many examples on YouTube).
- Ask the students to **explain** why males fight for a mate.
- Ask the students to **list** other things animals and plants might compete for. [O1]

Challenge and develop

- Show the class a beaker full of duckweed and a beaker with just a little duckweed in. **Discuss** as a class ways in which the population of each beaker will change over the next few days. **Discuss** reasons for the changes. [O1, O2]
- Students then complete the duckweed population questions on Worksheet 8.6 (split). If the students require practice at graph plotting they can **plot the points** before answering the questions. [O1, O2]
- Show students videos clips that show examples of animals competing. This could be courtship videos in birds, animals of the same species fighting over food or animals, or time-lapse photography of a plant growing in the rainforest competing for light. In each case ask the students to **state** if the competition is between different species or between organisms of the same species. [O3]

Explain

- **Explain** the term 'niche' as the role that a living thing occupies in a community. Two animals living in the same habitat eating the same food have a similar niche. Two animals living in the same habitat but eating different food, i.e. a carnivore and herbivore, are less likely to be in competition for food. [O2]
- **Explain** the terms 'interspecific competition' and 'intraspecific competition' using examples seen in the video clips. [O3]

Consolidate and apply

- Students research two animals, for example lions and hyenas that have a similar ecological niche; they then present their findings in some way. They should use Worksheet 8.6 to help them. [O1, O2, O3]
- Low demand students **research** and **produce** a poster using Worksheet 8.6.1. [O1, O2, O3]
- Standard demand students **research** and **produce** a poster using Worksheet 86.6.2. [O1, O2, O3]
- High demand students **research** and **produce** a poster using Worksheet 8.6.3. [O1, O2, O3]

Extend

Ask students able to progress further to:

- **think** back to the red and grey squirrels and write a paragraph about competition of these two species using all the following terms: 'niche', 'intraspecific competition', 'interspecific competition'. [O2, O3]

Plenary suggestions

Ask a student to volunteer to present their research to the rest of the class.

Answers to Worksheet 8.6

Duckweed populations

1. The duckweed population increases to a maximum and then starts to decrease slightly.
2. no light for photosynthesis
3. any two from: light, water, minerals, space
4. not enough space, water or minerals for any more plants
5. Population will start to decrease.

Lesson 7: Required practical: Measure the population size of a common species in a habitat

Lesson overview

AQA Specification reference

AQA 4.7.2.1

Learning objectives

- Describe a suitable method to investigate a population.
- Estimate the size of a population.
- Explain the effect of sample size.

Learning outcomes

- Plan an investigation to look at the effect of an abiotic factor on population size. [O1]
- Use results of an investigation to estimate the population. [O2]
- Evaluate data collected in terms of sample size. [O3]

Skills development

- WS 2.1 Use scientific theories and explanations to develop hypotheses.
- WS 2.2 Plan experiments or devise procedures to make observations, produce or characterise a substance, test hypotheses, check data or explore phenomena.
- WS 2.3 Apply a knowledge of a range of techniques, instruments, apparatus, and materials to select those appropriate to the experiment.
- WS 2.4 Carry out experiments appropriately having due regard for the correct manipulation of apparatus, the accuracy of measurements, and health and safety considerations.
- WS 2.5 Recognise when to apply a knowledge of sampling techniques to ensure any samples collected are representative.
- WS 2.6 Make and record observations and measurements using a range of apparatus and methods.
- WS 2.7 Evaluate methods and suggest possible improvements and further investigations.
- WS 3.5 Interpreting observations and other data (presented in verbal, diagrammatic, graphical, symbolic or numerical form), including identifying patterns and trends, making inferences and drawing conclusions.
- WS 3.7 Being objective, evaluating data in terms of accuracy, precision, repeatability and reproducibility and identifying potential sources of random and systematic error.
- WS 4.3 Use SI units (eg kg, g, mg; km, m, mm; kJ, J) and IUPAC chemical nomenclature unless inappropriate.

Maths focus

- 1d Make estimates of the results of simple calculations
- 2b Find arithmetic means

Resources needed Worksheet 8.7 (standard and high demand); Practical sheet 8.7 (low demand); Technician's notes 8.7

Digital resources

Key vocabulary correlation, mean, median, mode, quadrat, transect

Teaching and learning

Engage

- Show students two images, one showing plants growing in a flower meadow and the other showing plants growing under the shade of a tree. [O1]
- Ask students to **list** the abiotic factors that might affect the growth of the plants in the two areas. [O1]

- Ask them to suggest how they could **compare** the number of plants growing in the shade with those growing in full sunlight. [O1]

Challenge and develop

- **Introduce** the investigation by telling students they are going to plan an investigation into the effect of one abiotic factor on the growth of plants. Show them a quadrat and explain how it can be used estimate the population of a species. If students have not done any field work before, they will need support in planning their investigation. In some cases, this could be done as a class or they can be given Practical sheet 8.7. [O1]
- Low demand students should be told that they are investigating the effect of light intensity by looking at the population of one species growing in full sun or in shade. [O1]
- Other students can choose, but make it clear that they have to measure the abiotic factor so their choice may depend on available equipment, such as pH probes and soil moisture meters. [O1]
- In groups ask standard and high demand students to **plan** the investigation. The students use Worksheet 8.7 to help them plan. The investigation should involve sampling two different areas, one in shade and another in full sun. The students mark out the area and randomly place quadrats down. They should look for just one species, such as daises or dandelions. The object is to estimate the population in each area to see if the abiotic factor is having an effect. [O1]
- Low demand students can follow the instructions on Practical sheet 8.7; they will need to **identify** variables and **complete** a risk assessment, but the method is provided. [O1]

Explain

- Students **complete** their investigations in the school field. [O1]
- Low demand students **analyse** their results by completing the questions on Practical sheet 8.7. [O2]
- High and standard demand students **analyse** their results by completing Worksheet 8.7. [O2]

Consolidate and apply

- Low demand students **evaluate** their investigation by completing the questions on Practical sheet 8.7. [O3]
- High and standard demand students **evaluate** their investigation (Questions 1–3 on Worksheet 8.7). [O3]

Extend

Ask students able to progress further to:

- **complete** Question 4 in the evaluation section on Worksheet 8.7. [O3]

Plenary suggestions

Ask each group of students that worked together to **identify** one problem they had with their investigation and one improvement they could make.

Answers to Practical sheet 8.7

Investigation plan

1. light intensity
2. dandelion
3. light intensity (sunny or shady)
4. population or number of dandelions
5. spaced out along a transect line
6. few hazards; if grass is wet they might slip so wearing sensible clothing
7. light meter, tape or string, quadrat

Analysis / Evaluation

Depends on results – they may find that light intenisty has no effect as they have not taken into account biotic factors, such as length of surrounding grass and competition.

Lesson 8: Adapting for survival in animals

Lesson overview

AQA Specification reference

AQA 4.7.1.4

Learning objectives

- Recall why animals have adaptations.
- Explain some adaptations.
- Use surface area to volume ratios to explain some adaptations.

Learning outcomes

- State that organisms that are adapted to their environment are better able to compete and survive. [O1]
- Explain how different adaptations help animals survive in their environments. [O2]
- Explain the importance of surface area to volume ratios to heat loss in coping with environmental stress. [O3]

Skills development

- WS 1.2 Use a variety of models such as representational, spatial, descriptive, computational and mathematical to solve problems, make predictions and to develop scientific explanations and understanding of familiar and unfamiliar facts.
- WS 2.4 Carry out experiments appropriately having due regard for the correct manipulation of apparatus, the accuracy of measurements, and health and safety considerations.
- WS 2.6 Make and record observations and measurements using a range of apparatus and methods.
- WS 3.1 Presenting observations and other data using appropriate methods.
- WS 3.3 Carrying out and represent mathematical and statistical analysis.
- WS 3.5 Interpreting observations and other data (presented in verbal, diagrammatic, graphical, symbolic or numerical form), including identifying patterns and trends, making inferences and drawing conclusions.
- WS 3.6 Presenting reasoned explanations including relating data to hypotheses.

Maths focus

- 1c Use ratios, fractions and percentages
- 4a Translate information between graphical and numeric form
- 4c Plot two variables from experimental or other data

Resources needed A video clip of huddling penguins. Worksheet 8.8.1 (low demand); Worksheet 8.8.2 (standard demand); Worksheet 8.8.3 (high demand); Practical sheet 8.8; Technician's notes 6.8

Digital resources

Key vocabulary behaviour, camouflage, functional, structural

Teaching and learning

Engage

- Show students an image of a camel and **explain** that it lives in a hot, dry ecosystem. [O1]
- Ask students to work in small groups (could use mini-whiteboards) to **list** the adaptations of a camel. [O1]
- Groups report back and as a class **discuss** what they have written to find out what they know about adaptations. [O1, O2]

Challenge and develop

- **Explain** to students that adaptations are features that organisms have that help them survive and be successful in their habitat. **Explain** that these features can either be structural features, such as small ears to reduce surface area for heat loss, functional features, such as camels not sweating, or behavioural features, such as hibernation. [O1, O2, O3]

- Students **complete** the adaptation questions on Worksheet 8.8. [O1, O2, O3]

Explain

- Show students a video clip of huddling penguins and ask them to **explain** the behaviour. [O1, O2]
- Students then **investigate** this example of behavioural adaptation in animals (huddling in penguins) by following the instructions on Practical sheet 8.8. [O1, O2]

Consolidate and apply

- Low demand students should **complete** the card sort on Worksheet 8.8.1. [O1, O2, O3]
- Standard demand students should **complete** questions on investigating heat loss and surface area to volume ratios on Worksheet 8.8.2. [O3]
- High demand students should **complete** all the questions in section 2 of Worksheet 8.8.3. [O3]

Extend

Ask students able to progress further to:

- **write a summary** that **explains** the difference between behavioural, functional and structural adaptations using examples. [O1, O2, O3]

Plenary suggestions

Students should **draw** a diagram of an animal to match a written description of an imaginary environment.

For example: this animal lives in a forest that is very dark with tall trees. The animal is a carnivore, it hunts for food in the forest. The forest floor is very muddy so animals with small paws easily sink into the mud. The temperature is very low and it often freezes at night.

Their animal should have dark thick fur to stay camouflaged and keep warm. It may have small ears to reduce heat loss and large paws to stop it sinking into the ground. It will have eyes at the front of the head and sharp claws and teeth.

Answers to Worksheet 8.8

Adaptations

1. sharp teeth and claws for hunting, white for camouflage
2. so they have an energy store for the winter when there is less food
3. lack of food
4. vitamin D
5. temperature change, escape predators, escape flies
6. Isolated calves have more chance of being taken, in a herd they are protected by the large numbers of adults.
7. behavioural: hibernation, sun basking, migration, synchronised birth; functional: hedgehog's metabolism drops; structural: thick fur, blubber, small ears

Adaption card sort (low demand)

shark: sea – has gills and is streamlined – so it can swim fast and breathe under water; frog: pond – powerful back legs – so it can hop in and out of the water; penguin: Antarctic – wings that can be used as flippers – so it can swim instead of fly; blackbird: garden – sharp beak – so it can pull worms out of the soil

Surface area to volume ratio

1. 2, 2, 4, 4, 6
2. As surface area to volume ratio increases the change in temperature increases.
3. Large ears increase the SA:V ratio so the elephant can lose more heat.
4. (high demand): (a) Values for 4:1 and 5:1 are the same; (b) 4.4, 6.8, 8.7, 11.4,13.6; (c) Anomaly occurred because temperature at the start was much lower than in the other beakers.

Answers to Practical sheet 8.8

3. The students should find that the single test tube cools the quickest and the middle one the slowest. Penguins on the outside of the huddle cool faster, so they rotate so that no one penguin is left on the outside.

Lesson 9: Adapting for survival in plants

Lesson overview

AQA Specification reference

AQA 4.7.1.4

Learning objectives

- Identify some adaptations of plants and bacteria.
- Explain the importance of plant adaptations.
- Explain a range of plant adaptations.

Learning outcomes

- Identify different ways in which plants are adapted to survive. [O1]
- Understand that plant adaptations allow them to survive and reproduce in different environmental conditions. [O2]
- Explain the importance of surface area to volume ratios to water loss in plants. [O3]

Skills development

- WS 1.2 Use a variety of models such as representational, spatial, descriptive, computational and mathematical to solve problems, make predictions and to develop scientific explanations and understanding of familiar and unfamiliar facts.
- WS 2.4 Carry out experiments appropriately having due regard for the correct manipulation of apparatus, the accuracy of measurements, and health and safety considerations.
- WS 2.6 Make and record observations and measurements using a range of apparatus and methods.
- WS 3.3 Carrying out and represent mathematical and statistical analysis.
- WS 3.5 Interpreting observations and other data (presented in verbal, diagrammatic, graphical, symbolic or numerical form), including identifying patterns and trends, making inferences and drawing conclusions.

Maths focus

- 2b Find arithmetic means

Resources needed Image of a cactus; image of a cross section of a marram grass leaf; video clip of bacteria living in deep sea vents; image of a pine tree in the snow. Worksheet 8.9.1 (low demand); Worksheet 8.9.2 (standard demand); Worksheet 8.9.3 (high demand); Practical sheet 8.9; Technician's notes 8.9

Digital resources

Key vocabulary epiphyte, extremophile

Teaching and learning

Engage

- Show students an image of a cactus and **explain** that it lives in a hot, dry ecosystem. [O1]
- Ask students to work in small groups (could use mini-whiteboards) to **list** the adaptations of a cactus. [O1]
- Groups report back and as a class **discuss** what they have written to find out what they know about plant adaptations. [O1]

Challenge and develop

- **Explain** to students that plants are not just adapted to conserve water but have many different adaptations to help them survive and reproduce. [O1, O2]
- **Explain** to students the difference between organisms that are specialists and those that are generalists, using examples of cactus and grass. Ask the students to suggest an animal that is a specialist and another that is a generalist. [O1, O2]

- Ask students to **read** the Student Book pages 338–339 and use the information to **complete** the table on plant adaptation on Worksheet 8.9.1 (low demand: answers are given in the wrong order), Worksheet 8.9.2 (standard demand) or Worksheet 8.9.3 (high demand: students should also answer Question 2). [O1, O2]

Explain

- Students **investigate** how leaves are adapted to reduce water loss by following the instructions on Practical sheet 6.9. [O1, O2, O3]
- Use the results of the investigation to **discuss** why surface area to volume ratio in leaves is so important for both water loss and photosynthesis and how these two processes are in conflict. Leaves need a large surface area to absorb light but not too large so that they lose too much water. [O3]
- Ask students to **suggest** how plants get around the problem of needing a large surface area but also needing to reduce water loss. (They should think back to the lesson on leaf structure and the idea of waxy cuticles and reducing number of stomata.) [O2, O3]

Consolidate and apply

- Introduce the ideas of extremophiles as organisms that live in extreme environments. One example is bacteria living in deep sea vents where there is high pressure and no light or oxygen. [O3]
- **Show** students a video clip of bacteria living in deep sea vents. [O3]
- **Explain** that the bacteria are the producers in the food chain that survive around the vents. [O3]
- Ask students to **explain** how the bacteria have adapted to produce food without photosynthesis. [O3]
- Low demand students should **complete** questions 1–4 in the Student Book pages 338–339. [O1, O2, O3]
- Standard and high demand students should **complete** the questions on SA:V ratio on their worksheet. [O3]

Extend

Ask students able to progress further to:

- **look at** an image that shows the cross section of a leaf from marram grass. The image should show the stomata are sunken into pits. Ask the students to **explain** how the adaptation reduces water loss. They should use the words, diffusion, transpiration and humidity. [O1, O2]

Plenary suggestions

Show students an image of a pine tree growing in a snowy habitat. Ask them to **suggest** why the pine tree has needles rather than leaves, linked to the habitat they are found in. Are there any other adaptations they can think of that the pine tree has in order to survive?

Answers to Worksheet 8.9

1 Adaptations of plants

1. Flowers open petals every day in daylight but close them at night: flowers need to be open for pollination. Closing them protects the flowers at night. Deciduous trees lose leaves in winter: this protects the leaves from frost. A cactus plant has very long roots that spread just under the surface of the ground: to collect as much water as possible when it rains. Some plants have a thick waterproof cuticle on their leaves: reduces water loss by evaporation. Pine trees have needle-shaped leaves: reduces surface area for water loss.
2. (High demand) **specialists** are suited to one type of habitat. For example, cacti cannot survive in a wet climate.

   Generalists can survive in different conditions, e.g. grasses.

2 Surface area to volume ratio

1. B = 1.19; C = 33.3; D = 5.00
2. C, because it has the largest surface area to volume ratio and because the surface area is large so it may have more stomata.
3. D, because it has the smallest surface area and the smallest volume; pine needles are very small and thin.
4. Has a large volume, which suggests it stores water in its leaves.
5. fewer stomata, thick waxy cuticle

Lesson 10: Cycling materials

Lesson overview

AQA Specification reference

AQA 4.7.2.2; AQA 4.7.4.1

Learning objectives

- Recall that many materials are recycled in nature.
- Explain the stages in the water and decay cycles.
- Explain the importance of recycling materials.

Learning outcomes

- Describe the stages of the water cycle. [O1]
- Understand that materials are recycled when organisms decay. [O2]
- Understand that materials must be recycled in order to maintain the ecosystem. [O3]

Skills development

- WS 2.6 Make and record observations and measurements using a range of apparatus and methods.
- WS 3.8 Communicating the scientific rationale for investigations, methods used, findings and reasoned conclusions through paper-based and electronic reports and presentations using verbal, diagrammatic, graphical, numerical and symbolic forms.
- WS 4.1 Use scientific vocabulary, terminology and definitions.

Resources needed Time-lapse photography of a decaying animal. Worksheet 8.10.1 (low demand); Worksheet 8.10.2 (standard and high demand); Practical sheet 8.10 (demonstration); Technician's notes 8.10

Digital resources

Key vocabulary decomposers

Teaching and learning

Engage

- **Demonstrate** the water cycle to students using a bowl of warm water. See Practical sheet 8.10. [O1]
- Ask students to **describe** what is happening in the bowl. [O1]

Challenge and develop

- **Explain** that the demonstration is showing the water cycle. [O1]
- Ask the students to use the terms 'transpiration', 'precipitation', 'condensation' and 'evaporation' to **describe** the water cycle in nature. [O1]
- Students should then **draw** their own version of a water cycle, making sure they include the terms 'transpiration', 'precipitation', 'condensation' and 'evaporation'. [O1]
- Students should **add a summary** to their diagram to **explain** why the water cycle is so important. They can use the Student Book to help them. [O1]

Explain

- **Explain** that water is only one material that is recycled in nature. Ask the students if they can **name** any more. [O1, O2]
- **Show** the students time-lapse photography of a decaying animal. [O1, O2]
- Ask the students to **suggest** what causes the decay and why decay is so important to other organism in the ecosystem – link back to lesson on looking at energy loss where they were introduced to decomposers. [O1, O2, O3]

Consolidate and apply

- **Introduce** the role of detritivores in breaking down dead matter and increasing surface area for the action of decomposers. [O2, O3]
- Students **research** the role of earthworms in the soil by **completing** Worksheet 8.10.1 (low demand) or Worksheet 8.10.2 (standard and high demand). [O2, O3]
- Standard and high demand students should **complete** Worksheet 8.10.2, which has some extension questions included (Questions 4 and 5). [O2, O3]

Extend

Ask students able to progress further to:

- **explain** what would happen to an ecosystem if all the decomposers were removed. [O1, O2, O3]

Plenary suggestions

Ask a student to volunteer to come to the board and **draw** the outline of the water cycle with no words. Other students should volunteer to then add annotations until the cycle is complete.

Answers to Worksheet 8.10

Earthworms

1. e.g. nitrates, phosphates, potassium, magnesium
2. decomposed by bacteria and fungi
3. to add air to the soil
4. Earthworms also mix up soil layers.
5. Roots need the oxygen for respiration to provide energy for uptake of minerals by active transport.

Lesson 11: Cycling carbon

Lesson overview

AQA Specification reference

AQA 4.7.2.2

Learning objectives

- Recall that plants take in carbon as carbon dioxide.
- Explain how carbon is recycled.
- Interpret a diagram of the carbon cycle.

Learning outcomes

- State that plants take in carbon as carbon dioxide. [O1]
- Explain how carbon is recycled in nature using the terms photosynthesis, respiration and combustion. [O2]
- Understand that carbon can be locked up in limestone until it is released by weathering and that the ocean acts as a 'carbon sink'. [O3]

Skills development

- WS 2.6 Make and record observations and measurements using a range of apparatus and methods.
- WS 3.8 Communicating the scientific rationale for investigations, methods used, findings and reasoned conclusions through paper-based and electronic reports and presentations using verbal, diagrammatic, graphical, numerical and symbolic forms.
- WS 4.1 Use scientific vocabulary, terminology and definitions.

Resources needed Worksheet 8.11.1 (low demand); Worksheet 8.11.2 (standard and high demand); Practical sheet 8.11; Technician's notes 8.11

Digital resources

Key vocabulary carbon sink, combustion, photosynthesis, respiration

Teaching and learning

Engage

- **Demonstrate** the burning of leaves to show that a black substance (carbon) is formed.
  See Practical sheet 8.11. [O1]

Challenge and develop

- **Explain** that the carbon that forms came originally from the atmosphere. [O1]
- Ask the students to **explain** how it ended up as part of the leaves. [O1]
- **Discuss** with the students what would happen to the carbon if the leaves were eaten by an animal. [O1, O2]
- **Demonstrate** the burning of sugar to show that the carbon in plant material can be eaten and then released back into the atmosphere during respiration. [O1, O2]
- Low and standard demand students then **complete** the questions on Practical sheet 8.11. [O1, O2]
- High demand students should **write a summary** of what they observed in each demonstration and how the demonstrations show that carbon is transferred from the leaves to the atmosphere via animals. [O1, O2]

Explain

- Place the students in small groups of about four or five. Each group should be given an A3 piece of paper and coloured pens. Place a copy of the carbon cycle on the front desk (this can be the one in the Student Book) or if there is room, a few copies on the side benches. [O1, O2]
- Ask one student from each group to go to the copy of the carbon cycle and **observe** the diagram for 30 seconds. They return to their group and have 30 seconds to **describe** what they see while another member of the group **draws** it. The process is repeated, each time members of the group change roles. [O1, O2]
- After all members of the group have spent time observing, then **compare** the diagrams to see which group has got the most accurate drawing of the carbon cycle. [O1, O2]

Consolidate and apply

- **Explain** that carbon can become locked up in organisms for millions of years. Ask students to suggest where the carbon is held and how is it released. [O3]
- **Explain** the idea of carbon sinks as something that takes in carbon faster than it releases it, so storing the carbon. The oceans are one of the largest carbon sinks. [O3]
- Low demand students then **complete** the carbon cycle diagram on Worksheet 8.11.1. [O2, O3]
- Standard and high demand students **complete** the carbon cycle diagram and questions on Worksheet 8.11.2. [O3]

Extend

Ask students able to progress further to:

- **explain** how carbon atoms in the leaves that fall to the ground in autumn could be returned to the tree in spring. [O1, O2]

Plenary suggestions

Ask a student to **explain** why bacteria and fungi play an important part in the carbon cycle.

Answers to Worksheet 8.11

Carbon cycle (Q3 and 4: standard and high demand only)

1. top row: photosynthesis, respiration, respiration; bottom row: photosynthesis, respiration, decay
2. burning or combustion of fossil fuels
3. This acid rain weathers limestone and releases carbon dioxide that was originally in the shells of marine organisms contained as carbonates.
4. Carbon dioxide is absorbed by the ocean faster than it is released.

Answers to Practical sheet 8.11

Burning leaves

1. black
2. mainly carbon
3. carbon dioxide

Burning sugar

1. goes cloudy
2. It is carbon dioxide.
3. needs oxygen / releases energy / produces carbon dioxide / produces water

Lesson 12: Investigating decay

Lesson overview

AQA Specification reference

AQA 4.7.2.2; AQA 4.7.4.1

Learning objectives

- Recall the factors needed for decay.
- Describe how different factors affect decay.
- Explain extracellular digestion.

Learning outcomes

- State the factors needed for decay. [O1]
- Understand how different factors affect decay. [O2]
- Explain that extracellular digestion involves releasing enzymes then absorbing the digested food. [O3]

Skills development

- WS 1.2 Use a variety of models such as representational, spatial, descriptive, computational and mathematical to solve problems, make predictions and to develop scientific explanations and understanding of familiar and unfamiliar facts.
- WS 2.4 Carry out experiments appropriately having due regard for the correct manipulation of apparatus, the accuracy of measurements, and health and safety considerations.
- WS 2.6 Make and record observations and measurements using a range of apparatus and methods.
- WS 3.3 Carrying out and represent mathematical and statistical analysis.
- WS 3.5 Interpreting observations and other data (presented in verbal, diagrammatic, graphical, symbolic or numerical form), including identifying patterns and trends, making inferences and drawing conclusions.
- WS 4.1 Use scientific vocabulary, terminology and definitions.

Resources needed An image of Otzi the iceman (search 'Otzi the iceman') or a frozen woolly mammoth; a video or pictures of worm compost bins. Worksheet 8.12.1 (low demand); Worksheet 8.12.2 (standard demand); Worksheet 8.12.3 (high demand); Practical sheet 8.12; Technician's notes 8.12

Digital resources

Key vocabulary biogas, compost, detritivore, extracellular digestion, generator, secrete

Teaching and learning

Engage

- **Show** students an image of Otzi the iceman or a frozen woolly mammoth. [O1]
- Ask students to **suggest** why they are so well preserved after several thousand years. [O1]

Challenge and develop

- Ask students to work in small groups to try and **list** the conditions needed for decay. [O1]
- Discuss with the students the role of microorganisms in the decay process and the different factors that can speed up or slow down decay. [O1]
- **Show** students a set of the apparatus that has been left for a few days (Practical sheet 8.12) – the limewater should be cloudy in A while clear in B. [O1, O2]
- Ask students to explain the results. They then **complete** the questions on Practical sheet 8.12. [O1, O2]

Explain

- **Explain** that farmers and gardeners use the process of decay to produce compost. [O2]

- **Explain** to the students the role of detritivore in the process of forming the compost. This could be illustrated by showing the students a video or pictures of worm compost bins. [O2]
- **Discuss** with the students that the detritivores increase the surface area for the decomposers. [O2]
- **Explain** how the decomposers digest the dead plant matter in the compost by releasing enzymes. The enzymes digest the matter so that the decomposers can absorb the soluble food molecules. [O2]

Consolidate and apply

- Low demand students then **complete** the compost heap questions on Worksheet 8.12.1. [O1, O2, O3]
- Standard demand students should **complete** the first two sections of Worksheet 8.12.2. [O1, O2, O3]
- High demand students should **complete** the first two sections of Worksheet 8.12.3. [O1, O2, O3]
- Make sure that the students **recall** the term 'biomass' and that the energy in biomass can be released in different ways – by microorganisms breaking it down, by it being eaten or by burning the biomass.
- **Introduce** the idea of producing biogas from biomass using a generator. Explain that the decomposers use anaerobic respiration to produce methane. Make sure that students understand the term 'anaerobic respiration' and that there are different types; they may already know about the types that produce alcohol and lactic acid from Chapter 1. [O1, O2]
- Students **complete** the questions about biogas on their respective worksheets. [O2]

Extend

Ask students able to progress further to:

- **explain** why drying food, for example dried fruit, will preserve it for longer. [O1, O2]

Plenary suggestions

Ask students what advice they would give somebody with a compost bin about how to make it form compost faster.

Answers to Worksheet 8.12

Compost heap

1. to provide the microorganisms with water
2. to allow oxygen into the compost heap
3. Microorganisms are respiring, which releases heat.
4. bacteria and fungi
5. to digest the leaves

Answers to decomposing leaves

1. 10 mm all the organisms / 5 mm maggots, woodlice and microorganisms / 1 mm microorganisms only
2. bag with 10 mm holes
3. The more detritivores that can enter the bag, the more decomposition. This is because they increase the surface area for microorganisms to work on.

Answers to microorganisms and decay

1. It is slower, the temperature is lower therefore enzyme action is slower.
2. They release enzymes onto the food and then take up the partially digested food.

Answers to biogas questions

1. burning of trees, or anaerobic respiration of biomass by bacteria or yeast
2. methane and carbon dioxide
3. Because biomass can be grown to replace that which has been used.
4. Bacteria require anaerobic conditions.

Answers to Practical sheet 8.12

1. Soil in flask B has been heated.
2. A goes cloudy, B stays clear.
3. A contains microorganisms but B does not.
4. (a) It would take longer for the limewater in A to go cloudy; (b) it would take longer for the limewater in A to go cloudy.

Lesson 13: Required practical: Investigate the effect of temperature on the rate of decay of fresh milk by measuring pH change

Lesson overview

AQA Specification reference

AQA 4.7.2.3

Learning objectives

- Describe how safety is managed, apparatus is used and accurate measurements are made.
- Make and record observations and make accurate measurements.
- Evaluate methods and suggest possible improvements and further investigations.

Learning outcomes

- Able to use apparatus safely and measure volume accurately when investigating decay of fresh milk. [O1]
- Measure the pH of a reaction to show the rate of decay at different temperatures. [O2]
- Evaluate and suggest possible improvements to decay of fresh milk investigation. [O3]

Skills development

- WS 2.1 Use scientific theories and explanations to develop hypotheses.
- WS 2.4 Carry out experiments appropriately having due regard for the correct manipulation of apparatus, the accuracy of measurements, and health and safety considerations.
- WS 2.6 Make and record observations and measurements using a range of apparatus and methods.
- WS 2.7 Evaluate methods and suggest possible improvements and further investigations.
- WS 3.7 Being objective, evaluating data in terms of accuracy, precision, repeatability and reproducibility and identifying potential sources of random and systematic error.

Maths focus

- 1c Calculate rate changes in the decay of biological material (high demand)
- 4a Translate information between numerical and graphical form
- 4c Plot and draw appropriate graphs selecting appropriate scales for the axes

Resources needed Container of milk and a container of sour milk (milk that had been left out of the fridge for a few days). Worksheet 8.13.1 (low demand); Worksheet 8.13.2 (standard demand); Worksheet 8.13.3 (high demand); Practical sheet 8.13; Technician's notes 8.13

Digital resources

Key vocabulary lipase, phenolphthalein indicator

Teaching and learning

Engage

- **Show** students a container of fresh milk and a container of sour milk.
- Ask students to **discuss** the differences between sour and fresh milk and to **suggest** how the milk turned sour.

Challenge and develop

- **Discuss** with the students the idea that, when milk decays, enzymes from bacteria breakdown the food groups in the milk forming acids. It is the acids that make the milk turn sour. It is unusual to think of milk as decaying as we normally just say it has gone sour of 'gone off'. Explain that, as it is the result of bacteria breaking down the milk, it is similar to the decay of other foods, such as fruit.

- Ask students to **suggest** factors that increase the rate at which milk decays. They should link their ideas to the action of enzymes used by the bacteria in the process.
- Explain that, as the milk decays, acids are produced which change the pH of the milk. This change in pH can be used to monitor the rate of decay. [O1]

Explain

- Students then **investigate** the effect of temperature on the enzyme lipase as it breaks down lipids in the milk. Using Practical sheet 8.13, each group should be given a temperature to start at. If they complete that temperature, they just move onto the next one. In this way all the temperatures should be covered by the class at least once. [O1, O2]
- **Collect** together the class results. Students then **complete** the questions about their investigation on Worksheet 8.13.1 (low demand), 8.13.2 (standard demand) or 8.13.3 (high demand). [O3]

Consolidate and apply

- Ask students to **evaluate** their investigation by answering the evaluation questions on their worksheet. [O3]

Extend

Ask students able to progress further to:

- **complete** Questions 5–9 on page 349 of the Student Book; this will provide further practice at analysing and evaluation investigations. [O3]

Plenary suggestions

Select individual students to **identify** one problem with their investigation and **explain** how they could improve the investigation to help solve the problem. [O3]

Answers to Worksheet 8.13

1 Analysis

2. Should be 40°C
3. Low and high temperatures may not have worked; low temperature reaction is too slow to complete in the time; high temperature may denature lipase
4. As the temperature increases, the time to lose the pink colour / mixture to become acidic decreases (rate increases) until an optimum of around 40°C then a further increase in temperature causes the time taken to increase
5. As temperature increases, there are more collisions so time to react decreases (rate increases); at high temperatures the shape of the active site changes so the fat can no longer fit into it; lipase has become denatured

2 Evaluation

1. Depends on the results
2. Main problem will be maintaining the temperature
3. Several repeats are needed so that anomalies can be identified

Lesson 14: Changing the environment

Lesson overview

AQA Specification reference

AQA 4.7.2.4; AQA 4.7.3.1; AQA 4.7.3.2

Learning objectives

- Recall causes of environmental change.
- Describe the impact of environmental change.
- Explain the impact of an environmental change.

Learning outcomes

- State causes of environmental change. [O1]
- Describe the impact of environmental change on distribution of species. [O2]
- Explain the impact of an environmental change on distribution of species. [O3]

Skills development

- WS 1.4 Explain everyday and technological applications of science. Evaluate associated personal, social, economic and environmental implications, and make decisions based on the evaluation of evidence and arguments.
- WS 3.2 Translating data from one form to another.
- WS 3.8 Communicating the scientific rationale for investigations, methods used, findings and reasoned conclusions through paper-based and electronic reports and presentations using verbal, diagrammatic, graphical, numerical and symbolic forms.

Maths focus

- 4a Translate information between graphical and numeric form

Resources needed Video of migrating animals, such as caribou. Worksheet 8.14

Digital resources

Key vocabulary dredging, environmental change, global warming

Teaching and learning

Engage

- **Show** students a video of migrating animals such as caribou. [O1]
- Ask students to **suggest** why the animals are migrating. [O1]

Challenge and develop

- **Discuss** with the students the idea of seasonal changes and how they affect animals. Apart from migration, what other ways do animals cope with seasonal changes? [O1, O2]
- In groups students should **list ways** in which humans change the environment. They should start their list without the use of the Student Book and then add to the list using the Student Book. [O1, O3]
- They then **report back** on their group discussions in a whole class discussion. [O1, O3]

Explain

- Students then **complete** a spider diagram about environmental change. This is intended to introduce some of the topics that will be covered in more detail during the next few lessons. The activity is split into three steps of which low demand students should **complete** at least step 1, standard demand students at least steps 1 and 2 and high demand students must complete all three steps. [O1, O2, O3]
- The students can then swap their spider diagrams with another group to then **evaluate** them. [O1, O2, O3]

Consolidate and apply

- Students **complete** the questions on coral reefs on Worksheet 8.14. [O2, O3]

Extend

Ask students able to progress further to:

- **complete** Question 5 on page 351 of the Student Book. [O2, O3]

Plenary suggestions

Ask one student to come to the front of the class; they need to be able to defend one way humans cause environmental change. For example, they should talk for 1 minute on why humans should drill for oil in the Arctic despite it damaging the environment.

The rest of the class should then try and get them to change their mind by asking them questions, such as: Are there no alternatives to using oil as an energy source?

Answers to Worksheet 6.14

Coral reef

1. There is a decrease in percentage cover of coral reef from 1996 to 1999.
2. habitat destruction, decrease in coral population, less food for other organisms, less fish for local human population
3. saves money as there is no need to build coastal defences to stop coastal erosion by the sea; they have a big economic as well as environmental value, provide an income for the local people in tourism and fishing, help maintain biodiversity and protect the coast

Lesson 15: Learning about land use

Lesson overview

AQA Specification reference

AQA 4.7.3.3

Learning objectives

- Identify why land use has changed.
- Describe the effects of changing land use.
- Evaluate a change in land use.

Learning outcomes

- Understand that an increase in human population has led to changes in land use. [O1]
- Describe how the change in land use has caused pollution. [O2]
- Explain the advantages and disadvantages of one change in land use. [O3]

Skills development

- WS 1.4 Explain everyday and technological applications of science. Evaluate associated personal, social, economic and environmental implications, and make decisions based on the evaluation of evidence and arguments.
- WS 3.2 Translating data from one form to another.
- WS 3.8 Communicating the scientific rationale for investigations, methods used, findings and reasoned conclusions through paper-based and electronic reports and presentations using verbal, diagrammatic, graphical, numerical and symbolic forms.

Maths focus

- 4a Translate information between graphical and numeric form

Resources needed A video news clip about building houses on greenbelt land (search the news section at www.bbc.com); Internet access. Worksheet 8.15.1 (low and standard demand); Worksheet 8.15.2 (high demand)

Digital resources

Key vocabulary eutrophication, run-off

Teaching and learning

Engage

- **Show** students a video news clip about building houses on greenbelt land. [O1]
- Ask students to **debate** whether houses should or should not be built on greenbelt land. [O1]

Challenge and develop

- **Discuss** the need for more houses being linked to the increase in population; explain that this is a worldwide problem in that the human population is growing exponentially. [O1]
- In groups, students should **list** other resources that an increased population will need, apart from houses. [O1, O2]
- They then **report back** on their group discussions to produce a more complete list. [O1, O2]
- Use this list to **explain** that this need for resources has resulted in the change in land use by humans, and its impact on biodiversity. [O1, O2]
- Ask the students to **suggest** ways land use changes are linked to the need for resources. [O1, O2]

Explain

- Students **complete** the questions on population growth on Worksheet 8.15.1 (low and standard demand) or Worksheet 8.15.2 (high demand). [O1, O2, O3]
- High demand students should also **complete** the questions on eutrophication (Worksheet 8.15.2). [O2]

Consolidate and apply

- Students then **research** one change in land use, which they will then **evaluate**. They can take an example from the Student Book or the Internet. Instructions for the research are on Worksheet 8.15.1 and Worksheet 8.15.2. [O2, O3]

Extend

Ask students able to progress further to:

- **discuss** the connection between land usage and recycling bins. [O2]

Plenary suggestions

Ask students to explain the difference between cause and effect using examples from the lesson.

Answers to Worksheet 8.15

Population growth

1. population is increasing with time; population of underdeveloped countries is increasing faster than that of developed countries; sharp increase around 1950; population of developed countries has dipped / levelled off
2. exponential growth
3. Birth rate is greater than death rate.
4. farming / industrial revolution, better medical care
5. any three from: farming, building, quarrying, dumping waste
6. Ideas should include habitat destruction, loss of food sources, pollution of soil and water.

Eutrophication

Fertilisers or sewage pollute water, algae grows rapidly.

Algae die, rot and fall to bottom.

Bacteria decompose algae and use up oxygen.

Animals die from the lack of oxygen.

Lesson 16: Changing the landscape

Lesson overview

AQA Specification reference

AQA 4.7.3.3; AQA 4.7.3.4

Learning objectives

- Identify the reasons for deforestation.
- Describe the impact of peat bog destruction and deforestation.
- Evaluate the destruction of peat bogs and forests.

Learning outcomes

- State the reasons for deforestation. [O1]
- Understand the impact of peat bog destruction and deforestation. [O2]
- Evaluate the destruction of peat bogs and forests. [O3]

Skills development

- WS 1.3 Appreciate the power and limitations of science and consider any ethical issues which may arise.
- WS 1.4 Explain everyday and technological applications of science. Evaluate associated personal, social, economic and environmental implications, and make decisions based on the evaluation of evidence and arguments.
- WS 3.8 Communicating the scientific rationale for investigations, methods used, findings and reasoned conclusions through paper-based and electronic reports and presentations using verbal, diagrammatic, graphical, numerical and symbolic forms.

Resources needed A video clip showing deforestation; a video showing the effects of deforestation on orangutans or another species; Internet access; sheets of A3 plain paper. Worksheet 8.16.1 (low demand); Worksheet 8.16.2 (standard and high demand)

Digital resources

Key vocabulary deforestation, peatlands

Teaching and learning

Engage

- **Show** students a video clip showing deforestation. [O1]
- Ask students to **discuss** what they understand by the term 'deforestation'. [O1]

Challenge and develop

- In groups students should **list** reasons for deforestation. [O1]
- They then **report back** on their group discussions to produce a more complete list. [O1]
- **Show** students a video clip about the impact of deforestation on the population of one species, for example orangutans. (Search 'effects of deforestation on orangutans'.) [O1, O2]
- **Discuss** with the students ways in which deforestation affects both animal and human populations. [O1, O2]

Explain

- **Explain** to students what peat bogs are and why they are being destroyed. [O1, O2]
- Low demand students **complete** the summary activity on Worksheet 8.16.1. [O1, O2]
- Standard and high demand students **write** a short summary of what deforestation is and why it is happening; they should do the same for peat bogs. [O1, O2]

Consolidate and apply

- Split the class into groups. Then divide the groups into two sets. One set **researches** deforestation of rainforests, and the other researches destruction of peat bogs. The information they need to **complete** the task is in the Student Book if there is no access to the Internet. [O2, O3]
- They then **produce** a storyboard for a TV advert encouraging people to help save the rainforest or peat bog. They use Worksheet 8.16.1 (low demand) or Worksheet 8.16.2 (standard and high demand) to help them. Low demand students have been given a help sheet, which is a very basic idea using the example of greenbelt land being used for building. If students use this, make it clear that they cannot just copy it, as it is not about deforestation or peat bogs. [O2, O3]

Extend

Ask students able to progress further to:

- **debate** the practicalities of the statement 'All countries should be forced to protect all the forests in their countries and to stop all logging'. [O3]

Plenary suggestions

Students present their storyboards to the class and accept feedback.

Answers to Worksheet 8.16.1

Changing the landscape summary

Deforestation

Deforestation involves **removing** trees.

Huge areas of tropical forest are being destroyed to provide land for growing **crops**.

Deforestation **decreases** biodiversity.

Growing rice in cleared land increases levels of **methane**.

Peat bogs

Peat bogs take **thousands** of years to form.

They are areas of very **wet** land.

Peat is used as a **fuel** source.

Burning peat increases **carbon dioxide** levels.

Lesson 17: Thinking about global warming

Lesson overview

AQA Specification reference

AQA 4.7.3.5

Learning objectives

- Recall what global warming is.
- Describe the causes of global warming.
- Explain how global warming impacts on biodiversity.

Learning outcomes

- State that global warming is the idea that the average global temperature of the Earth and its atmosphere is increasing. [O1]
- Understand that global warming is caused by increased levels of carbon dioxide and methane in the atmosphere. [O2]
- Understand that global warming could lead to a decrease in biodiversity. [O3]

Skills development

- WS 1.4 Explain everyday and technological applications of science. Evaluate associated personal, social, economic and environmental implications, and make decisions based on the evaluation of evidence and arguments.
- WS 3.5 Interpreting observations and other data (presented in verbal, diagrammatic, graphical, symbolic or numerical form), including identifying patterns and trends, making inferences and drawing conclusions.
- WS 3.8 Communicating the scientific rationale for investigations, methods used, findings and reasoned conclusions through paper-based and electronic reports and presentations using verbal, diagrammatic, graphical, numerical and symbolic forms.

Resources needed Video clip about global warming (e.g. from http://video.nationalgeographic.com); Internet access. Worksheet 8.17

Digital resources

Key vocabulary global warming

Teaching and learning

Engage

- **Write** the term 'global warming' on the board. [O1]
- Ask students to **write down** one fact they already know about the global warming on a sticky note. [O1, O2, O3]
- The students then put their sticky notes onto the main board. [O1, O2, O3]

Challenge and develop

- **Discuss** as a class what they have written down on their sticky notes; check for any confusions with other issues, such as acid rain or ozone depletion. [O1, O2, O3]
- **Produce** a class list of facts they know about global warming by choosing sticky notes with correct facts on and eliminating any incorrect ideas or duplicates. [O1, O2, O3]
- **Show** students a video clip about global warming. [O1, O2, O3]
- Having watched the video ask students if they can **add** anymore facts to the list of sticky notes. [O1, O2, O3]

Explain

- **Introduce** the idea of a debate to the students: explain that they are going to **decide** if they, as an individual, could help stop global warming. [O1, O2, O3]
- Ask students to **consider** the question at the start of Worksheet 8.17. They should then take a vote; **record** the results of the vote in a copy of the table on the main board. [O1, O2, O3]
- Students then use the Internet to **research** the ideas on Worksheet 8.17. Most of the ideas can be found in the Student Book if no Internet is available. [O1, O2, O3]

Consolidate and apply

- Split the class into groups of those that initially voted yes and those that initially voted no. The 'don't knows' should be split between the two groups to even them up. Alternatively, split them into the two groups randomly. The groups then **prepare** their argument. [O1, O2, O3].
- Each group **presents** their argument to the others and then they take a new vote. [O1, O2, O3]
- Standard and high demand students should them **complete** the conclusion section on Worksheet 8.17. [O1, O2, O3]

Extend

Ask students able to progress further to:

- **explain** why research into alternative energy sources is so important. [O2]

Plenary suggestions

Ask a student who changed their mind to explain why they did so.

Lesson 18: Looking at waste management

Lesson overview

AQA Specification reference

AQA 4.7.3.2

Learning objectives

- Describe how waste production is linked to human population growth.
- Describe the impact of waste on ecosystems.
- Explain how waste impacts on biodiversity.

Learning outcomes

- Understand how waste production is linked to human population growth. [O1]
- Understand the impact of waste on ecosystems. [O2]
- Understand how waste impacts on biodiversity. [O3]

Skills development

- WS 1.4 Explain everyday and technological applications of science. Evaluate associated personal, social, economic and environmental implications, and make decisions based on the evaluation of evidence and arguments.
- WS 3.8 Communicating the scientific rationale for investigations, methods used, findings and reasoned conclusions through paper-based and electronic reports and presentations using verbal, diagrammatic, graphical, numerical and symbolic forms.
- WS 4.1 Use scientific vocabulary, terminology and definitions.

Resources needed Video clip about bioaccumulation; Internet access. Worksheet 8.18

Digital resources

Key vocabulary bioaccumulation, sewage

Teaching and learning

Engage

- **Show** students an image of a landfill site. [O1]
- Ask students to **discuss** where all the rubbish has come form and why we need more and more landfill sites. [O1]
- Ask the students to **suggest** some types of pollution that might be caused by household waste. [O1, O2]

Challenge and develop

- Ask students to **recall** the term 'eutrophication' from lesson 8.15 'Learning about land use'; in that lesson they discussed eutrophication caused by fertilisers. Can they **explain** why sewage would have the same effect if it got into rivers? [O2, O3]
- **Show** students a video clip about bioaccumulation. **Discuss** with the students why the toxins build up in the food chain. [O2, O3]
- Ask the students to **state** what they know about acid rain. **Discuss** its cause and its effect on aquatic life. [O2, O3]

Explain

- Low and standard demand students should **answer** Questions 1–4 in the Student Book (pages 358–359). [O1, O2, O3]
- High demand students should **write** a summary that compares acid rain, pollution from sewage and bioaccumulation. [O1, O2, O3]

Consolidate and apply

- Split the class into groups to **research and produce** a presentation on pollution. They should follow the instructions on Worksheet 8.18. [O1, O2, O3].
- High demand students should also include the points in part 2.

Extend

Ask students able to progress further to:

- **explain** the difference between acid rain and global warming. [O2, O3]

Plenary suggestions

Ask the students to **feed back** to the class what they have found out about pollution.

Lesson 19: Investigating pollution

Lesson overview

AQA Specification reference

AQA 4.7.3.2

Learning objectives

- Identify pollution levels using indicator species.
- Explain how indicator species measure pollution.
- Compare different methods of measuring pollution.

Learning outcomes

- Understand that pollution can affect the distribution of organisms living in a particular habitat. [O1]
- Interpret data on indicator species. [O2]
- Identify the advantages and disadvantages of different methods of measuring pollution. [O3]

Skills development

- WS 2.4 Carry out experiments appropriately having due regard for the correct manipulation of apparatus, the accuracy of measurements, and health and safety considerations.
- WS 2.5 Recognise when to apply a knowledge of sampling techniques to ensure any samples collected are representative.
- WS 2.6 Make and record observations and measurements using a range of apparatus and methods.
- WS 3.2 Translating data from one form to another.
- WS 3.4 Representing distributions of results and make estimations of uncertainty.
- WS 3.5 Interpreting observations and other data (presented in verbal, diagrammatic, graphical, symbolic or numerical form), including identifying patterns and trends, making inferences and drawing conclusions.

Maths focus

- 1a Recognise and use expressions in decimal form
- 3a Understand and use the symbols: =, <>, >, ∝, ~
- 3d Solve simple algebraic equations
- 4a Translate information between graphical and numeric form

Resources needed Short video clip about indicator species. Worksheet 8.19.1 (low and standard demand); Worksheet 8.19.2 (high demand); Practical sheet 8.19; Technician's notes 8.19

Digital resources

Key vocabulary indicator species

Teaching and learning

Engage

- **Show** a short video clip about indicator species (search online for 'animals as indicator species'). [O1]
- **Discuss** with the students why these animals are so useful. [O1]

Challenge and develop

- **Explain** to students that monitoring the levels of indicator species can help determine water pollution levels. [O2,3]
- Ask students to **recall** some of the ways water can become polluted; discuss that some organisms will be found in polluted water but not others due to their ability to tolerate the pollution. [O2, O3]
- Ask the students to **suggest** other ways water pollution levels could be measured. [O2, O3]

Explain

- Divide the class into small groups and provide them with water samples from different places. Ask students to **investigate** the use of indicator species by following the instructions on Practical sheet 8.19. [O1, O2]

Consolidate and apply

- Collate the results for the class and then students should **complete** the biotic index calculations and questions on Worksheet 8.19. Low demand students may need help with this calculation. [O1, O2]
- Students should then **complete** the 'direct methods' questions on Worksheet 8.19. [O3].
- High demand students should then **complete** the questions on biotic indicators on Worksheet 8.19.2. [O1, O2]

Extend

Ask students able to progress further to:

- **suggest** how they could use different species of lichen to investigate levels of air pollution in different habitats. [O2, O3]

Plenary suggestions

Call out the names of some animals found in a water sample. Ask the class to decide if the sample is polluted or not.

Answers to Worksheet 8.19

Biotic index

1. Depends on results.
2. Depends on results.
3. This will depend on how many species were found; students may suggest their sample size is not big enough.

Direct methods

1. pH probe / universal indicator solution
2. living – take samples from different areas and identify living organisms present – use these to determine if the samples are high or low in oxygen; advantage: does not rely on technology that can go wrong; disadvantage: may harm non-living; oxygen probe – advantage: measures exact values; disadvantage: relies on technology that can go wrong, i.e. batteries needed

Biotic indicators (high demand)

1. increases until 3 km then starts to fall
2. more nutrients to feed on
3. oxygen
4. should follow line of oxygen down then up again
5. after 5 km as they only live in water with plenty of oxygen
6. e.g. bloodworms, as they can live in water with low oxygen levels

Lesson 20: Maintaining biodiversity

Lesson overview

AQA Specification reference

AQA 4.7.3.6

Learning objectives

- Describe some conservation measures.
- Describe the impact of breeding programmes.
- Explain how habitats are regenerated.

Learning outcomes

- Describe conservation programs such as recycling, breeding programs, regeneration of habitats, re-introduction of field margins and reducing deforestation. [O1]
- Understand why conservation programmes are needed. [O2]
- Understand that there are advantages and disadvantages to conservation measures. [O3]

Skills development

- WS 1.3 Appreciate the power and limitations of science and consider any ethical issues which may arise.
- WS 3.8 Communicating the scientific rationale for investigations, methods used, findings and reasoned conclusions through paper-based and electronic reports and presentations using verbal, diagrammatic, graphical, numerical and symbolic forms.
- WS 4.1 Use scientific vocabulary, terminology and definitions.

Resources needed Short video clip about coral reefs (e.g. from http://video.nationalgeographic.com); Internet access. Worksheet 8.20.1 (low demand); Worksheet 8.20.2 (standard demand); Worksheet 8.20.3 (high demand)

Digital resources

Key vocabulary conservation, monocultures, regeneration, sustainable

Teaching and learning

Engage

- **Show** a short video clip about coral reefs. [O1]
- **Discuss** with the students why damaging the coral affects all the species in the coral reef. [O1]

Challenge and develop

- Use the video to **explain** to students that efforts are being made to regenerate the coral reefs, this was mentioned in the video, can they **recall** how this is being done. [O1, O2]
- **Explain** that regeneration is just part of the conservation programme to protect the biodiversity of coral reefs. Ask students to **describe** any other conservation programmes linked to other habitats or species they know about. Can they **suggest** why these conservation programmes are needed? [O1, O2]

Explain

- Ask the students to **read about** the conservation programmes on page 363 of the Student Book. They should then **write a summary** of each one in their own words. Low demand students can complete the table in section 1 of Worksheet 8.20.1 instead of writing the summary. [O1, O2]

Consolidate and apply

- Working individually or in groups, students **produce** a leaflet about conservation. They should follow the instructions on Worksheet 8.20. [O1,2,3]
- Low demand students should **complete** the instructions on Worksheet 8.20.1. [O1, O2]
- Standard demand students should **complete** the instructions on Worksheet 8.20.2. [O1, O2, O3]
- High demand students should **complete** the instructions on Worksheet 8.20.3. [O1, O2, O3]

Extend

Ask students able to progress further to:

- **debate** the statement 'There is a food shortage around the world yet we waste valuable farming land reintroducing hedges and field margins'. Do they agree with statement, if so why? Can they think of reasons why people would disagree with the statement? [O2, O3]

Plenary suggestions

Ask the students to feed back to the class what they have found out about the conservation programme they researched.

Answers to Worksheet 8.20.1

Correct table

conservation programme	description
regeneration of rare habitats	Coral reefs, mangroves, and heathland are just some of the ecosystems that are being restored.
re-introducing field margins and hedgerows	Only one type of crop is grown on a lot of farmland. This will increase the biodiversity around the farmland.
reducing deforestation and carbon dioxide emissions	This can be done using sustainable strategies including replanting trees that are cut down and finding alternative energy sources.
recycling resources	This can reduce the need for large areas of landfill.

Lesson 21: Learning about food security

Lesson overview

AQA Specification reference

AQA 4.7.5.1

Learning objectives

- Identify factors affecting food security.
- Describe how different factors affect food security.
- Interpret data to evaluate food security.

Learning outcomes

- State factors affecting food security. [O1]
- Understand how different factors affect food security. [O2]
- Interpret data to evaluate food security. [O3]

Skills development

- WS 1.4 Explain everyday and technological applications of science. Evaluate associated personal, social, economic and environmental implications, and make decisions based on the evaluation of evidence and arguments.
- WS 3.4 Representing distributions of results and make estimations of uncertainty.
- WS 3.5 Interpreting observations and other data (presented in verbal, diagrammatic, graphical, symbolic or numerical form), including identifying patterns and trends, making inferences and drawing conclusions.
- WS 3.8 Communicating the scientific rationale for investigations, methods used, findings and reasoned conclusions through paper-based and electronic reports and presentations using verbal, diagrammatic, graphical, numerical and symbolic forms.

Maths focus

- 2c Construct and interpret frequency tables and diagrams, bar charts and histograms
- 3a Understand and use the symbols: =, <>, >, ∝, ~
- 4a Translate information between graphical and numeric form

Resources needed Short video news clip about food shortages (search 'world food shortage' on BBC news); news clip about food wastage; paper and pens for poster. Worksheet 8.21.1 (low demand); Worksheet 8.21.2 (standard demand); Worksheet 8.21.3 (high demand)

Digital resources

Key vocabulary food security

Teaching and learning

Engage

- **Show** students a video news clip about food shortages to highlight the issue. [O1]
- Ask students to work in groups and **produce** a list of reasons why food shortages may occur in parts of the world. [O1]

Challenge and develop

- **Compare** the list each group has made with the one in the Student Book. [O1]
- Working in the same groups, students take the list in the Student Book and **discuss** why each situation affects food security. They should try and do this without reading on in the Student Book. [O1, O2]
- Groups should then **feed back** to a class discussion about the factors that affect food security. As a class can they **suggest** any solutions to the problems? [O1, O2]

Explain

- Students then **interpret** data about food security by answering the questions on Worksheet 8.21. [O1, O2, O3]
- Low demand students should **complete** section 1 of Worksheet 8.21.1. [O1, O2, O3]
- Standard demand students should **complete** section 1 of Worksheet 8.21.2. [O1, O2, O3]
- High demand students should **complete** section 1 of Worksheet 8.21.3, which includes extension questions. [O1, O2, O3]

Consolidate and apply

- **Explain** to students that food shortage is not just a problem in developing countries. As well as the factors they have already looked at, food wastage is a big issue. **Show** the students a news clip about food wastage. [O2]
- Students then **produce** a poster encouraging people to not waste food, the poster should link to the factors affecting food security they discussed at the start of the lesson. Worksheet 8.21 section 2 has instructions for what to include. [O1, O2]

Extend

Ask students able to progress further to:

- **explain** why protecting biodiversity can help ensure future food security. [O2]

Plenary suggestions

Ask the students to **suggest** ways they as individual people could help secure future food security.

Answers to Worksheet 8.21

Food security

2. Ethiopia and Madagascar
3. very good – USA, Norway, Spain, UK, Chile; good – Mexico, Brazil, Egypt; moderate – Indonesia, Kenya; poor – Ethiopia, Madagascar
4. European countries are the ones with very good food security; African or developing countries are poor or moderate; South American countries are good.
5. Reasons should be based on their biological knowledge of factors that can affect food shortages not their knowledge of the country; this is good practice for the exam where they might be asked about a country they have never heard of but they will be provided with data they can apply their knowledge to.
   Reasons could include developed countries have a better economy so can import food; African countries have problems with growing food in harsh habitats; South American countries – growing conditions may be better than in Africa, may have lower populations.

Food insecurity (high demand)

1. Overall food insecurity has gone down from 1999 to 2008; in urban areas it rose in 2005 but then went down in 2008. It is always higher in rural areas than urban but the percentage went down each time in the rural areas.
2. Overall fall may be to do with increased economy, better farming methods; increase in 2005 in urban areas could be due to economy, or influx of people to the cities and towns increasing the population. Higher in rural areas as rely more on growing own food than buying food so affected by drought conditions.

Lesson 22: Maintaining food security

Lesson overview

AQA Specification reference

AQA 4.7.5.2

Learning objectives

- Describe some intensive farming methods.
- Explain ethical issues related to intensive farming.
- Evaluate modern farming techniques.
- Describe methods to maintain sustainable fisheries.

Learning outcomes

- Understand that factory farming is an example of intensive farming. [O1]
- Describe ways in which fish stocks can be maintained and the use of biological control. [O2]
- Understand the ethical issues associated with intensive farming. [O3]

Skills development

- WS 1.4 Explain everyday and technological applications of science. Evaluate associated personal, social, economic and environmental implications, and make decisions based on the evaluation of evidence and arguments.
- WS 3.4 Representing distributions of results and make estimations of uncertainty.
- WS 3.5 Interpreting observations and other data (presented in verbal, diagrammatic, graphical, symbolic or numerical form), including identifying patterns and trends, making inferences and drawing conclusions.
- WS 3.8 Communicating the scientific rationale for investigations, methods used, findings and reasoned conclusions through paper-based and electronic reports and presentations using verbal, diagrammatic, graphical, numerical and symbolic forms.

Maths focus

- 1d Make estimates of the results of simple calculations

Resources needed Images of battery chickens and free range chickens; video about one example of biological control (search 'cane toads biological control'). Worksheet 8.22

Digital resources

Key vocabulary biological control, factory farming, sustainable fisheries

Teaching and learning

Engage

- **Show** students images of chickens in large huts (take care with the image you choose, this is not meant to be a shock tactic about battery farming) and others roaming free in fields. [O1]
- Ask students to **use their knowledge** of energy transfer to **explain** which type of farming results in more energy loss from the food chain. [O1]

Challenge and develop

- **Discuss** with the students the different farming methods of factory farming, fish farming and hydroponics. Video clips showing the different methods can be used, but take care when choosing them as many have a political message. [O1]
- **Explain** to students that one of the reasons for farming fish is the low numbers in the seas due to overfishing in the past. Ask students to **suggest** other ways apart from farming that we can help preserve fish numbers. Use their ideas to lead into a discussion about sustainability and fish stocks. [O2]
- Write the word pesticides on the board and ask student to **state** what they know about pesticides. Can they **suggest** any other ways to remove pests or weeds for crops? [O2]

- **Show** the students a video about one example of biological control and how it can sometimes go wrong.

Explain

- Students then complete the questions on Worksheet 8.22. Low demand students should at least **complete** the questions on factory farming and fish quotas. [O1, O2]
- Standard demand should at least **complete** the questions on factory farming and tomatoes. [O1, O2]
- High demand students should **complete** all the sections. [O1, O2, O3]

Consolidate and apply

- **Discuss** with the students some of the ethics involved in intensive farming. Make it clear to students that not all intensive farming methods are on the same scale. [O3]
- **Ask** the students if they ever think about where their food comes from; do they understand the concept of organic food? They may have noticed the 'red tractor' symbol on food packets, which means the food came from a farm in the UK that has to meet certain standards – for example, allowing only a certain number of chickens in a certain area. [O3]
- Low demand students should **produce** a table that shows the advantages of intensive farming in one column and the disadvantages in the other. They should use the Student Book to help them. [O3]
- Show standard and high demand students this question 'Should we ban factory farming of animals?'. They should then **write a balanced answer** to the question making their **own decision** based on facts from both sides of the argument. High demand students should try to do this without the Student Book. [O3]

Extend

Ask students able to progress further to:

- **explain** why organic farming methods would not be viable in countries like Madagascar. [O2]

Plenary suggestions

Ask the students to **vote** on whether or not factory farming should be banned. After the vote, ask some of the students to explain their choice.

Answers to Worksheet 8.22

Factory farming and fish stocks

1. what it is – idea of intensive farming of animals by keeping lots of them in one place to restrict movement and feeding them high energy food sources

   why it is done – to produce food quickly and cheaply

   how it reduces energy loss from the food chain – less movement means more food intake is used to make biomass, less energy lost as heat to keep warm if sheds or barns are heated
2. Protein is needed for growth so will increase biomass quickly.
3. overfishing
4. maintaining fish stocks by not taking too many
5. fishing quotas to conserve stocks at sustainable levels, control of net size, fish farms

Growing tomatoes

1. hydroponics
2. barren conditions where there is little soil
3. minerals
4. for respiration in roots, to release energy for active transport for uptake of minerals
5. disadvantage – lack of support; advantage – insecticides can be added and the minerals can be monitored and adjusted if necessary

Cane toads

1. No effective pesticide was available until five years later.
2. 300–400 years at the present rate but will actually be a lot less due to the ever-increasing toad population.
3. used a secure trial/pilot to see whether the toads ate the cane beetles and to make sure they didn't cause any other problems
4. introduce sterile toads to reduce offspring, contain advance and destroy them, try to discover a specific poison or disease

Lesson 23: Using biotechnology

Lesson overview

AQA Specification reference

AQA 4.7.5.4

Learning objectives

- Describe some uses of biotechnology.
- Explain the advantages of some uses of biotechnology.
- Evaluate some uses of biotechnology.

Learning outcomes

- Describe some uses of biotechnology to include mycoprotein, insulin production and GM crops. [O1]
- Understand the advantages of some uses of biotechnology. [O2]
- Evaluate some uses of biotechnology in terms of food security and effect on the environment. [O3]

Skills development

- WS 1.4 Explain everyday and technological applications of science. Evaluate associated personal, social, economic and environmental implications, and make decisions based on the evaluation of evidence and arguments.
- WS 3.8 Communicating the scientific rationale for investigations, methods used, findings and reasoned conclusions through paper-based and electronic reports and presentations using verbal, diagrammatic, graphical, numerical and symbolic forms.
- WS 4.1 Use scientific vocabulary, terminology and definitions.

Resources needed Images of unusual genetically engineered organisms (use 'genetically engineered animals' or 'genetically engineered plants' as search terms); image of a fermentation vat (search 'fermentation tanks') used to grow bacteria. Worksheet 8.23.1 (low demand); Worksheet 8.23.2 (standard and high demand)

Digital resources

Key vocabulary biotechnology, mycoprotein

Teaching and learning

Engage

- **Show** students images of unusual genetically engineered organisms and ask them to decide which ones are real. [O1]
- Explain to the students that the organisms have been genetically engineered. Ask students to work in small groups (perhaps using mini-whiteboards) to **write down** what they know about genetic engineering. [O1]

Challenge and develop

- **Explain** that genetic engineering is just one example of biotechnology – using living organisms to make a product to improve the quality of life. In most cases it involves enabling large quantities of microorganisms to be cultured in industrially controlled vats for food. [O1]
- **Show** the students an image of a fermentation vat used to grow bacteria, explain that the bacteria can be engineered to produce a required substance, such as insulin; they are then put in the vats to multiply and produce the chemical. Ask students to **suggest** the conditions needed inside the tank for optimum growth. [O1]
- **Describe** the examples of biotechnology; the students need to know about mycoprotein, production of human insulin and GM crops such as golden rice. [O1]

Explain

- Students then **read** about the different examples in the Student Book. They should then **write a summary** of each one in their own words; the summary should include a description of each method and the benefits associated with it. Low demand students can complete the sentences on Worksheet 8.23.1 instead of writing the summary. [O1, O2]

Consolidate and apply

- **Discuss** with the students some of the ethics involved in GM crops and other uses of biotechnology. [O3]
- Low demand students should **complete** the questions on Worksheet 8.23.1. [O1, O2, O3]
- Standard demand and high demand students can **produce** a PowerPoint presentation about biotechnology following the instructions on Worksheet 8.23.2. [O1, O2, O3]

Extend

Ask students able to progress further to:

- **explain** why there is no need to produce golden rice for the UK food market. [O2]

Plenary suggestions

Ask a student to volunteer to talk for 1 minute about any aspect of biotechnology without repeating any words. Other students can challenge when they hear a repeated word, and then they take control of the game until the time is up.

Answers to Worksheet 8.23

Biotechnology

1. golden, Asian, A, blindness, carrots A; hormone, diabetes, inject, bacteria, purified; fermenters, *Fusarium,* anaerobic, dried

Genetic engineering

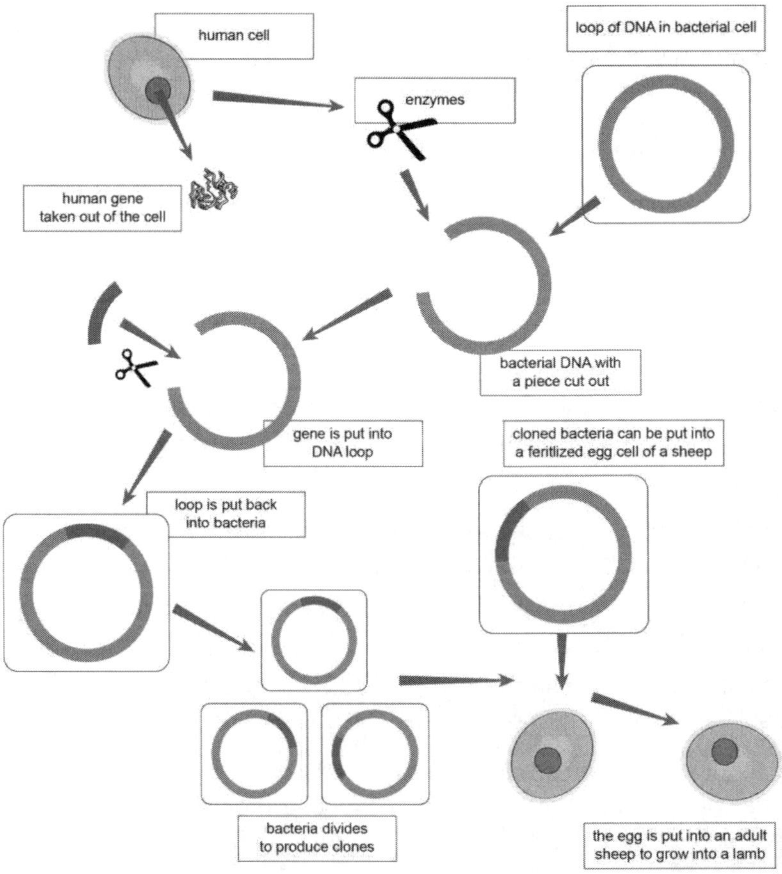

1. Food can be produced that contains required vitamins, grown in areas where other crops will not grow, be resistant to disease.
2. Risk of unexpected harmful side effects, danger genes could get into plants growing in the wild.
3. Can produce large quantities of food inside fermenters so not dependent on weather.

Lesson 24: Maths skills: Using graphs to show relationships

Lesson overview

Learning objectives

- To recognise direct proportionality in a graph.
- To calculate reaction rates in linear graphs.
- To use the gradient of a graph to calculate the rate.

Learning outcomes

- Able to recognise direct proportionality in a graph. [O1]
- Understand how to calculate reaction rates in linear graphs. [O2]
- Understand how to use the gradient of a graph to calculate the rate. [O3]

Skills development

- WS 3.1 Presenting observations and other data using appropriate methods.
- WS 3.2 Translating data from one form to another.
- WS 3.3 Carrying out and representing mathematical and statistical analysis.
- WS 3.5 Interpreting observations and other data (presented in verbal, diagrammatic, graphical, symbolic or numerical form), including identifying patterns and trends, making inferences and drawing conclusions.

Maths focus

- 4a Translate information between graphical and numeric form
- 4c Plot two variables from experimental or other data
- 4d Determine the slope and intercept of a linear graph

Resources needed Image of the World population growth curve. Worksheet 8.24

Digital resources

Key vocabulary directly proportional, gradient, intercept, linear rate

Teaching and learning

Engage

- Ask students to **look** at the two graphs on page 370 of the Student Book.
- In groups the students should **list** information that the graphs shows. The groups should then report back to the class.

Challenge and develop

- **Explain** what is meant by a linear graph. Use the the two graphs on page 370 of the Student Book to **explain** the terms 'directly proportional' and 'intercept'. [O1]
- **Explain** how rate can be calculated from the gradient using the values on the *x* and *y* axis. [O2]
- Ask students to **calculate** the gradient of the graphs on page 370 of the Student Book. [O2]

Explain

- **Discuss** the idea of how the rate often changes over the course of a graph. They can see the change in rate by looking at how the slope of the graph changes. **Explain** that the steeper the gradient the faster the rate. They may be familiar with this from physics where they will have learnt about distance time graphs. [O3]
- Students then **complete** questions 3–5 on page 371 of the Student Book. [O3]

Consolidate and apply

- All students then **answer** the questions on Worksheet 8.24. [O1, O2, O3]
- Low demand students may need help with the scale for the graph. This could be done by asking each group to record one set of results and then complete the calculations together with a class set of results. [O1, O2, O3]

Extend

Ask students able to progress further to:

- (high demand) **complete** questions 6 and 7 on page 371 of the Student Book. [O3]

Plenary suggestions

Show the students an image of the World population growth curve.

Ask the students to **describe** how the rate of growth has changed. Can they **suggest** reasons for the changes (especially the change in the 1900s)? Can they predict how the rate of growth may change in the future; will it get faster, slower or stay the same?

Answers to Worksheet 8.24

1. time on *x* axis, number of bacteria on *y* axis
2. S-shaped, rises slowly to start with then sharp increase and then starts to levels off
3. 40/40 = 1 bacteria/minute
4. (156 – 40)/40 = 2.9 bacteria/minute
5. rate has almost tripled, idea that you have more bacteria dividing at any one time so more can be produced
6.
   a. 10/20 = 0.5 bacteria/minute
   b. Rate of growth has dropped.
   c. Space or food is limited.

When and how to use these pages: Check your progress, Worked example and End of chapter test

Check your progress

Check your progress is a summary of what students should know and be able to do when they have completed the chapter. Check your progress is organised in three columns to show how ideas and skills progress in sophistication. Students aiming for top grades need to have mastered all the skills and ideas articulated in the final column (shaded pink in the Student book).

Check your progress can be used for individual or class revision using any combination of the suggestions below:

- Ask students to construct a mind map linking the points in Check your progress
- Work through Check your progress as a class and note the points that need further discussion
- Ask the students to tick the boxes on the Check your progress worksheet (Teacher Pack CD). Any points they have not been confident to tick they should revisit in the Student Book.
- Ask students to do further research on the different points listed in Check your progress
- Students work in pairs and ask each other what points they think they can do and why they think they can do those, and not others

Worked example

The worked example talks students through a series of exam-style questions. Sample student answers are provided, which are annotated to show how they could be improved.

- Give students the Worked example worksheet (Teacher Pack CD). The annotation boxes on this are blank. Ask students to discuss and write their own improvements before reviewing the annotated Worked example in the Student Book. This can be done as an individual, group or class activity.

End of chapter test

The End of chapter test gives students the opportunity to practice answering the different types of questions that they will encounter in their final exams. You can use the Marking grid provided in this Teacher Pack or on the CD Rom to analyse results. This shows the Assessment Objective for each question, so you can review trends and see individual student and class performance in answering questions for the different Assessment Objectives and to highlight areas for improvement.

- Questions could be used as a test once you have completed the chapter
- Questions could be worked through as part of a revision lesson
- Ask Students to mark each other's work and then talk through the mark scheme provided
- As a class, make a list of questions that most students did not get right. Work through these as a class.

Marking Grid for End of Chapter 8 Test

	Getting Started [Foundation Tier]				Going Further [Foundation and Higher Tiers]		More Challenging [Higher Tier]			Most demanding [Higher Tier]		Total marks	Percentage
Student Name	Q. 1 (A01) 1 mark	Q. 2 (A01) 2 marks	Q. 3 (A02) 2 marks	Q. 4 (A03) 2 marks	Q. 5 (A01) 1 mark	Q. 6 (A01, AO2, AO3) 7marks	Q. 7 (A01) 2 marks	Q. 8 (A01) 1 mark	Q. 9 (A02) 4 marks	Q. 10 (A01, AO2) 6 marks	Q. 11 (A02) 4 marks		

Check your progress

You should be able to:

identify abiotic and biotic factors that affect ecosystems →	explain the impact of introducing a new species or pathogen →	explain how a gradual change in an abiotic factor can affect the distribution of species
describe how changes in the population of one organism can affect other organisms →	explain how the size of a predator population will affect the numbers of prey and vice versa →	explain why the cycles of predator and prey populations are out of phase with each other
explain what a trophic level is →	construct food chains →	construct accurate pyramids of biomass
describe how to use some sampling techniques →	describe how to use a transect line →	explain how to perform accurate sampling
recall that plants take in carbon as carbon dioxide →	explain how carbon is recycled →	interpret a diagram of the carbon cycle
recall causes of environmental change →	describe the impact of environmental change →	explain the impact of environmental change
identify factors affecting food security →	describe how different factors affect food security →	interpret data to evaluate food security
describe some uses of biotechnology →	explain the advantages of some uses of biotechnology →	evaluate some uses of biotechnology

Worked example

Some students investigated the average number of plant species growing in a habitat. They investigated different sized areas of the habitat. Look at the graph.

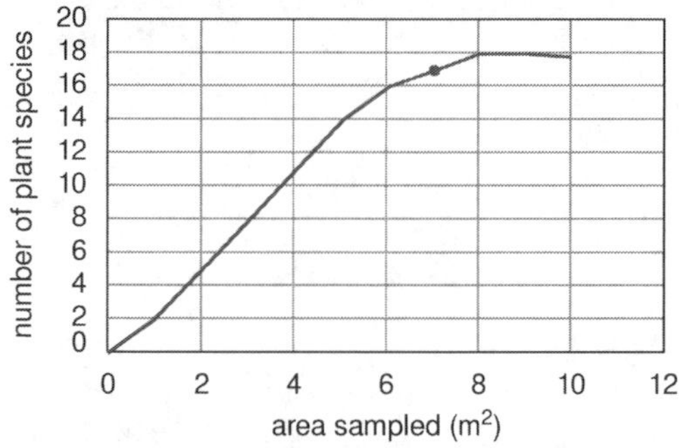

1 Describe what the graph shows.

The number of different plants found increases from 0 to 18

Maximum number of plants found identified.

More detailed responses include: as area sampled increases the number of species found increases; the ideal area to sample is 8 m^2.

2 Amy and Laura decide to investigate how the number and type of plant species growing in a salt marsh change with distance from the sea.

What equipment should the use to do this?

A quadrat

A more detailed answer would include a measuring instrument, e.g. tape measure.

3 Describe how Laura and Amy should use this apparatus.

They make a straight line and place the quadrat every metre along it. Then count the different species and their numbers.

4 Why will the results of Amy's and Laura's investigation be an estimate of the population sizes?

They have only sampled the habitat

Correct answer identified. A more detailed response would include that it is impossible to count every plant.

5 Laura and Amy repeat the process in two other locations. Explain why.

To improve the accuracy of their results

This is a correct response.

Overall, there is some good understanding of the topic shown but the answers need to be more detailed.

Chapter 1: Cell biology

Lesson 1.1 Looking at cells

1 cell membrane, nucleus (and mitochondria and ribosomes)

2 a chloroplasts, permanent vacuole

b (nucleus) - controls the activities of the cell

(cell membrane) - controls the passage of substances into and out of the cell

3 cell wall

4 0.00004 m/4 x 10^{-5} m

5 10 – 100 µm

6 0.007 µm

7 1 (=x10)

8 Oxygen and glucose pass in to supply muscle with reactants for respiration to make the energy needed for muscles to contract. Carbon dioxide passes out as this is a waste product of respiration.

Lesson 1.2 The light microscope

1 x600

2 x20

3 x1250

4 the magnification of an image in a book will give an accurate indication of true size; the size of an image on a computer will be different on different computers, depending on screen size and resolution. It is of little value, therefore, to show the magnification.

5 0.2 µm/200 nm

6 around x1500.

7 Pros: Easy and quick to use, inexpensive, able to view live specimens.

Cons: high magnification not possible so cannot see very small structures, resolution restricted to 200nm, only 2D, not 3D.

Lesson 1.3 Looking at cells in more detail

1 <1 nm

2 non-living

TEM is used to look at internal structure in fine detail. SEM is used to look at the surface of cells and small organisms.

3 It has enabled us to study cells in greater detail.

4 ribosome

Mitochondria *are* visible when viewed with the light microscope.

5 protein synthesis

6 A Scanning Electron Microscope (SEM) as this uses electrons that bounce of the surface of the specimen to reveal the surface shape of structures such as very small organism and cells

Lesson 1.4 Required practical: Using a light microscope to observe and record animal and plant cells

1 Sterile cotton swabs – to rub inside of cheek to get cells

Glass microscope slides – to wipe swab to transfer cheek cells onto

Methylene Blue solution – to stain cells

Plastic pipette or dropper – to drop MB solution onto cells

Plastic cover slips – to protect cells and microscope, keeps cells flat and retains liquid

Paper towels or tissue – to remove excess MB solution

2 Cells are less likely to expand and burst in saline

3 a Red blood cell; Nucleus

b Frog RBC contain a nucleus

4 Easier to locate cells; can see structure of tissue; can see different cell types and numbers of each

5 From middle diagram, top label – region of cell elongation; mid – meristem and bottom label – root cap

6 Difficult to say, as not looking at whole cells (have been sliced) and small image

7 Assume p/m is at the same scale as the top image, so 10mm represents 10 micrometres. Any scale between 1 – 5cm with correct micrometres. But suggest that this isn't actually the case – text needs to state magnification of image

8 Measure paramecium (10mm) Read off value in mm from scale drawn in (7) Paramecium usually 100 – 350 micrometres

Lesson 1.5 Primitive Cells

1 prokaryotic cells are much smaller/eukaryotic much larger.

prokaryotic cells have no nucleus – the DNA is free in the cytoplasm/eukaryotic cells have a nucleus

2 as a single loop/chromosome free in the cytoplasm and in one or more structures called plasmids

3 Archaea, Bacteria and Eukaryota

4 Eukaryota

5 Accept example of extreme environment such as hot spring, salt lake, deep ocean etc.

6 chemical nature is different from Bacteria (chemicals in cell membranes of Archaea are unique to them; nucleic acid [of ribosomes] is closer to eukaryotes)

7 It has only been since we have had the technology and awareness of how to analyse DNA of different organisms that it has become evident that Archaea and genetic makeup distinct from bacteria and eukaryotes.

Lesson 1.6 Cell division

1 46 or 23 pairs

2 In pairs, in order of decreasing size. The sex chromosomes are displayed last.

3 growth; when cells need to be replaced, e.g. dead cells, cells that are damaged

4 same number of chromosomes as the parent cell, i.e. 46; two daughter cells

5 they have doubled their DNA (before dividing). So there are two molecules of DNA per chromosome

6 One-eighth/12.5%. Working out: 45°/360°

7 15 minutes. Working out: (45/360 x 120) minutes

8 Repair and replacement of cells at site of wound or injury, during growth spurts, during production of red blood cells in bone marrow (happens all the time).

Lesson 1.7 Cell differentiation

1 sperm cell – tail and mitochondria for swimming to fertilise the egg; sperm nucleus carries genetic material to combine with that of the egg

muscle cell – protein fibres that move over each other and cause the muscle to contract; mitochondria to release energy.

nerve cell – long, with extensions, to communicate with other nerve cells and muscles and sensory structures; gaps in the myelin sheath enable rapid conduction of impulses

2 have many ribosomes for protein synthesis; many mitochondria will supply the energy for protein synthesis

3 (smallest) cell tissue organ system human body

4 cell – red blood cell, white blood cell; tissue – blood

5 The biconcave shape increases the surface area which is important for the increases the rate of diffusion for oxygen and carbon dioxide , both of which are transported around the body by RBCs

Lesson 1.8 Cancer

1 when cells divide uncontrollably (and do not fulfil their normal function, and spread throughout the body)

2 leukaemia

3 benign, malignant

4 to supply it with food and oxygen

5 secondary

6 of the cervix/cervical

7 The graph shows a positive correlation. As animal fat intake increases the breast cancer death rate increases. The graph suggests there is a link between animal fat intake and breast cancer. The higher a person's fat intake the greater their risk of breast cancer.

Note this is only a theory as it is impossible to establish a causal link between the two variables without further research

Lesson 1.9 Stem cells

1 replace body cells lost through injury or disease.

2 embryonic

3 Alzheimer's disease, diabetes, other disease resulting from cells that have been destroyed, injured, or have degenerated

4 chemotherapy destroys bone marrow, where blood cells are produced, so these patients need stem cell transplants to be able to produce new blood cells

5 (embryonic) stem cells have to be removed from an embryo

6 requires growth of an embryo, and removal of cells from the embryo

The embryo is destroyed in the process

7 Benefits: Potential role in treating injuries where certain body cells degenerate, treatment of alzheimers, diabetes, cancer, burn victims, etc

Drawbacks: Moral and ethical questions raised particularly for the use of embryonic stem cells. Costs of stem cell research and therapies

Lesson 1.10 Stem cell banks

1 an identical twin, twin, close member of the family (in that order).

2 Advantages – from the patient, so there will be no problem with rejection, or ethical issues.

Disadvantages – cost, problems with storage (possible changes to cells; keeping conditions stable).

3 the production of an embryo from a patient's body cell, to generate stem cells for therapy.

The nucleus from the body cell is removed and inserted into an egg cell that has had its nucleus removed.

4 after 4-5 days.

5 safety – have there been any mutations; have the cells been transformed into cancer cells?

is it morally/ethically correct to produce an embryo with the intent of destroying it?

could embryos just become a resource for research?

could people be exploited for eggs?

could an embryo be implanted to produce a cloned human?

6 Risks – We don't know how safe or successful therapeutic cloning will be.

Benefits: If successful therapeutic cloning could produce stem cells that could be used to treat patients.

Ethics – should we be creating an embryo with the intent of then destroying it? When does life start?

Lesson 1.11 Key Concept: Cell development

1 skeletal muscle

2 They have no sub-cellular structures to perform normal cell functions.

They have only nucleic acid for reproduction.

3 cell replacement; asexual reproduction (eukaryotes); cancer cells; cloned cells

4 cells producing gametes

5 embryonic stem cells cannot differentiate into cells of the placenta

6 Stem cells give scientists a more accurate prediction of a drugs potential toxicity than rat cells.

Lesson 1.12 Cells at work

1 drive chemical reactions needed for them to live, cell division, movement, to keep warm, active transport, transmit nerve impulses, maintaining a constant environment

2 drive chemical reactions needed for them to live, cell division, transport, active transport (mineral ions and opening and closing stomata), maintaining a constant environment

3 to provide energy

4 to keep warm/maintain a constant body temperature.

5 Respiration takes place in all cells all the time

6 have many mitochondria.

7 Respiration produces heat energy and during exercise. Respiration increases as the muscles need more energy to contract.

Lesson 1.13 Living without oxygen

1 glucose → ethanol + carbon dioxide (+energy released)

2 The yeast divide so rapidly that the oxygen runs out and therefore the yeast switch to anaerobic respiration.

3 for the yeast to be able to respire/a respiratory substrate
4 because of the carbon dioxide produced (you can see where bubbles of the gas have been in the baked bread)
5 It evaporates off in the oven.
6 They have to grow down through the tissue of the female reproduction system, where there will be little oxygen.
7 $C_6H_{12}O_6 \rightarrow 2C_2H_5OH + 2CO_2$
(+energy released)
8 glucose → lactic acid (+energy released)
$C_6H_{12}O_6 \rightarrow 2C_3H_6O_3$ (+energy released)
Note that you do not need to know the *symbol* equation for this process.
9 Reactants for aerobic respiration are glucose and oxygen. Reactants for anaerobic respiration are only glucose.
Products for aerobic respiration are carbon dioxide and water. Products for anaerobic respiration are ethanol and carbon dioxide.
Both reactions produce energy, but aerobic respiration produces more than anaerobic respiration.

Lesson 1.14 Growing microorganisms

1 to study the organism under investigation, and not have results affected by microorganisms that could disrupt the cultures environment
2 inoculating loop
3 6 h 40 min/400 min (to produce 1 048 576 bacteria)
4 1.05×10^{-6} g/1.05 μg
Possible shortage of food; build-up of waste products.
5 Exponential growth is ever increasing growth. An example is binary fission.
6 the food has run out; build-up of toxic waste products.
7 Resources would no longer be scarce so the bacteria would begin to divide again and grow exponentially until the food supply runs low at which point the stationary phase would happen a second time, followed by the death phase.

Lesson 1.15 Testing new antibiotics

1 heated, so as not to transfer bacteria from the environment; cooled, so as not to kill the bacteria being tested
2 it is important that bacteria are present, but have not begun to grow
3 agar, beef, milk protein (not blood – that is an added extra)
4 the clear zones may overlap each other and make measurement impossible
5 a sample that is typical of the whole.
A representative sample is important to ensure the characteristics and behaviour of the bacteria sample are the same as the rest of the population.
6 choosing a colony on the culture plate that looks identical to – and not different from – the others.
7 MRSA shows antibiotic resistance and is therefore difficult to treat.

Lesson 1.16 Required practical: Investigating disinfectants

1 independent variable – concentration of disinfectant
dependent – size/diameter/area of clear zone
2 size of filter paper disc, volume of disinfectant, species of bacterium, size of agar plate, number of discs applied, depth of agar, duration of incubation.
3 learner answer
4 Mueller-Hinton agar would grow pathogens
Again, if the plate was incubated at body temperature, pathogens may grow
5 Using the formula πr^2
6 500, 550, 570, 580, 580, 580 (note that these are all to three significant figures)
7 470 (2.0 g/dm^3, Test 1)
It is not consistent with the other two results.
8 4.5 g/dm^3
9 student answer
10 collect more data in the area of the graph where there is a sharp increase in clear zone diameter observed in the graph; at appropriate, e.g. 0.1 g/dm^3 intervals of concentration
11 Collect a sample from a clear zone (with a sterilised inoculating loop) and transfer it to a fresh agar plate. If any bacteria grow, they were inhibited from growing by the antibiotic; if none grow, the bacteria must have been killed by the antibiotic.

Lesson 1.17 Maths skills: Size and number

1 Place a transparent ruler on the stage of the microscope on lower power.
2 It is quicker.
Giving a quick indication of the size of cells could be important in picking up abnormal cells, or if examining cell types that we haven't examined before.
3 182 μm.
Calculation:
1 division ≡ 2.8 μm
∴ 65 divisions
≡ (65 × 2.8) μm = 182 μm
4 12.5 divisions.
Calculation:
2.8 μm ≡ 1 division
$\therefore$ 35 μm $\equiv \frac{1}{2.8} \times 35 = 12.5$
5 megakaryocyte (1.5×10^{-4} m)
Macrophage (5.0×10^{-5} m)
neutrophil (1.2×10^{-5} m)
lymphocyte (7.5×10^{-6} m)
6 20
Calculation:
$$\frac{1.5 \times 10^{-4}}{7.5 \times 10^{-6}} = 0.2 \times 10^2 = 2.0 \times 10^1 = 20$$
Note that this is not an order of magnitude calculation; the question says simply, 'How many times larger is a megakaryocyte than a lymphocyte'

End of chapter questions

1 a A: i (cell membrane) B: iii (cytoplasm) (1)
b iii (DNA) (1)
2 Two correct answers from:
Cell wall (1)
Chloroplasts (1)
Large vacuole (1)
[Your answer will be rejected if you have said that it's a 'permanent' vacuole. We cannot tell this from the diagram.]
3 (from smallest to largest:)
sperm cell root hair cell hen's egg nerve cell (1)
Note that all the answers have to be in the correct order for the mark.
4 8
5 a Two correct answers from:
Wear goggles/eye protection (1)
Wear gloves (1)
Use in well-ventilated area (1)
[Reject answers related to using a respirator]
b Chromosomes (1)
Are being pulled/moving to opposite ends of the cell (1)
[Reject answers related to there being two cells already]
6 Mitochondrion – respiration
Ribosome – protein synthesis
Note that both need to be correct for (1) mark.
7 glucose + oxygen → carbon dioxide + water (+ energy)
Note that both need to be correct for (1) mark.
8 Two correct answers from:
Uses tail to swim (1)
Large number of mitochondria for energy (1)
Acrosome/enzymes to dissolve egg membrane (1)
Nucleus carries genetic information (1)
9 a Four correct answers from:
For both investigations (1)
Control temperature (1)
Same (starting) concentrations of respective sugars (1)
Other reasonable factor (1)
Use of controls/distilled water instead of sugars and yeast (1)
b Two correct answers from:
Yeast respires using sucrose (1)
Yeast unable to respire using lactose (1)
Answers relating to trends in sucrose data:
Little use of sucrose at the beginning (1)
Use of sucrose/respiration levels off at the end/after 100 min. (1)
10 Benign and malignant/cancerous (1) *[Accept cancer]*
Note that both need to be correct for one mark.
11 <u>Therapeutic</u> cloning (1)
[Reject answers related to cloning alone]
12 One mark related to growth/development/replacement of cells:
Growth (1)
Development (1)
Replace worn out/damaged cells (1) [accept repair]
One mark for:
(ensures daughter) cells are genetically identical (1)
Maximum (2)
13 Keeps culture (under investigation) pure/uncontaminated (1)
Prevents release/escape (into environment) (1)
14 Correct axes (1)
All five points plotted correctly (2); 1-3 points plotted correctly (1)
Points connected using appropriate line of best fit (a smooth curve) (1)
15 Two correct answers from:
Aerobic uses oxygen, anaerobic does not (1)
Products are ethanol and carbon dioxide/lactic acid (in anaerobic) (1)
Instead of carbon dioxide and water (in aerobic) (1)
Less energy produced (in anaerobic respiration) (1)
16 The answer should cover:

- a clear description of the applications
- an appreciation of the benefits and technical difficulties of the procedures involved
- ideas about reproducibility of results across scientific researchers in finds of other types of stem cells.

Four correct scientific points from:
<u>Adult</u> stem cells in cord blood;
Can differentiate into cells of the blood/vascular system;
(May be) useful in later life in cases of leukaemia/cases where bone marrow has been destroyed so the patient is no longer able to produce own blood cells;
No issues with rejection as stem cells are from the patient;
Storage (for many years) may be costly;
(Possible) technical difficulties with preserving stem cells that will remain unchanged for years/don't become cancerous;
Some scientists make claim for the presence of embryonic/stem cells able to differentiate into but results cannot be repeated;
Further repeatable/reproducible peer-reviewed studies necessary.
17 Two marks relating to ideas about timing of the cell cycle:
Cell cycle in normal cells is shorter for normal cells/70% of time (1)
Length of time in mitosis is 78 min. in normal cells and 56 min. in cancerous cells/length of time not dividing is 522 min. in normal cells and 364 min in cancerous cells (1)
(So) <u>proportion</u> of time spent not dividing and dividing is the same in normal and cancerous cells (1)
Mitosis is shorter in cancerous cells as cell division is uncontrolled (1)
Two marks for ideas relating to evaluation of data:
Cells observed and counted must be representative of sample (1) Data limited from just two patients (1) Scientist needs to check whether data are reproducible or words to that effect (1).

Chapter 2: Photosynthesis

Lesson 2.1 Explaining photosynthesis

1 Raw materials: carbon dioxide and water
Products: glucose and oxygen

2 a The leaf did not contain starch
b the plant was kept in the dark

3 The control was unchanged. It is used as a comparison with Tubes 1 and 2.
Tube 1: The indicator changed from orange to yellow as conditions became more acidic
Tube 2: The indicator changed from orange to red as the conditions became less acidic

4 Tube 1: more carbon dioxide was in the tube because the leaf was respiring and producing carbon dioxide. The foil blocked light from reaching the leaf and prevented it from photosynthesising so no carbon dioxide was being used. This caused a net increase in carbon dioxide in the tube which dissolved in the solution, turning the indicator yellow.
Tube 2: the leaf was still respiring, but because light was present it was using up the carbon dioxide produced during photosynthesis. Oxygen gas was released as a waste product, making the solution less acidic so the indicator became red.

5 Light provides the energy needed for the reaction
Chlorophyll absorbs the light energy for the reaction

6 It would change from yellow to red as the light would allow photosynthesis to take place which uses carbon dioxide. The reduction of carbon dioxide dissolved in the water would make it less acidic therefore turning red.

Lesson 2.2 Looking at photosynthesis

1 The increase in the mass of the willow tree was not just because of minerals in the soil

2 Take two similar plants, put one in a dark place and one in a bright light place. Leave for 24 hours and then test a leaf from both plants for starch.

3 Plants respire all of the time to produce the energy they need for the chemical reactions in cells

4 1. For respiration
2. To produce insoluble starch for storage
3. To produce cellulose for cell walls
4. combine glucose with nitrate ions and other minerals to produce amino acids for protein synthesis
5. To produce fats and oils for storage

5 Plants use carbon dioxide to photosynthesise and produce oxygen. They use some of the oxygen for respiration and produce carbon dioxide (for photosynthesis)

6 Plants do release oxygen during photosynthesis but it is a waste product of the reaction and plants do not photosynthesise so humans can breathe. They photosynthesise to make glucose a source of energy.

Lesson 2.3 Investigating leaves

1 Broad: large surface area to absorb light
Thin: carbon dioxide only needs to diffuse a short distance

2 Support the leaf and transport water to the leaf and glucose away from it

3 Palisade mesophyll cells photosynthesise and spongey mesophyll cells allow the diffusion of gases

4 Roots are underground and receive no light energy so are unable to photosynthesise and need no chloroplasts. However, leaf cells can receive light energy and photosynthesise so they have chloroplasts

5 To reduce water loss by evaporation as the underside of the leaf is more protected from direct sunlight and wind.

6 The cells are arranged end on so that more cells can fit along the upper surface of the leaf and the chloroplasts are more concentrated towards the upper surface so that they can absorb the maximum amount of light energy possible

7 The cuticle is thin to allow light to pass to the mesophyll. It is waxy to provide protection from insects and to stop water loss. The wax is transparent to allow light through for photosynthesis.

Lesson 2.4 Required practical: Investigate the effect of light intensity on the rate of photosynthesis using an aquatic organism such as pondweed

1 When the lamp is closest to the pondweed, at a distance of 10cm.

2 Increasing light intensity will speed up the rate of photosynthesis.

3 As the rate of photosynthesis increases, more oxygen will be produced.

4 Temperature of the water; volume of water; size of the pondweed; angle of lamp.

5 As the light intensity increases more bubbles of oxygen will be produced because the rate of photosynthesis increases.

6 The results in test 1 are all higher, than in the the subsequent tests, expect for the first reading at 10cm, which was much lower.

7 In the first test, the apparatus may not have been set up correctly; or consistently; the number of bubbles was mis-counted; the water was warmer; the lamp was closer (distance not accurately measured).

8 If they identified that problem with Test 1 results was experimental error, they should consider not using these results. It would be best to repeat the investigation a 3rd time to make sure.

9 To identify anomalous results and experimental error; to increase reliability.

10 As a line graph, because photosynthesis is a continuous process.

11 Independent variable (lamp distance) on x axis, number of bubbles on y axis. Axes labelled, including units. Sensible scale used on both axes. Points plotted accurately. Points joined together.

Lesson 2.5 Increasing photosynthesis

1 Tropical forest: very warm and plenty of rain
Tundra: extremely cold and ice most of the year
Desert: very hot days (cool nights) and little rain

2 Tropical forest: ideal conditions for maximum rate of photosynthesis

Tundra: low temperature and liquid water availability will result in little photosynthesise most of the time
Desert: very high temperatures and low water availability will result in little photosynthesis most of the time

3 The temperature, carbon dioxide availability and the amount of chlorophyll

4 Early morning/sun rise/6am: temperature and light availability
Noon: carbon dioxide availability
Evening/ sunset/6pm: carbon dioxide availability, (temperature) and light availability
Midnight: temperature and light availability

5 Higher ratio of leaves to roots increases the proportion of the plant that can photosynthesise
Thinner leaves with larger surface area and more chloroplasts to maximise the rate of photosynthesis

6 Advantage of having needles: Less water loss through transpiration as less surface area and fewer stomata than broad, flat leaves. Stomata in pine needles are also very deep.
Disadvantage: Less surface area for absorbing light compared to a broad flat leaf.

Lesson 2.6 Increasing food production

1 The sun rises earlier and sets later; temperatures are warmer.

2 It is much warmer in the greenhouse than outside; plants are protected from frost/snow; plants germinate faster.

3 To maintain carbon dioxide levels and protect plants from excessive heat

4 To increase levels of carbon dioxide and maintain an optimum temperature

5 Conditions in the greenhouses can be controlled to optimise limiting factors and maximise the rate of photosynthesis

6 Benefits: can maximise photosynthesise throughout the year to produce large yields quickly.
Drawbacks: cost of heaters, extra lighting etc. and computer technology needs to be balanced against increased crop yield and income to ensure it is economically viable.

7 Conditions in Spain and Abu Dhabi are not ideal for growing crop plants. There is too little water, a lack of fertile soil and very high temperatures during the day. Abu Dhabi can also have very cold nights. Using large greenhouses means that conditions can be controlled and large crops produced. Even though this is at a comparatively high cost, these crops can still be sold for profit as they are cheaper than the cost of imported fresh produce.

8 It is true that light is a limiting factor for photosynthesis but in the height of summer too much light raises the temperature in greenhouses beyond what plants can tolerate so blinds are needed to provide some shade when temperatures rise too much.

Lesson 2.7 Key Concept: Diffusion in living systems

1 It will increase the speed of diffusion

2 Temperature; surface area of the cell

3 As cells photosynthesise the concentration of oxygen in cells increases so it diffuses across the cell membrane into the spongey mesophyll. The concentration of oxygen in the leaf is then higher than outside the leaf. This results in a concentration gradient, and oxygen diffuses out of the leaf

4
- They have a large surface area with many guard cells and stomata on the underside to allow gases to diffuse in and out of the leaf
- They are very thin so gases only have a short distance to diffuse
- Spongey mesophyll has air spaces to allow diffusion to occur

These adaptations maximise the concentration gradient to facilitate diffusion

5 As the temperature rises the molecules gain energy and move faster, so the rate of diffusion will increase

6 Partially permeable membranes allow smaller molecules to diffuse through the gaps along a concentration gradient, but block larger molecules. Look at Figure 5.21

7 It will appear to stop when an equilibrium is reached. The particles will always be moving randomly but there is no change in concentration

Lesson 2.8 Looking at stomata

1 The movement of water through a plant

2 Water enters the root hairs, passes into the roots and travels up the stem to the leaves, where it escapes through the stomata.

3 High temperatures, increased light intensity, increased wind and lower humidity

4 Stomatal density can depend on the environment where the plant species is found. Plants found in hot dry environments have less stomata to decreases water loss by evaporation. Plant species found in humid/wet environments have more stomata as there is no need to conserve water

5 Leaves have stomata to allow gas exchange during photosynthesis. When stomata are open for gas exchange, water loss occurs too.

Because leaves are thin for rapid diffusion of gases, water only has a short distance to diffuse too.

6 Marram grass leaves are cured and rolled. The stomata are sunk in pits inside the roll making diffusion more difficult. They have interlocking hairs which hold water vapour; this increases humidity which decreases the concentration gradient, slowing down diffusion even more. They also have very thick cuticles to prevent water loss

Lesson 2.9 Moving water

1 Water is a reactant in photosynthesis

2 Xylem transports water and dissolved minerals from the roots up through the plant; phloem transports sugars from the leaves around the plant

3 Translocation is the movement of sugar around the plant

4 Roots have a large surface area and have root hair cells to increase the surface area even more for absorption of water. There is only a short distance for the water to diffuse across to reach the xylem

5 Xylem long cells reinforced by lignin to make them strong and they are waterproof to prevent water escaping before it reaches the leaves. The internal cell walls and contents breakdown to leave a hollow tube

6 To prevent wilting, stomata close and photosynthesis stops

7 Waterlogged soil contains little air which means that the root cells do not receive the oxygen they need for respiration. Therefore, they die if waterlogging lasts a long time.

Lesson 2.10 Investigating transpiration

1 Leaves in the dark will not be photosynthesising so the stomata will be closed and no transpiration will occur. These photosynthesising leaves kept in the light can be compared to these leaves

2 Mass before and after the investigation to calculate percentage weight loss
Surface area of each leaf to calculate percentage water loss for a given area

3 The leaves put in the light will begin to show signs of wilting. This is because leaves put in the light will photosynthesis so the stomata will open to allow carbon dioxide in which will also increase transpiration and therefore water loss. Leaves put in the dark will show less signs of wilting because there stomata will close which in turn will reduce transpiration and water loss.

4 Set up two shoots in potometers. Put one in front of a fan (wind) and one in an area with no wind for a set period of time and measure the distance moved by the air bubble

5 As humidity decreases transpiration increases

6 Most stomata are on the underside of the leaves; they have a waxy, cuticle; leaves maybe spiny, needles or rolled; stomata may be in sunken pits

7 Stomata on the underside of leaves are protected from direct sunlight and wind to reduce water loss; the waxy cuticle is waterproof to prevent water loss from the leaf; spiny, needles or rolled leaves all have a reduced surface area to reduce water loss; rolled leaves and stomata in pits trap water vapour, to increase humidity and reduce water loss

8 Some water is used for chemical reactions in the cells

9

a. As temperature decreases transpiration will not change at first, but then it will decrease rapidly until at 0°C it almost zero too.

b. As the wind increases the rate of transpiration increases.

c. As light intensity decreases transpiration will not change at first, but then it will decrease rapidly until at 0°C it almost zero too.

d. As humidity increases transpiration decreases.

10 Plants grown in cooler cloudy and humid conditions will lose less water because the evaporation concentration gradient will be lower. However cool temperatures may limit the rate of photosynthesis. Cloud cover may reduce light intensity which could also limit the rate of photosynthesis.

Lesson 2.11 Moving sugar

1 Carbohydrates and proteins are insoluble bur amino acids and sucrose are soluble

2 The movement of sugar around the plant

3 Diagram might show glucose being:
Stored as starch, converted to cellulose, used to make fruits/seeds, converted into lipids and proteins, used for growth and repair.

4 Phloem cells have end walls with sieve plates, cytoplasm but no nucleus and a companion cell with a nucleus next to them.

5 Phloem cells have end walls with sieve plates but xylem are hollow tubes. Phloem cells contain cytoplasm but xylem cells do not

6 Phloem are living cells with no nucleus; the companion cells have a nucleus to control the phloem cells activities and lots of mitochondria to supply the energy needed for translocation to take place in the phloem cells

7 Sieve plates allow the movement of soluble substances in the phloem

Lesson 2.12 Maths skills: Surface area to volume ratio

1 11 full squares + 3 halves = 12.5cm^2

2 a 30 x 3 = 90cm^2

b 50 x 8 = 400mm^2

3 a 10 x 15 x 3 = 450m^3

b 2 x 3 x 4 = 24cm^3

4 The branch is immersed in the tank, all of the spilled water is caught and the volume of this is measured. This is the same as the volume of the branch. Or, the volume of the water tank is known, the water tank is bigger and not full, the rise in the level of the water is recorded, when the branch is immersed. The original volume of the water is subtracted from this new volume and the difference is the volume of the branch.

5 Surface area = surface of side x number of sides (cm^2)
Volume = height x width x breadth (cm^3)
Surface area: volume = SA/V
Correct units – SA cm^2; V cm^3; SA: V none, answers given in that order below:
1cm cube = 6; 1; 6/1
2cm cube = 24; 8; 24/8 = 3
3cm cube = 54; 27; 54/27 = 2
4 cm cube = 96; 64; 96/64 = 1.5
6 cm cube = 216; 216; 216/216 = 1
8 cm cube = 384; 512; 384/512 = 0.75

6 A - 40/16 = 2.5
B - 64/32 = 2
C – 136/64 = 2.125

a B, because it has the smallest SA:V ratio (ie small surface area compared with volume)

b A, because it has the largest SA:V ratio – ie larger surface area compared with small volume

7 Plastic bag, fill with water – measure capacity of bag in cm^3. Empty and completely dry the bag, breath out normally into the bag. Use a range of bags of different sizes.
Or use an empty plastic 'pop' bottle, mark volumes up the side (by either adding specific, known amounts of water, or dividing the length of the bottle into half and half again etc, if the total volume is known). Put some water in a bowl, fill

the bottle with water and invert it in the bowl – making sure none of the water from the bottle escapes. Take a piece of plastic tube (or a flexible straw). Put one end up into the bottle and breathe out down through the straw/tube. Measure/read off how much air you have put into the bottle from your lungs.

End of chapter questions

1

- **a.** To produce glucose to provide the plant with energy (1)
- **b.** Chloroplasts (1)
- **c.** Iodine (1)
- **d.** 0.06 (1)

2

- **a.** any three methods (3) with the correct explanation (3) (total 6)
  - can increase carbon dioxide levels (1) to increase photosynthesis (1)
  - can increase the temperature (1) to increase photosynthesis (1)
  - can increase availability of light using a lighting system (1) to increase photosynthesis (1)
  - install a watering system (1) to supply sufficient water to maximise photosynthesis (1)
  - supply nitrates (1) for protein synthesis (1)
  - supply magnesium (1) for chlorophyll (1)

3

- **a.** 7°C (1)
- **b.** B (1)

4 Any three from

- For respiration
- To produce insoluble starch for storage
- To produce cellulose for cell walls
- combine glucose with nitrate ions and other minerals to produce amino acids for protein synthesis
- To produce fats and oils for storage (3)

5 Any from

- cost of heating, lighting etc.
- need to add fertilisers
- need to supply water regularly
- pest infestations can spread quickly (1)

6

- **a.** $14cm^3$ (1)
- **b.** $25.8cm^3$ (1)
- **c.** The leaves in the shaded area had a larger mean surface area because they need to be able to absorb as much sunlight (1) available to them for photosynthesis (1)

7

- **a.** between 5-6pm (1)
- **b.** As the transpiration rate increases and more water is lost from the leaves, the rate at which the roots take water in starts to increase to replace the water lost

  or

  as the transpiration rate decreases and less water is lost from the leaves, the rate at which the roots take water in starts to decrease to replace the water lost (1)

8

- **a.** Description of method (2 marks)
  - Measure the amount of oxygen release by counting the number of bubbles or measuring volume of gas being released released from the pondweed using a syringe
  - Take measurements at a range of distances from lamp/light source.
  - and using a syringe to capture and measure the oxygen released (2)
- **b.** 2 marks for suitable diagram of equipment (one for correct labelling of equipment and one for correct set up).

9 The leaf is very thin(**1)**
This allows carbon dioxide to diffuse rapidly to the cells inside the leaf as there is only a short distance for the carbon dioxide to diffuse from the air to the photosynthesising cells inside the leaf.(1)
The leaf has lots of tiny openings called stomata (1)
These allow carbon dioxide diffuse from the air into the cells inside the leaf (1)

10 Any 2 from:

- The waxy cuticle helps prevent water loss through transpiration/ evaporation – especially important as pine trees are often found in areas of low rainfall.
- Provides a waterproof layer preventing water entering or leaving through the leaf (also repels water so water tends to drop of the leaves and not weigh them down)
- It provides some protection from insects or any other substances e.g. pollutants/UV light that might otherwise penetrate the leaf by providing a barrier (2)

11 (1)

$$6CO_2 + 6H_2O \longrightarrow C_6H_{12}O_6 + 6O_2$$

12 Temperature, humidity and amount of wind (3)

13

- **a.** At low light intensity/carbon dioxide concentration as the light intensity/carbon dioxide concentration the rate of photosynthesis, increases proportionately (1)
  As light intensity is increased further, however, the rate of photosynthesis is eventually limited by some other factor. (1)
- **b.** At A, the rate of photosynthesis increases proportionately to the increase in the light intensity/carbon dioxoide concentration. This is because the light intensity or carbon dioxide levels are limiting factors. (1)
  At B, the rate plateaus/stops increasing because the rate of photosynthesis is limited by some other factor. (1)
- **c.** The graph would be a similar shape but the rate of reaction would be higher (1) because the temperature would not be a limiting factor (1)

Chapter 3: Moving and changing materials

Lesson 3.1 Explaining water movement

1 Water moves in and out of cells by osmosis: diffusion of water molecules from a dilute solution to a concentrated through a partially permeable membrane

2 Boiling potatoes destroys the cell wall and the cells swell as water enters them quickly by osmosis making them softer

3 They both occur down a concentration gradient.

Dissolved solutes/gases and gases in air diffuse; only water moves by osmosis

Osmosis needs a partially permeable membrane

4 Animal cells in a solution less concentrated than in the cell take in water by osmosis, swell and burst

Animal cells in a solution more concentrated than in the cell lose water by osmosis, shrink and shrivel.

5 When water enters plant cells by osmosis, it fills the vacuole which pushes against the inelastic cell wall making the cell turgid but preventing it from bursting.

If water moves out by osmosis the vacuole shrinks and the cell becomes flaccid, but if too much water leaves the cell it becomes plasmolysed, but it is a reversible process

Plants have a cellulose cell wall, when water enters the cell, it fills the vacuole and makes the cell turgid. Animals don't have a cell wall and their vacuoles are much smaller, so when water enters an animal, the cells swells and may burst (if too much water enters).

6 Osmosis is the movement of water from a region of relatively high concentration (of water) to a region of lower water concentration across a partially permeable membrane, thereby attempting to equal out the amount of water on either side of the membrane

Lesson 3.2 Required practical: Investigate the effect of a range of concentrations of salt or sugar solutions on the mass of plant tissue

1 Concentration inside the cell is higher than the water/concentration outside the cell

2 Concentration of salt solution is higher than the concentration inside the cell

3 Water will move out of the cell

4 Water will flood the cell when it is put is water and will leave the cell when it is put in a salt solution due the difference in concentrations inside and outside the cell, causing the water to move by osmosis. The stronger the salt solution, the greater the water loss.

5 Independent variable – salt concentration. Dependent variable – mass of potato cylinders

6 So that the excess water was not weighed

7 b No.

c Depends on possible reason for anomaly, best to repeat investigation. If reasons are found for anomaly, can possibly disregard and explain in write up of investigation

d Because the potato cylinders were not all of the same mass and so it would be difficult to make a meaningful comparison

e Potato cylinders placed in water and weak salt solutions take up water (and increase in mass). The stronger the salt solution the potatoes are placed the more water they lose.

8 The stronger the salt solution, the greater the water loss.

9 Proved (upheld, is more correct)

10 Water is moving by osmosis from a region of relatively high concentration (the cell) to an area of relatively lesser concentration of water (the stronger salt solutions) across the cell membrane (partially permeable) and so the potato in these stronger solution lost mass

Lesson 3.3 Learning about active transport

1 Plants absorb minerals by active transport into the root hair cells

2 Active transport needs energy and it is against the concentration gradient

3 Cells that are adapted for active transport have many mitochondria to supply the energy needed

4 It is important that plants select the minerals they need so that they do not suffer from mineral deficiencies and to use the energy they have efficiently

5 Molecules to be transported are attached to the transport protein. The transport protein rotates in the cell membrane and then releases the molecule inside the cell. The transport protein rotates back, ready to move another molecule.

6 Osmosis and diffusion occur down a concentration gradient and do not energy but active transport occurs against a concentration gradient and needs energy.

Diffusion and active transport move dissolved solutes but osmosis moves water.

Osmosis needs a partially permeable membrane.

Active transport needs a transport protein

Lesson 3.4 Key Concept: Investigating the need for transport systems

1 The larger the cell the more chemical activity takes place

2 Cells take absorb glucose and oxygen for respiration and produce carbon dioxide and water.

Other correct examples include DNA replication, protein synthesis and digestion

3 Surface areas and volume both increase as the size increases, but surface area increases more than volume

4 SA:V decreases as the size increases

5 Smaller SA:V make it more difficult to absorb enough of the substances the cell needs to stay alive and to get rid of the waste products made

6 Efficient exchange surfaces have a large surface area, thin membranes and a transport system

7 In smaller organisms, nutrients, oxygen and waste substances can quickly diffuse in and out.

Transport systems are needed in larger organisms so that cells deep inside the organism:

- get the oxygen and nutrients they need for chemical reactions
- are able to get rid of the waste products of chemical reactions and any excess heat generated

Lesson 3.5 Explaining enzymes

1 An enzyme is a chemical hat speeds up a reaction without being used up itself

2 Enzymes catalyse/speed up the chemical reactions in cells e.g. respiration, protein synthesis and photosynthesis.

Accept:

- break down large molecules into smaller ones
- build large molecules from smaller ones
- change one molecule into another molecule

3 40°C is the optimum temperature for many enzymes to work at so they can breakdown proteins in stains

At 60°C the enzymes are denatured/ the active site changes shape and cannot catalyse reactions

Do not accept: the enzymes are killed

4 Each enzyme has a differently shaped active site that only one particular substrate can fit into, so each enzyme catalyses just one type of reaction.

5 Enzymes hold reactants in place so that there is an increase in collision rate in a given time which increases the rate of reaction. Once the reactants start to run out, the amount of product made starts to level off as collisions start to decrease again.

Lesson 3.6 Required practical: Investigate the effect of pH on the rate of reaction of amylase enzyme

1 a $10cm^3$ measuring cylinder or a calibrated dropping pipette

b Glass rod or dropping pipette

2 Drops should be done at a consistent rate, each drop equates to about 10 seconds of reaction time

3 Care over use of chemicals – iodine (stains irritant); powered amylase is harmful (for making the solution); buffers – maybe an irritant; use of safety goggles; care with heated water

4 Original orange colour of iodine

5 To catch the point at which all the starch has been 'digested'/converted ie the end point of the reaction

6 To ensure that the concentration he was using were sufficient to see the reaction – it wasn't going too fast or too slow

7 Every 10 seconds (5 is Ok, but errors may occur as this is quite quick to do, 20 seconds isn't often enough)

8 Line graph

9 a To make his results more valid

b pH 9, test 2, 90 s, much lower than tests 1 and 3

c Think about why it might be anomalous, if due to his poor/inaccurate technique, he could consider ignoring it

10 pH5 156.7
pH6 73.3
pH7 43.3
pH8 73.3
pH9 120 (or 135 if Test 2 disregarded)

11 Line graph (continuous process), independent variable (pH) on x axis, time on y axis. Axes labelled, including units. Sensible scale used on both axes. Points plotted accurately. Points joined together

12 Rate of reaction is 1/time

13 Reaction is fastest at pH 7 and slows when the pH is either more or less than this. This is the pH found in the body

Lesson 3.7 Learning about the digestive system

1 Digestion changes insoluble nutrients into soluble products that can be absorbed and transported to all cells

2 Absorption is when the soluble products of digestion pass through the intestine wall

3 Emulsification is when fats are broken up into smaller pieces of fat resulting in a larger surface area for digestion to occur

Digestion is when the fats are broken down into fatty acids and glycerol

4 They are still too large to pass through the capillaries

5 Digestion breaks down large insoluble molecules to small soluble molecules. These can then be absorbed into the bloodstream and transported to every cell in the body where they are used for cellular metabolism

6 The small intestine is an effective exchange surface because it:

- is very long to allow time for absorption
- the villi and microvilli increase the surface area for absorption
- it has a thin, permeable membrane for easy diffusion
- it has many blood capillaries to transport products of digestion and maintain the concentration gradient

Lesson 3.8 Explaining digestion

1 Physical digestion breaks the food down into smaller pieces to allow it to be swallowed and increases the surface area to speed up chemical digestion

2 Chemical digestion uses enzymes and breaks the food down into different products

3 Proteases are produced in the stomach, pancreas, small intestine. They catalyse the breakdown of proteins into amino acids

4 If digestive enzymes worked inside cells they would break down the cell which is made of protein etc.

5 The optimum pH of pepsin is about pH 2.5-3.0/acidic but the optimum pH of trypsin is about pH7-8/which is very slightly alkaline

The range that pepsin works at is pH0-about 5.5

The range that trypsin works at is pH4-about 11.5

6 as the temperature increases, the rate of enzyme action increases because enzymes have more (kinetic) energy/collide more (frequently)/reaches a

maximum activity rate at about 38°C (its optimum temperature)

decreases rapidly because enzymes denature above 40°C

Lesson 3.9 Required practical

1 a Pestle and mortar

b Dropping pipette or glass rod

c Measuring cylinder or a calibrated dropping pipette

2 To ensure maximum surface area is exposed to the test solutions/ molecules released from cells

3 Transfer the filtrate of ground food and distilled water to test tube. Add the same amount of sodium hydroxide (NaOH) to the filtrate, mix carefully. Add a few drops of 1% copper sulfate, CuSO4, do not shake the mixture. The solution turns purple if protein is present (from blue).

4 Care with grinding food; safety googles for Biuret's NaOH and $CuSO_4$; heat with a water bath (not a Bunsen); do not eat food in laboratory; check for allergies

5 To make her results more valid

6 To check that the procedure does not have an effect/ to make a fair test

7 Table with RHC headed: food type, rows to be labelled with the 3 food types (A< B & C) and the control and each one tested twiceA1, A2, B1, B2, C1, C2, control 1, control 2. Text in book says to include units – but we are recording colour changes, so no units required. Other column headings: starch, protein and fat. Ravi can tick the appropriate box if the food group in question is detected, or he can write the final colour
Allow table to be other way round

8 a Food A had no protein – solution remained blue (purple indicates presence of protein)

b Foods A and C had Glucose – solution turned red

c Food A had lipids – red layer at top of tube

Lesson 3.10 Looking at more exchange surfaces

1 Fish use gills, insects use trachea and amphibians use skin and lungs for gas exchange

2 Tadpoles have external gills and use their skin for gas exchange but fish only use internal gills

3 Skin: has thin membrane for diffusion, is moist for dissolving gases and organisms usually a large SA:V

Trachea: have spiracles that can close to prevent evaporation to keep exchange surfaces moist, tracheae have many branches to increase surface area, insects pump air in and out of tracheae,

tracheae are stiffened to prevent collapse, there is only a short diffusion distance

Gills: have feathery projections to increase the surface area, very thin walls for diffusion and water is pumped over the gills

Lungs: the alveoli have thin membranes, millions of alveoli to provide large surface area, moist for dissolving gases, constant ventilation to maintain the concentration gradient

4 When they are being more active and need more oxygen more quickly or if their skin becomes dry and the oxygen cannot dissolve to be absorbed through the skin.

5 When you breathe out, the concentration of oxygen left in the lungs and the air pressure is low compared to outside, so oxygen (in the air) is drawn into the lungs

Lesson 3.11 Learning about plants and minerals

1 Root hairs help to anchor the plant and absorb water and minerals from the soil

2 Water is absorbed through the root hair cells carrying mineral ions with it

3 Root hair cell adaptations for osmosis are:

- they have a large surface area for absorption of water
- have no cuticle, just a thin membrane to allow absorption
- have a thin wall to reduce the distance for osmosis
- have a large permanent vacuole to absorb as much water as possible
- are close to the xylem so that materials can be moved around the plant

4 Plants need minerals to make proteins and other chemicals needed for the plant to grow

5 Enzymes catalyse the chemical reactions in plants cells e.g. respiration to supply the energy needed for active transport

6 Fertiliser is dissolved in the soil water and absorbed into the root hair cells by active uptake. The mineral ions then move by transpiration through the xylem up through the plant. This happens more quickly on windy days because the water evaporates more quickly off the leaves increasing the rate of transpiration.

Lesson 3.12 Investigating how plants use materials

1 Fertilisers replace the essential mineral nutrients that plans have used up from the soil

2 Gardeners add fertiliser to water before watering their plants so that the fertiliser dissolves and the mineral ions can be quickly absorbed

3 Plants with different mineral deficiencies show specific symptoms. Accept a named example, e.g. plants with nitrate deficiencies have yellow leaves and poor growth

4 Magnesium deficiency symptoms are yellow leaves; plants with phosphate deficiency have discoloured leaves and poor buds and flowers

5 Organic and inorganic fertilisers both contain essential plant nutrients

Organic fertilisers come from animal or plant matter; take time to break down in the soil and slowly release the minerals; can reduce soil erosion and improve water retention in the soil.

Inorganic fertilisers are man-made; come from concentrated sources of mineral;, are used in small amounts and can be absorbed immediately; do not smell, are easy to apply and store.

6 Advantages: Increase in crop yields; crops receive all minerals they need so more likely to be healthy; farmer makes more money due to higher yield.

Disadvantages:

High cost of buying a applying fertiliser ;may pollute nearby water and cause eutrophication

Lesson 3.13 Learning about the circulatory system

1 A double circulation is where the blood goes from the heart to the lungs, back to the heart, to the body and back to the heart again.

2 Closed circulation systems need a heart to pump the blood under high pressure around the system

3 Blood from the heart travels under high pressure in arteries with thick walls to the body. At body tissues the blood travels in capillaries. The pressure gradually falls as blood flows through the capillaries to the veins. Veins carry blood under low pressure back to the heart.

4 Arteries: thick elastic walls to withstand pressure; small lumen increases blood flow speed

Veins: have valves along their length to prevent the backflow of blood; large lumen gives least flow resistance

Capillaries: capillary networks have a large surface area for diffusion; walls are one cell thick and permeable to allow substances to diffuse in and out easily

5 Double circulation is an advantage to active animals because:

- blood pressure is higher
- there is a higher blood flow to body tissues
- oxygenated blood is separate from deoxygenated blood.

Lesson 3.14 Exploring the heart

1 The function of the heart is to pump blood around the body

2 Blood from the lungs enters the heart at the left atrium. It passes into the left ventricle and is pumped out to the body.

Blood from the body enters the heart at the right atrium passes into the left ventricle and is pumped to the lungs

3 Atria pump blood to the ventricles in the heart

Ventricles pump blood to the rest of the body

4 Blood enters the heart at the atria. The ventricles relax whilst the atria contract and push blood into the ventricles. The valve between the atria and ventricles closes and the atria relax whilst the ventricles contract to push blood out of the heart towards the body.

5 The left atrium receives oxygenated blood from the lungs and contract to push it through the open valve into the left ventricle. At the same time, the right atrium receives deoxygenated blood from the rest of the body and contract to push it through the open valve into the right ventricle. The valves between the atria and ventricles then close.

The left ventricle contracts to send oxygenated blood through the open valve to the rest of the body. At the same time, the right ventricle contracts to send the deoxygenated blood through the open valve to the lungs. These valves then close

6 A blocked coronary artery would prevent oxygen and glucose reaching some of the heart so the muscle cells would not be able to respire and produce energy to contact. If the muscle stopped contracting then the heart might not be able to pump the blood effectively around the body.

Lesson 3.15 Studying blood

1 Red blood cells are coloured red

White blood cells are coloured purple

Platelets are coloured green

2 Red blood cells transport oxygen from the lungs to body cells

White blood cells help to protect the body against infection.

3 Plasma transports substances around the body e.g. carbon dioxide

4 The adaptations of red blood cells are:

- they are tiny so they can pass through narrow capillaries
- they have a large surface area to increase diffusion of oxygen into and out of the cell.
- contain **haemoglobin** to bind to oxygen to transport it
- have no nucleus, increasing the space for haemoglobin

5 Haemoglobin binds with oxygen to form bright red oxyhaemoglobin when there is a high concentration of oxygen. The bonds between the haemoglobin and oxygen are weak, and oxyhaemoglobin dissociates to haemoglobin and oxygen in low oxygen concentrations.

oxygen + haemoglobin ⇌ oxyhaemoglobin

6 The biconcave disc shape of red blood cells gives them a large surface area to volume ratio to increase the rate of diffusion of oxygen into and out of the cell

7 Haemoglobin binds to oxygen and transports it around the body. A person with sickle cell anaemia has abnormal haemoglobin therefore they can't transport oxygen around the body with the same efficiency as normal red blood cells. As a result they tend to feel tired and easily get breathless when exercising.

Lesson 3.16 Investigating gas exchange

1 Air is filtered, warmed and moistened in the mouth and nasal passages, passes into the trachea, through one of the bronchi into the lungs. In the lungs, air passes the bronchioles to the alveoli.

2 Gas exchange is the absorption of oxygen and the release of carbon dioxide across a gas exchange/specialised surface e.g. the alveoli.

3 Air entering the alveoli has a greater oxygen concentration than the deoxygenated blood flowing through the lungs. This steep concentration gradient allows the oxygen to diffuse into the blood. Deoxygenated blood contains a greater concentration of carbon dioxide, so it diffuses out of the blood and into the alveoli to be breathed out.

4 Breathing is taking oxygen into the lungs and expelling carbon dioxide out of the lungs

Respiration the chemical reaction between glucose and oxygen to release energy. Carbon dioxide and water are waste products.

5 Alveoli are efficient exchange surfaces because they:

- are spherical with a large surface area to volume ratio, resulting in efficient diffusion of gases
- have very thin walls so the diffusion distance is very short.
- are surrounded by blood capillaries to ensure a good blood supply. Oxygen is constantly moved into the blood and carbon dioxide is constantly taken to the lungs to be removed. This means that gas exchange happens at the steepest concentration gradients possible.
- have moist surfaces so gases dissolve to allow efficient diffusion

6 Blood is pumped with continuous fresh supplies of oxygen from the lungs to all body cells. Capillaries around the alveoli are efficient transfer surfaces, there is a steep concentration gradient and oxygen from the lungs is quickly diffused into the bloodstream and carried to cells deep in the body. Again, capillaries around the cells are efficient transfer surfaces and oxygen quickly and continuously diffuses down the concentration gradient into the body cells and carbon dioxide diffuses into the blood to be carried back to the lungs. At the lungs the capillaries allow the rapid diffusion of carbon dioxide down the concentration gradient into the lungs to be breathed out.

Lesson 3.17 Learning about coronary heart disease

1 The symptoms of coronary heart disease are a build-up of fatty material inside the coronary artery, reducing blood flow so less glucose and oxygen reach the heart cells for respiration. Less energy is available for the heart to contract. If cells are starved of nutrients, they can die and a heart attack may happen.

2 It is caused by gender, age, diet and genetic factors

3 It is treated by the insertion of stents to open up blocked arteries or bypass surgery

4 Artificial valves are mechanical valves that are used to replace damaged valves and they control the blood flow through the heart

5 Artificial valves: no rejection problems and no blood supply needed but can damage red blood cells and need to take anti-clotting drugs. Good option especially if there is no damage to RBCs

Artificial pacemakers: no surgery is needed but the immune system can reject them and they may need replacing. Good option as, even though they may need replacing, it is not major surgery like the heart transplant

Heart transplant: does give a better quality of life but it is major surgery, anti-rejection drugs need to be taken (greater infection risk) and there is a shortage of donors. Probably a final option for most people as major surgery carries great risks and finding a suitable donor s hard due to the shortage of suitable organs

6 Statins are being increasingly prescribed to reduce levels of cholesterol produced by the liver to prevent coronary arteries from becoming blocked

Lesson 3.18 Maths skills: Extracting and interpreting information

1 a 6 units

b It remained constant (approximate)

2 a 9 units, the rate no longer increase as light intensity increased

b The rate increased to a maximum light intensity (9 units) but did not increase any further with increasing light intensity

3 A good, positive correlation – as BMI increases so does total cholesterol

4 As carbon dioxide concentration increased, the rate of photosynthesis increased to a maximum (13-15 % concentration) but did not increase any further with increasing carbon dioxide concentration

End of chapter questions

1 By diffusion through the villi in the small intestine (1)

2 To pump the blood around the body (1)

3 Respiration is the chemical reaction that takes place in cells to produce energy from glucose and oxygen and breathing is the movement of air in and out of the lungs. (1)

4 In 2015, 175000 people from a population of 64,000 000 suffered from a heart attack. Calculate the percentage of the population that suffered a heart attack. Give your answer to 2 decimal places (1)

175000/64000000 x 100 =0.27%

5 30x16.1=483 KJ (1)

6 The chest muscles are needed to move the ribs which changes the pressure of air in the lungs to move air in and out (1). If the chest is paralysed, we wouldn't be able to breath and the body would become starved of oxygen. (1)

7 a pH 6 (1)

b pH 8 (1)

8 Emulsifies fats (1)

9 The shape of the active site is changed (it can no longer bind with the molecules to catalyse the reaction) (1)

10 Any two from:

- It is very long,
- very thin, permeable membrane
- many small villi, with microvilli,
- many blood capillaries for transport (2)

11 Water molecules will continually move by osmosis (1) from the dilute solution outside the amoeba (1) to the more concentrated solution inside (1) through the partially permeable membrane (1)

12 a 39.1 (test 1 at 50°C) (1)

b As the temperature increases so does the respiration in yeast. This happens up until 40°C at which point it begins to decrease. (1)

13 Between meals the concentration of dissolved food molecules in the blood can be higher than in the intestine and diffusion will not work so active transport is used. (1)

14 Make sure the blood flows in just one direction(1)

15 Water enters cells by osmosis (1) fills the vacuole and pushes against the inelastic cell wall making it turgid (1)

16 Any two from:

- large surface area for absorption of water
- a thin membrane to allow absorption
- a thin wall to reduce the distance for osmosis
- large permanent vacuole to absorb as much water as possible
- close to xylem so materials can be moved around the plant(2)

17 70x74x10 / 1000 = 51.8L (1)

18 Any 3 from:

- they are spherical. And have a large SA:V, causing efficient diffusion of gases
- alveoli wall is very thin so diffusion distance for gases is very short.
- each alveolus has many blood capillaries so oxygen is constantly moved into the blood and carbon dioxide is constantly taken to the lungs to be removed. So gas exchange occurs at the steepest concentration gradients possible.
- alveoli surfaces are moist and gases dissolve for efficient diffusion (3)

19 Break up of fat molecules into smaller pieces to increase the surface area o increase the rate of digestion (2)

20 If too much water leaves a plant cell the cytoplasm moves away from the cell wall (1)

21 The smaller organism (1) because it has a larger SA:V (1)

22 Rate of exchange of particles with the surrounding environment may be too low for diffusion to occur efficiently (1) so need exchange surfaces/ transport system (1)

23 From the top:

In pure water and very weak sugar solution the core increased in length (1) because water moved by osmosis / from the dilute solution outside the cell to the more concentrated solution inside making them turgid. (1)

In chip 4 there is no (net) osmosis / same amount of water in and out and the chip stays the same (1)

Chips 5 and 6 water moves out the chip from the dilute solution in the cell to the more concentrated solution outside /by osmosis so the potato cells become flaccid and shrink (1)

Chapter 4: Health matters

Lesson 4.1 Learning about health

1 Health is the physical and mental well-being of an individual

2

- disease
- diet
- stress
- life situations.

3 Communicable: e.g. measles; HIV etc.

Non-communicable: e.g. cancer; cardiovascular disease etc.

4 Moving home, family death, divorce etc., cause stress which causes depression, high blood pressure, risk of obesity, heart disease etc.

5 For example:

Immune system diseases, such as AIDS, can cause an increased risk of contracting infectious diseases.

Immune reactions initially caused by a pathogen can cause allergies, such as skin rashes. Severe viral respiratory infections in early childhood can trigger asthma as children grow.

Severe physical ill health, for example obesity or cancer, can

cause depression and other mental illnesses.

6 Any correct analysis, e.g.

- Most common cancer in men is oropharynx and in females is in the cervix
- Most cases of cervical cancer can be attributed to HPV (approximately 10200 out of 11300 etc.

7 For example:

- Patient may get depression which affects family
- Loss of earnings if it is an adult
- Loss of quality time with other family members
- Travelling expenses and time for treatment etc.

Lesson 4.2 Key concept: Looking at risk factors

1 Genetics, diet/obesity and exercise

2 Not taking drugs; stop drinking alcohol; eating healthily; losing weight.

3 Men: number of cigarettes smoked per person increased from 1900 until approx. 1970 and then plateaued. Lung cancer incidences increased from 1920 and then plateaued from 1980

Women: started smoking after men; number of cigarettes smoked per person slowly increased from mid-1920's – late-1930's but has increased rapidly since; lung cancer incidences slowly increased from early 1930's – early 1940's and then the rate has continually increased and still shows an upward trend in 1990

4 Drinking alcohol, smoking cigarettes and taking drugs, poor diet and lack of exercise, being over or under weight contributes to certain diseases e.g. type-2 diabetes, cardiovascular disease, cancer etc. and can affect foetal development

5

- smoking can cause serious health problems e.g. cardiovascular disease, lung disease, cancer, low birthweight of babies and premature birth; possible death, dangers of passive smoking to non-smokers
- cost of cigarettes from monthly income, loss of family income due to illness or death and a depleted workforce affecting local, national and world economies
- financial burden of healthcare costs on local, national and global economies.

Lesson 4.3 Exploring non-communicable diseases

1 Uncontrolled cell division

2 a UV light, age

b smoking

3 Benign tumours divide slowly, do not spread and are harmless.

Malignant tumours divide quickly, spread and cause cancer

4

- not smoking
- staying out of the sun and using sun screen
- drinking alcohol in moderation
- healthy diet and exercise.

5 Cigarette consumption showed an increasing trend from 1900 to early 1960's and as slowly decreased until 2000.

Cancer deaths in men was always higher than women until 2000

Cancer deaths in men rose rapidly from 1930 – 1990 (peaked at 90) and then started to decline (to mid 70's).

Cancer deaths in women rose very slightly from 1930 to mid 1960's (<500) and then then rapidly (but not as much as I men) to late 1990's and then started to slow down slightly (just over 40).

6 Cancer debilitates individuals leading to a poor quality of life and even death.

Families lose quality time due to the patient being incapacitated and time for treatment etc., have increased expenses caused by hospital visits (and medical bills in some countries) etc., may lose family income affecting living conditions, diet etc.

Nations lose financial contributions to the economy and have the financial burden of medical costs

7 Financial cost to individuals and families and nation may include cost of treatment and loss of earnings due to being unable to work (patient and family members having to take time off work). Cost to nation also in terms of investment in research and development of cancer treatments and financing cancer care.

Lesson 4.4 Analysing and evaluating data

1 The more cigarettes smoked the lower the average birth weight.

2 Vaccination uptake was at around 93% from 2000-2004 then dropped to 83% in 2005 and staying at around this level since

Whooping cough increased slowly from 2000 – 2004 then there was a rapid rise in 2005 before falling back down to previous levels from 2007-2009

3 Smoking habits of men increased from about 750 cigarettes smoked per capita per year in 1900 to about 5000 in 1990. The incidence of lung cancer also increased in this time from below 5 per 100000 per year in 1920 to approximately 93 per 100000 in 1990.

4 The data strongly suggests that there could be a link between smoking and lung cancer as the incidence of lung cancer increased in line with the number of cigarettes smoked.

5 Increases steadily from 150 in 1940 to 600 in late 1960's then falls steadily to 150 around 2005. The female trend follows the same general pattern but is always lower than the male trend and peaks at about 300

6 Large sample taken over many years so data should be reliable if there is no bias but there is no information about who completed the research, any personal interests etc.

The trends follow the same pattern

Lesson 4.5 Studying pathogens

1 Any correct examples e.g. rose black spot fungus, salmonella, malaria protist, measles virus

2 Direct contact, by water or through the air

3 They release toxins

Effects can be reduced by taking painkillers, having vaccinations etc.

4 Spread can be prevented by:

- simple hygiene: covering mouth when coughing, using a handkerchief when sneezing and washing hands after using the toilet
- isolation of infected individuals
- destroying vectors, e.g. killing mosquitoes with insecticides
- vaccination.

5

- reducing the risk of contact with infected animals – hard to monitor and control as bats fly around
- wearing protective clothing – effective in hospitals but too expensive to provide for all unaffected individuals due to cost
- washing hands frequently – everyone can do this but must have education programmes to ensure people know the importance of this
- isolation of infected people – can be difficult if people do not report cases
- safe burials of the dead – can enforce for reported cases but may go against local customs
- travel restrictions – difficult to do but is possible especially air travel to and from countries with epidemics.

6 An epidemic is when disease spreads rapidly to many people. If a disease spreads globally, it is a pandemic.

Lesson 4.6 Learning about viral diseases

1 Genetic material replicates inside host cells

2 Measles: fever and a red rash over the skin.

HIV: initially causes a flu-like illness

TMV: leaves have discoloured 'mosaic' patterns

3 HIV is spread by sexual contact or exchange of body fluids such as blood. This can occur if drug users share needles.

Control: use condoms, not sharing needles, taking antiretroviral drugs

4 TMV: virus gains access through a break in the skin or plant

Epidermis. Controlled by removal of infected plant material,controlling pests, sterilising tools (using heat), washing hands after handling infected plants

5 HIV virus enters the lymph nodes and attacks the body's immune cells. AIDS, occurs when the body's immune system is no longer able to deal with other infections or cancers

6 Bacteria are living cells that survive outside body cells. They can be treated with prescribed drugs. Viruses are only found inside host cells, where they are protected from drugs

7 Measles spreads through droplet infection (talking, coughing, sneezing so tiny droplets of the virus are inhaled by others) whereas HIV is only spread by direct contact (sexual contact or exchange of body fluids such as blood). Measles is considered more infectious.

Lesson 4.7 Studying bacterial diseases

1 Salmonella and Gonorrhoea

2 Salmonella: fever; abdominal cramps; vomiting; diarrhoea

Gonorrhoea: a thick yellow or green discharge from the vagina or penis; pain when urinating

3

- ingesting food contaminated with Salmonella bacteria
- preparing food in unhygienic conditions.

4

- treatment with antibiotics e.g. penicillin, but in recent years many resistant strains have appeared
- use of a barrier method of contraception, e.g. a condom, to prevent contact.

5 There are more cases in:

- summer months – due to BBQs and food left out in warm rooms
- November-December: Christmas parties/dinners with undercooked food, food left in warm rooms etc

6 If we know how diseases are spread, measures can be taken to prevent these situations e.g. if we know salmonella is passed by:

- poor personal hygiene, people can be informed, facilities provided etc.
- undercooked food – food can be prepared according to instructions and cooked for required time
- poor storage of food – food can be stored in the correct conditions etc.

7 Bacterial diseases caused by living bacterial cells that release toxins and are easy to control by

using antibiotics. Some bacteria are useful. Bacteria are larger than viruses

Viral diseases caused by virus particles living inside host cell and destroying them; they cannot be affected by drugs so you need to be vaccinated

Lesson 4.8 Looking at fungal diseases

1 Causes black or purple spots on upper surface of leaves; leaves can be yellow around the spots; leaves can drop early; can form black scab-like spots on stems

2 Warm, wet conditions

3 Produces spores released in wet conditions and spread by wind. Spores overwinter on shoots and survive in soil. Spread by infected leaves, which should not be composted

4 Removing infected leaves – removes spores

Pruning and burning infected stems – removes spores

Using fungicides – kills fungus

Putting manure/mulch around plants stops spores reaching stems

Not composting infected leaves – stops spores spreading

5 Blackspot fungus grows in the mesophyll. This causes irreversible structural changes in affected cells and cells cannot photosynthesise to make food needed for growth.

Lesson 4.9 Learning about malaria

1 A protist called Plasmodium

2 recurrent episodes of fever; sweats and chills; muscle pains; headaches; diarrhoea; cough.

3 An organism that spreads disease without causing it

4 Malaria is spread by female mosquitoes, which feed on blood.

They suck infected blood from someone with malaria and then pass the protists on when they feed on a new person.

5 Mosquitoes breed and lay eggs in still water. The eggs hatch and each larva develops into a pupa in the water. When an adult mosquito hatches, it rests on the water surface to let its body dry and harden so it can fly.

6

- spraying pools of water with insecticide to kill mosquito eggs/larvae.
- draining stagnant water pools – means there are no places for mosquitoes to lay eggs
- spraying pools with oil to prevent the larvae from breathing – kills larvae

All these measures mean that once mature adults die the mosquito population will die out with no new adults to replace them and continue the cycle.

- using mosquito nets to avoid bites – prevents infection of protected individuals only.
- taking drugs to kill the protist in the blood – effective in the short term but can be re-infected if no other measures taken.

7 Because we know that mosquitoes lay eggs on or near water these areas can be targeted by spraying them with insecticide to kill the eggs/larvae

Lesson 4.10 Protecting the body

1 Skin is a barrier and produces antimicrobial secretions.

The nose traps particles and pathogens.

Your trachea and bronchi secrete mucus to trap pathogens.

The stomach produces acid, which kills pathogens

Platelets help to form clots which dry to form scabs to seal wounds

2 Stomach acid (and possibly skin)

3 When you cough or sneeze, thousands of tiny drops of liquid containing pathogens are sprayed into the air.

4 Platelets in blood are exposed to the air at the wound site. They make protein fibres that form a mesh and platelets and red blood cells get caught in the fibres to form a clot.

Scabs protect the body by forming a barrier and preventing pathogens entering the body through the site of the injury whilst the skin is healing.

5 Pathogens cannot easily penetrate the outer layer of dead cells.

Sebaceous glands produce antimicrobial oils.

6 Hairs in the nose trap larger microbes and dust particles. The trachea and bronchi have a ciliated epithelium.

Goblet cells in the epithelium produce mucus; mucus traps smaller dust particles and microbes. Cilia beat together to waft mucus to the back of the throat, where it is swallowed.

7 Damage and paralysis of the cilia result in mucus that trap dust and microbes not being moved to the back of the throat so they are more likely to cause infection in the respiratory tract.

Lesson 4.11 Exploring white blood cells

1 They ingest pathogens, make antitoxins and antibodies

2 It is ingested, killed and digested by enzymes

3 Antitoxins combine with toxins to make safe chemicals.

Antitoxins are specific to a particular toxin.

4 Lymphocytes quickly reproduce when they detect a pathogen to make lots of antibodies. If the same type of pathogen enters the body again, lymphocyte cells recognise it and immediately make lots of antibodies. The person is then immune to that disease.

5 Phagocytosis destroys pathogens to stop them producing toxins and antitoxins neutralise any toxins that are produced to stop illness happening.

6 Each lymphocyte has a specific antibody to attack a specific pathogen. Pathogens carry chemicals called antigens on their surface. The appropriate antibodies lock onto the matching antigens, sticking the pathogens together todestroy them.

7 HIV damages white blood cells, meaning that they cannot make antibodies or kill other infected cells. When HIV has destroyed sufficient white blood cells, the body cannot make appropriate immune

responses because they have fewer or no lymphocytes to recognise simple infections and release antibodies

Lesson 4.12 Using antibiotics and painkillers

1 Medicines contain useful drugs that alter how the body works

2 Similar: alleviate symptoms of disease

Different: painkillers do not affect the pathogen but antibiotics kill the pathogens

3 When you are injured, sensory nerve endings send pain messages to your brain. Painkillers stop these nerve impulses, so you feel little or no pain

4 Painkillers alleviate symptoms but do not cure the disease whilst antibiotics kill the pathogens and cure the disease

5 Antibiotics cure bacterial infections but do not affect viral infections

6 a insufficient antibiotic so more bacteria are being produced than being killed

b a prescribed course of antibiotics has the correct amount to kill the bacterial pathogen completely. If the course is not finished, remaining bacteria will reproduce and the infection will return

Lesson 4.13 Building immunity

1 A vaccination introduces a small quantity of an inactive or dead form of a pathogen into the body.

2 They protect us from viral diseases

3 a Lymphocytes recognise the pathogen and the shape of the antigen so quickly make many specific antibodies.

b If a person has had the infection s/he will have made their own antibodies and are unlikely to become ill again (or if they do, the infection will not be as severe and they will recover more quickly)

4 They increase the number of people who are immune to a pathogen, making it difficult for the pathogen to pass to people who are not immunised. If a large proportion of the population is immune to a pathogen, its spread is very much reduced.

5 The flu virus mutates into new strains very quickly and the old vaccines will not have the correct shape for the new antigens

6 Many countries have eradicated certain diseases that are still found in different continents/countries. Global vaccination programmes can eradicate these diseases word wide or eradicate new diseases in specific areas (e.g. Ebola) and prevent them from rapidly spreading again

Lesson 4.14 Making new drugs

1 Digitalis – from foxgloves; aspirin – from willow bark; penicillin – from Penicillium mould

2 To ensure that they are effective, stable and safe

3 Stages are: preclinical testing in the lab and animals followed by clinical trials using volunteers: phase I tests for dosage and safety; phase II to monitor efficacy and side effects; phase III to monitor long term effects

4 Patients have a weaker immune system and these tests ensure that the drugs will be safe, the dose will be suitable, they will fight the infection and major side effects will have been determined.

5 So that the trials cannot be influenced by the people involved in them, (patients, doctors or employees of pharmaceutical companies)

6 When people feel better just because they think they will if they take a medicine

7 The doctor or researcher may give away clues to the volunteers/patients without realising it. This is called observer bias and can make the results unreliable.

Lesson 4.15 Investigating monoclonal antibodies

1 cells that are cloned from one cell.

2 Pregnancy tests, research, treatment of diseases and laboratory tests for hormones etc.

3 To stimulate lymphocytes that make specific antibodies

4 To break down the cell membranes to help them fuse

5 The hybridoma cells all come from the one specific lymphocyte cell so all produce the same antibody

6 Possibilities are endless as large amounts of antibodies can be produced very quickly. They are very specific and will not damage unaffected cells in e.g. in cancer treatment. BUT because they are produced in mice immune reactions occur due to the foreign proteins so the manufacturing technique needs to be improved.

Lesson 4.16 Looking at plant diseases

1 viral, bacterial and fungal pathogens, as well as by pests such as insects.

2 Magnesium is needed to make chlorophyll for photosynthesis and nitrates are needed to make amino acids to make proteins for growth.

3 Symptoms are mottling or discoloured leaves, curled leaves, stunted growth, yellow streaks or spots on leaves. Identification of TMV is difficult because symptoms are similar to other plant diseases. May need lab analysis

4 Magnesium is needed to make chlorophyll and If plants do not have sufficient chlorophyll they cannot produce the glucose they need for respiration to release the energy needed for chemical reactions in cells

5 mABs are produced by injecting e.g. a rabbit with the plant virus or an antigen of the virus to produce the antibodies. The mABs are then used to identify plant diseases.

Lesson 4.17 Learning about plant defences

1 They prevent pathogens from entering the epidermis

2 They prevent pests/pathogens from entering the living cells underneath

3 layers of dead cells around stems and bark prevent pests from entering the living cells underneath

tough waxy leaf cuticles prevent pathogens from entering the epidermis and reaching internal cells

cellulose cell walls prevent pathogens from entering cells.

4 Thorns impale insects and prevent egg laying and stop herbivores grazing

Hairs prevent larvae reaching the epidermis to feed.

Some leaves droop or curl when touched to protect them

Mimicry – features trick animals into not feeding or not laying eggs. Some grasses have anthers that look like aphids and hollyhocks have stem markings that look like aphids; both help prevent aphid attacks.

5 Antibacterial chemicals are produced when the plant is attacked by herbivores or pathogens

Production of poisons which taste bad to deter herbivores.

6 Physical defences only prevent entry by small insects, fungi, bacteria and viruses. If leaves are damaged in any way, no matter how small the wound, these pests can all enter through the wounds.

Chemical defences attack the pathogens once they have gained entry to the leaves

The two systems work together to protect the plant

Lesson 4.18 Maths skills: Sampling and scientific data

1 Prevent pathogens entering the epidermis; prevent pests eating the leaves; reduce excessive water loss

2 Prevent pests entering the living cells underneath; give support; prevent excessive water loss

3 Thorns impale insects and prevent egg laying; hairs prevent larvae feeding

4 Leaves droop or curl when touched, due to changes in water pressure (loss of water from vacuoles)

5 Antibacterial chemical production – increased when plant attacked by herbivore or pathogens; production of poisons to deter herbivores (taste bad)

6 What defences a plant has evolved are due to the random nature of evolution and genetic 'mistakes'. It depends on where they are – climatic defences; the pressure of herbivory they are subject to and how the herbivores and pathogens have in turn evolved in the face of the evolving plant defences.

End of chapter questions

1 Poor diet, smoking or lack of exercise (2)

2 Fungus (1)

3 Phagocytes/white blood cells (1)

4 There will not be sufficient phagocytes to ingest pathogens, or lymphocytes to produce large enough quantities of antitoxins and antibodies (1)

5 Cilia cannot trap dust particles and pathogens and waft them out of the trachea so they enter the lungs (1)

6 It makes it very difficult for the aphids to pierce the cuticle to suck the cell contents out (2)

7 a 1946 in England and Wales and 1949 in Bristol (1)

b 17 years in England and Wales, 11 years in Bristol (1)

8 Malaria (1)

9 Inactive or dead form of the pathogen (1)

10

- layers of dead cells around stems (bark on trees) that prevent pests from entering the living cells underneath
- tough waxy leaf cuticles that prevent pathogens from entering the epidermis
- cellulose cell walls which prevent pathogens from entering cells (2)

11 Painkillers to relieve the headache by stopping these nerve impulses, so you feel little or no pain. (2)

The correct antibiotics to kill the bacterial pathogen causing the sore throat (2)

12 Light purple – inactive; lilac – moderate activity; deep lilac – very active (1)

Obesity causes a shorter life expectancy, but physical activity reduces some of the harm. People who were obese and inactive had a life expectancy that was between five to seven years shorter (depending on their level of obesity) than people who were normal weight and moderately active. (1)

13 They spread malaria but do not cause it (2)

14 Bacteria are living cells that antibiotics are able to kill

Viruses replicate inside cells – antibiotics cannot reach them and they are able to carry on replicating (2)

15 a $452mm^2$ (1)

b It is a growth medium and contains the nutrients needed for bacteria for grow. (1)

c 483/568 x 100 =85.03% (1)

d Patients aged 18-25 are just a sub-set of the population and therefore not representative of the entire population. (1)

16 Every mental health condition is higher, some very significantly, in cannabis users than in non-cannabis users so it does appear to show a causal mechanism (2)

17 A mouse is injected with a specific antigen to stimulate lymphocytes to make particular antibodies (1)

The lymphocytes are extracted and combined with a tumour to make a hybridoma cell which can divide and make more antibodies (1)

The single hybridoma cells are cloned to produce many identical cells that all make the same antibody (1)

The antibodies are collected and purified.(1)

18 Main causes of disease are diet, stress and life situations.

Having poor nutrition, being overweight, smoking, drinking, taking drugs, lack of exercise and exposure to carcinogens all increase the risk of disease (Maximum 2)

Eating a balanced diet, losing weight, not smoking or taking drugs, limiting alcohol, exercising and reducing exposure to carcinogens all reduce the likelihood of becoming ill. (Maximum 2)

19 A big advantage is that unlike radiotherapy or chemotherapy (typically used to treat cancer) monoclonal antibodies only target specific cells so healthy cells are unaffected by treatment (1)

Monoclonal antibodies do not have the unpleasant side effect of hair loss as in traditional cancer treatments, although using monoclonal antibodies has resulted in more side effects than previously expected such as itchy rashes, nausea, headaches and breathlessness. (1)

Monoclonal antibodies are produced in mice so they have foreign antigens which can produce an immune response in humans (1)

Monoclonal antibodies have not been as widely used and tested as other forms of cancer treatment so there could be other long term effects (1)

Accept any other valid points

Chapter 5: Coordination and control

Lesson 5.1 Homeostasis

1 it is the optimum temperature for enzyme action (and other cell functions).

Organs such as the brain are sensitive to changes in temperature.

2 water; blood glucose concentration.

3 regulation of internal conditions within the body, to keep them as constant as possible.

4 nervous; endocrine.

5 receptors. Sometimes, these are grouped together in sense organs.

6

	Nervous system
Response	Rapid and precise
Nature of message	Nerve impulse – electrical
Action	Carried in nerves to specific location, e.g. muscle

	Endocrine system
Response	Slower but acts for longer
Nature of message	A **hormone** – chemical
Action	Carried in blood to all organs, but has an effect on the **target organ** only

Lesson 5.2 The nervous system

1 it enables us to detect our surroundings, and respond to stimuli, and coordinate our bodies and behaviour.

2 central nervous system; peripheral nervous system/nerves.

3 neurone.

4 (stimulus) → receptor → coordinator → effector (→ response)

5 The transmission speed of the nerve would be slower and the nerve impulses would jump around, so the impulses would not go where was intended

6 neurones are long, extended cells with extensions that enable them to communicate with other neurones; the myelin sheath of neurones enables the nerve impulse to jump from gap to gap and speed up transmission.

Lesson 5.3 Reflex actions

1 a rapid, automatic response to a stimulus

spinal reflexes that do not rely on the brain to produce the action.

2 they prevent us from getting hurt.

3 reflex arc.

4 the receptor detects the stimulus;

the sensory neurone carries the impulse to the central nervous system;

the long motor neurone connects the CNS with a muscle and brings about the response;

the short relay neurone, in the CNS, connects the two;

the communication of neurones through synapses and not physical connections enables the neurones to connect with other nerves, so the brain, for instance, is aware of what happened to cause the reflex action.

5 synapse.

6 chemical transmitter (neurotransmitter) molecules.

7 Flow diagram to include:

1. stimulus
2. receptor
3. sensory neuron
4. relay neuron
5. motor neuron
6. effector
7. response

Lesson 5.4 The brain

1 cerebral cortex.

2 cerebral cortex – concerned with consciousness, intelligence, memory and language.

cerebellum – coordination of muscular activity.

medulla – unconscious actions such as the heart rate and breathing.

3 observing the effect of damage to certain areas of the brain;

electrically stimulating regions of an exposed brain (for instance, before or during an operation).

4 (MRI scanners) produce very detailed images of the brain (and other regions of the human body); they're often used in preference to other techniques as they're non-invasive and don't use ionising radiation.

Functional MRI (fMRI) is used to measure brain activity (by detecting changes associated with blood flow) and not just brain structure.

5 removing a tumour; draining excess fluid; adding an implant to modify brain function, e.g. to help to control Parkinson's disease.

6 student answer should include ideas about:

sometimes treatment is not possible, e.g. if the nervous system is damaged, or it might cause further damage, owing to the delicate nature of nervous tissue and complex arrangement of neurones;

surgery may be necessary in the case of a tumour to save/prolong life; implants can help brain function; but risks brain damage;

destruction of tumours or cancer cells by radio- or chemotherapy can, again, save or prolong life, but these techniques damage normal cells also;

modern techniques such as monoclonal antibodies and gene therapy offer hope in the treatment of cancer/tumour treatment, while stem cell therapies offer help in repairing damaged nervous systems.

7 Investigations and treatment of the brain is very risky due to the intricate nature of the brain tissue and risk of infection or stroke

Lesson 5.5 Required practical: Investigating reaction time

1 student answer.

A model Risk assessment proforma could include the following:

Name of hazard, e.g. chemical, material, biological material or activity that could be a hazard; type of hazard; risks associated with hazard; steps taken to minimise risks; (optional) emergency procedure

The hazards listed in the Risk Assessment could include the following, where appropriate:

- Handling/carrying/catching metre rulers.
- Caffeine. The caffeine is likely to be in coffee, so care must be taken with hot drinks.

The hazard for caffeine is indicated with a warning label.

It should not be drunk in a laboratory (use a dining room or classroom), where there is risk of contamination with chemicals or microorganisms. Caffeine may have adverse effects on some students and intake must be limited for all students.

2 Test 5 for Experiment 1.

3 before drinking coffee: 116 mm; 112 mm with the anomalous result removed

after drinking coffee: 93 mm.

4 before drinking coffee: 4.86 s/ 4.77 s ignoring the anomalous result.

After drinking coffee: .26 s

Calculations:

Before drinking coffee:

$$t = \sqrt{\frac{2d}{a}} = \sqrt{\frac{2 \times 116}{9.81}} = \sqrt{\frac{232}{9.81}} = \sqrt{23.6} = 4$$

s

Ignoring the anomalous result:

$$t = \sqrt{\frac{2d}{a}} = \sqrt{\frac{2 \times 112}{9.81}} = \sqrt{\frac{224}{9.81}} = \sqrt{22.8} = 4$$

s

After drinking coffee:

$$t = \sqrt{\frac{2d}{a}} = \sqrt{\frac{2 \times 93}{9.81}} = \sqrt{\frac{186}{9.81}} = \sqrt{19} = 4.36$$

s

5 if taking in caffeine through a drink of coffee, the intake of caffeine needs to be standardised, i.e. the caffeine must be of the same concentration and the volume drunk must be the same for each student.

A further improved way of administering the caffeine would be for a set mass (e.g. 1-3 mg) per kilogram of student body mass.

If the caffeine intake were not standardised, the results obtained from different students could not be compared.

6 student histogram. Note that bars should not be separated, and the bars are not all the same width.

7 Median – occurs most often: 301-325
Modal – mid range of ranked data): 301-325

8 the student will have no indication of when the computer screen might change, e.g. change colour, whereas the student may be able to tell when his/her partner is about to drop the ruler;

The measurement of time is more accurate;

The resolution of the computer timer is higher.

Lesson 5.6 The eye

1 retina.

2 iris – regulates the amount of light entering the eye.

ciliary muscles and suspensory ligaments – control the size of the pupil and therefore regulate the amount of light reaching the retina.

optic nerve – carries impulses from the retina to the brain.

3 cones.

4 each cone connects with a neurone; there is a high concentration on the visual axis of the eye.

5 (in low light intensities,) radial muscles in the iris contract, and the pupil becomes larger.

6 rods are very sensitive to low light intensity (1000 times more sensitive than cones).

Lesson 5.7 Seeing in focus

1 the bending of light rays as they travel from one medium to another.

2 the cornea and the lens.

Note that the fluid and gel that fill the chambers – the aqueous humour and the vitreous humour – also refract light, but the question refers to *structures*.

3 distant object – the light rays are almost parallel.

near object – the light rays are diverging.

4 the lens is thicker/becomes fatter.

5 accommodation.

6 Table to include:

Focusing on nearby object:

Light rays diverge

Thick lens refracts light rays strongly

Light rays brought into focus on the retina.

Focusing on distant object:

Light rays are nearly parallel

Thin lens so only slight refraction of light rays needed

Light rays brought into focus on the retina.

Lesson 5.8 Eye defects

1 the lens is too long; eyeball is too long for the strength of the lens; the cornea is too steeply curved.

2 (a concave lens) causes the light rays to diverge before they reach/enter the eye.

3 the lens is too weak; the eyeball is too short; the cornea is not sufficiently curved.

4 (a convex lens) causes the light rays to to converge before they reach/enter the eye.

5 hard lenses are more rigid and durable; soft lenses are made from a gel and therefore softer and more comfortable. They are freely permeable to gases (though note that modern hard lenses are also gas permeable).

6 a laser is used to change the shape of the cornea.

7 Advantage: If successful, sight corrected and no need for contact lenses or glasses again.

Disadvantage: Risk of surgery not being successful, risks associated with surgery such as infection or eye damage, cost.

Lesson 5.9 Controlling body temperature

1 the thermoregulatory centre in the brain.

2 blood vessels in the skin become wider (vasodilation); sweating increases.

3 when blood vessels in the skin become narrower, so that less blood flows through the skin, and less heat is lost through it.

4 shivering/(involuntary) contraction of skeletal muscles.

5 heat transfer is increased through the skin, by the (increased) evaporation of sweat and vasodilation.

6 blood flow through her skin will be reduced, so less heat energy is transferred from her body to the environment (so her skin will turn blue); she will shiver, so that heat released by respiration of her muscles will help to warm her body.

Any sweating will also stop.

Lesson 5.10 The endocrine system

1 to produce hormones, or chemical messengers (that produce an effect on target organs).

2 target organs/effectors.

3 pancreas.

4 ovaries.

5 it acts on other glands, causing them to secrete hormones.

6 growth hormone.

Lesson 5.11 Controlling blood glucose

1 insulin.

2 it causes glucose to move into our body's cells.

3 it increases for a time (as it is absorbed into the bloodstream from the gut), then falls as insulin is secreted.

4 60 minutes.

5 70-100 mg of glucose per 100 cm^3 of blood.

It is important to maintain blood glucose as it is needed for cells for respiration to make energy. If blood glucose goes too high kidney failure or death can result. Similarly if blood glucose decreases too much a person can fall unconscious and die.

6 a combination of the actions of the hormones insulin and glucagon.

when blood glucose concentration increases, insulin is secreted by the pancreas, which causes glucose to be taken up by the cells;

the glucose is then used for respiration or converted to glycogen in the cells of the liver and in muscles;

when blood glucose falls, glucagon is secreted by the pancreas, which causes glycogen to be broken down in the liver and released into the bloodstream.

Lesson 5.12 Diabetes

1 when the pancreas is unable to produce enough, or any, insulin.

2 the body's cells are unable to take up glucose; the glucose concentration in the blood increases; glucose is excreted in the urine; fat and protein are used for energy; the person loses weight; if left uncontrolled, the kidneys will fail, and the person will die.

3 the person's body cells lose their sensitivity/no longer respond to insulin;

caused by lifestyle of high energy, 'fast' food and an inactive life.

4 the person fasts for 8-12 hours, and their blood glucose is measured; the person then takes in glucose, and their blood glucose tested 2 hours later; their blood glucose concentration will indicate how well they are able to regulate the glucose taken in.

5 The non-diabetic eats meals at around 8am, 12 noon and 4.30pm, insulin is released and helps return blood-glucose levels to normal (normal amount of glucose in the blood)

The type 2 diabetic with treatment eats at the same times, the treatment means they are able to regulate the levels of glucose in their blood to near normal levels

The type 2 diabetic without treatment also eats at the same times, but they have lost their sensitivity to the insulin being released and so the level of glucose in their blood does not return to normal levels in between meals.

6 type 2 diabetes tends to cluster in families; so genetics will have an effect; but the people in a family are also likely to have a similar diet, i.e. an environmental effect; a western diet is known to be a cause.

7 there is a (strong) correlation between body weight/BMI and Type 2 diabetes; the relative risk of diabetes increases markedly as the person becomes overweight and obese; the relative risk in women is over double that in men.

Lesson 5.13 Diabetes recommendations

1 insulin injections.

2 to manage the condition by modifying their lifestyle; involves increasing exercise and a carbohydrate-controlled diet.

3 African Americans (also Mexican Americans, American Indians, Native Hawaiians, Pacific Islanders and Asian Americans)

4 a small alcohol intake *reduces* the risk of Type 2 diabetes (<around 59 g per day in men; <48 g per day in women); the risk then increases up to around 80 g per day in men – at around 1.16 the relative risk; 53 g per day in women – at around 1.19 the relative risk; above these levels, further increase in alcohol consumption leads to no further increase in risk.

5 improve health qualities of food; if manufacturers don't, have a sugar tax imposed.

6 risk of diabetes increases with social deprivation; most deprived most likely to have poor lifestyle/diet/eat fast foods.

Lesson 5.14 Water balance

1 0.4 dm^3/400 cm^3.

2 lungs.

3 the cells become dehydrated; the cells are unable to function properly as water is required for the chemical reactions that go on inside the cells/cell metabolism.

4 The recommendation is that the workforce should have regular access to drinking water because it is good for general health and is vital for the efficient functioning of our cells, thus reducing the amount of sickness generally. When we are working in warm environments, we lose more water through sweating and from our lungs and so will need to drink more water to maintain a safe level of water in our body. If we are thirsty, we may already have already lost 2% water by body weight and are dehydrated – we need to have access to drinking water before we feel thirsty, otherwise we can lose concentration and this may be dangerous in the workplace. If we lose as much as 5% we lose our ability to work and this would have a significant impact on production.

5 we cannot store excess amino acids.

6 the amino group of the amino acid is removed as ammonia by deamination; ammonia is quickly converted to urea.

Lesson 5.15 The kidneys

1 filter waste substances, and substances the body doesn't need, from the blood, for removal from the body/excretion.

2 small molecules, including water, urea, glucose and ions.

3 proteins.

4 water; glucose.

5 detected by body; increase in the release of ADH by the pituitary gland; ADH acts on kidney tubules; more water is reabsorbed by the kidney tubules.

6 If less ADH is produced, less water is reabsorbed back into the blood from the kidney tubules and dilute urine is produced.

Lesson 5.16 Negative feedback

1 increased basal metabolic rate; thin/unexplained weight loss; rapid heart rate.

2 controls rate of metabolism/metabolic rate; controls growth and development.

3 thyroid-stimulating hormone (TSH).

4 thyroxine and adrenaline.

5 when the desired effect/action of a hormone is reached, the secretion of the hormone is switched off.

6 (with a heating system/thermostat, the heating system is switched off by the effect, i.e. the temperature having been reached);

the switching on and off of thyroxine secretion is not regulated by the action of the thyroxine on the thyroid gland itself; the thyroxine acts on the pituitary gland, which responds by producing another hormone (TSH) that acts on the thyroid gland.

Lesson 5.17 Kidney failure

1 the filtration of wastes from a patient's blood.

blood is removed from the patient's arm and circulated through the dialysis machine; wastes filter from the blood through a partially permeable membrane; the blood is returned to the arm.

2 when a person has lost their kidney function.

3 close relative, i.e. an identical twin is best; a brother, sister, father, mother, son or daughter may be suitable.

4 the patient's tissue type and blood group will already be on record and can be matched (or not) quickly when a possible donor kidney becomes available.

5 it is complex operation – the kidney needs to be connected to an artery and vein and also to the bladder;

risk of rejection;

Note that the usual risks associated with surgery apply – the possibility of blood clots and infection.

6 the patient needs to take immunosuppressant drugs (for the rest of their life).

Lesson 5.18 Dialysis or transplant?

1 a sudden fall in blood pressure; risk of infection.

2 some people, e.g. the elderly, are too weak; conditions such as heart disease may prevent a patient from having the operation.

3 immunosuppressant drugs must be taken; the person is able to have a much more varied diet; but lifestyle changes (stopping smoking, limiting alcohol and restricting weight) must be made to minimise risk.

4 50-60%.

5 being an organ donor; relatives overruling donor's wishes; offering organs for sale; breeding other animals specifically for organ donation.

6 these offer the possibility of genetically modifying the pigs so that the organs to be transplanted will not be rejected by the (human) patient/recipient of the organ.

Note that in the question, genetically modified is italicized. This question refers to this aspect, and not the suitability of pigs as donors, though these are suitable, owing to being readily available, and providing they're raised in a clean environment, the possibility of infection may be low.

Genetically, primate donors would offer the closest match, but their close relationship increases the likelihood of transferring a virus liable to cause infection in humans. There are also ethical issues with the use of primates and the practicality of breeding primates in sufficient numbers.

7 Stop smoking, limit alcohol, healthy diet, healthy BMI.

Lesson 5.19 Human reproduction

1 males – testosterone.

females – oestrogen and progestogens, e.g. progesterone.

2 development of secondary sexual characteristics

3 follicle stimulating hormone (FSH); luteinising hormone (LH), oestrogen and progesterone.

4 oestrogen and progesterone.

5 oestrogen and progesterone.

6 oestrogen and progesterone – negative feedback on FSH secretion.

progesterone – negative feedback on LH secretion.

oestrogen – positive feedback/stimulation of LH secretion.

Lesson 5.20 IVF

1 FSH – causes eggs to mature in the ovaries.

LH – triggers ovulation.

2 (to monitor) the number and size of developing follicles.

3 when eggs are removed from the mother and fertilised, in the laboratory, with sperm collected from the father.

4 Two from: young mother; mother has been pregnant previously; has a BMI in the range of 19-30; has a low alcohol and caffeine intake, and does not smoke.

5 a single sperm could be collected and injected into an egg. The procedure is carried out microscopically.

6 the IVF process, from stimulation of the ovaries to implantation of an embryo.

7 Similarities:

Both involve treatment with hormones FSH and LH to stimulate the production of eggs from the ovaries.

Both used for couples wanting to conceive.

Both can be emotionally quite stressful.

Differences:

In IVF the eggs are collected and then mixed with the father's sperm in a lab. Successful embryos are then placed in mother's uterus

In fertility treatment there is a risk of multiple births because the fertility drugs can stimulate more than one egg to be released.

Lesson 5.21 IVF evaluation

1 NICE provides guidelines; the final decision is made by the woman's local NHS.

2 to prepare the couple for the process – success or failure.

3 learner answer.

it can be estimated by increasing the success rate – from 32.2% in 2010 – by 1% each year, according to the year when the student makes the estimation.

4 Low success rate; possibility of multiple births; accompanying risk to mother and babies; there are reports of higher incidences of premature births, stillbirths, low birth-weights and infant deaths (but rates are higher still in babies born to couples with infertility problems who eventually manage to conceive).

5 student answer, related to IVF not being a natural process; production of many embryos, and the fate of those that are unused.

6 selecting the characteristics of a baby to be born/eugenics.

Lesson 5.22 Systems working together

1 adrenaline.

2 it is diverted from other areas (the gut and the skin).

3 the stimulus is detected by the nervous system (the sense organ, the eye, and perhaps other senses); nerve impulses transmitted to and through the brain; nervous message is sent to the adrenal medulla – an endocrine gland; the hormone adrenaline is secreted.

4 glucagon.

5 *insulin* promotes the conversion of glucose into glycogen.

glucagon and *adrenaline* promote the conversion of glycogen to glucose.

Thyroxine and cortisol also affect glucose metabolism, but you do not need to know how.

6 *thyroxine* affects our mental and physical development, in the embryo, infant and child.

oestrogen and *testosterone* control our sexual development, including the development of secondary sexual characteristics.

growth hormone controls our growth through its effects on bone and muscle and other cells.

7 Concept map to link words with correct explanations for links.

Lesson 5.23 Contraception

1 hormonal methods; non-hormonal methods, which tend to use barriers to sperm reaching the egg.

2 time in the cycle, at around 14 days; a slight increase in body temperature; a thinning of mucus secreted from the cervix.

3 condom; diaphragm.

4 with other contraceptives, such as the diaphragm.

it is not reliable used on its own, and should not be used with a condom (it is not necessary to enable the condom to work effectively, and they are now known to *not* reduce infections such as gonorrhoea, HIV or chlamydia; in fact, they may increase the risk of infection).

5 use reproductive hormones; oestrogen and progestogens inhibit the release of pituitary hormones that control egg maturation and release.

6 an intrauterine device – it prevents a fertilised egg from implanting in the uterus.

plastic IUDs also release a progestogen.

some IUDs have copper wound around them; copper is toxic to sperm.

7 IUDs prevent implantation of an egg in the uterus so conception can occur. For those that consider life to begin at conception, IUDs could be considered unethical. However, others consider life begins at implantation and the IUD prevents this so it could be argued not to be unethical.

Lesson 5.24 Which contraceptive?

1 there are computerised devices that monitor hormonal changes and therefore reduce the possibility of error.

2 IUDs and oral contraceptives (>99% effective).

3 oral contraceptives are hormonal contraceptives taken in through the mouth and are absorbed through the gut. if a person vomits or has diarrhoea, the contraceptive may not have been absorbed into the bloodstream before the vomiting or diarrhoea occurs.

the hormonal contraceptives in injections, implants or patches are injected or are absorbed into the bloodstream and are therefore not affected by factors that affect absorption through the gut.

4 Advantage – works immediately when it's inserted; can stay in place for several years

Disadvantage – insertion may be uncomfortable; periods may be longer and more painful.

5 contraceptive pill.

6 student answers. these should include religious factors; economic and lifestyle factors such as when a couple could afford to start a family, how frequently would contraception need to be used, how soon would the women like to become pregnant after contraception ceases.

7 Advantages of sterilisation:

Works immediately, very reliable, does not affect hormone levels, does not affect sexual health, long term affordable

Disadvantages:

Not easy or impossible to reverse, risks associated with surgery, no protection again STIs.

Lesson 5.25 Auxins

1 tropism.

2 grow towards the light.

3 negatively gravitropic/grow upwards.

4 as the shoot begins to grow (from the lowest point), it will change direction and grow upwards; a the root begins to grow (from the uppermost point), it will change direction and grow downwards.

5 auxin produced in the shoot tip diffuses towards the shaded side; causes cells on the shaded side to grow/elongate; so more growth on the shaded side causes the shoot to grow towards the light.

6 owing to gravity, auxin accumulates on the lower side; the auxin promotes growth, so there is increased growth on the lower side; the shoot grows upwards.

Negative gravitropism ensures plant shoots grow upwards and therefore towards sunlight.

Note that this is the reverse of the response in roots (positively gravitropic), where the auxin is inhibitory, so there is increased growth on the upper side; the root grows downwards.

Lesson 5.26 Application of auxins

1 in meristems;

(at the tip of the shoot; towards the tip of a root [in the root, the very tip is covered by the root cap, which protects the tip as it is growing through the soil]).

2 to produce clones of plants (quickly).

3 auxin and other types of plant hormones, nutrients, agar, water.

4 to promote cell division and cell enlargement.

5 roots: approximately 5×10^{-9} mol/dm^{-3}.

shoots: approximately 2×10^{-4} mol/dm^{-3}.

6 At the concentration used in auxin-based weedkillers, plants with broad leaves are more sensitive than narrow-leaved plants such as grass and cereal crop therefore farmers can use auxin-based weedkiller on wheat and barley. They can also be used on lawns to kill broad-leafed weeds such as dandelion without damaging the grass.

Lesson 5.27 Required practical: The effect of light and gravity on the growth of germinating seeds

1 Amount of water
Carbon dioxide levels
Temperature
Amount of light
Supply of nutrients
Make sure all roots are horizontal
Make sure plants are the same age and size

2 Result after 20 minutes (accept 10 mins) maybe anomalous, because it is different from all the others (and unexpected)

3 Geotropism or gravitropism

4 After 30 mins

5 Diagram should include stem and leaves, and the bending towards the light and an indication of the direction of the light source

6 Auxins are produced in the shoot tip and travel down the shoot, if you cut off the tip, then the auxin cannot be produced, so it is better to cover the tip with foil.

7 Line graph (continuous process), independent variable (time) on x axis, curvature on y axis. Axes labelled, including units. Sensible scale used on both axes. Points plotted accurately. Points joined together

8 As the shoot grows (time passes) the greater the mean angle of curvature

9 Auxin is produced in the shoot tip and travels down the shoot. Auxin makes plant grow more (where the auxin is) The auxin is destroyed by sunlight, so there is less growth on this side of the shoot. The shaded side still have auxin and the

shoot grows more on this side. The result is that the shoots bends towards the light

10 Repeat the investigation for gravitropism (geotropism) on page 224, but as well as keeping the seedlings in normal light (control), also germinate 3 additional sets of seedling and put one set with a blue filter in the canister, one set with a red filter and one set with a green filter, using the canister from the investigation on page 225, make sure the seedlings are placed with their roots horizontal and keep all other conditions the same.

Lesson 5.28 Other plant hormones

1 speed up/promote germination; promote flowering, e.g. of apples; increase fruit size, e.g. in cherries, (seedless) grapes.

2 increasing the concentration of gibberellins increases the average mass of the grapes; the graph levels off above around 30 ppm.

3 the banana produces ethene, which causes the tomato to ripen.

4 ethene production increases up to around 4.5 days, and then levels off, with production at around 4 nanolitres/g/h. There may be a slight decrease at 8 days, when the tomato is fully ripe.

5 so that they can be at the optimum degree of ripeness in the supermarket/not overripe by the time they reach the supermarket.

6 student answer. The answer should include reference to adequate ventilation.

7 Bananas produce ethene gas which ripens fruit so any fruit placed next to the banana will ripen.

Lesson 5.29 Maths skills: The spread of scientific data

1 12.5 °C.

The range bars are shorter.

2 the data collected at 5 days (for 12.5 °C).

3 student plots.

4 medians of the data are identical.

the treated plot has a much wider spread of data – interquartile ranges and maximum and minimum values.

but there are no outliers.

It is preferable to make comparisons using data with a narrower range of values than the treated set.

5 0.4/100

6. 80th percentile.

End of chapter questions

1.

a. B (1)

b. C (1)

2 Lead a more active lifestyle (1)

Change diet/eat healthier diet/eat less fast food/carbohydrate/suitable alternative (1)

3 Two correct answers from:
(the plant is growing towards the light)
because auxin (synthesised in the tip of the shoot) moves/diffuses to the shaded side/away from the light (1)
leading to more growth on the shaded side (1)
so the plant grows towards the light (1)

4 Three correct answers from:
so plant shoot would grow horizontally/not grow upwards/not change direction of growth (1)
auxin would not accumulate on a particular side of the shoot and root (1)
no differential/extra growth on a particular side (1)

5 Two correct answers from:
recommended water intake increases with increasing temperature (1)
to replace water lost (1)
because sweating increased (1)
to keep body temperature constant (1)
rest period also recommended at temperatures above 32 °C (1)
because the body may overheat/respiration also releases heat (1)

6 Type of message: nervous – electrical/nerve impulse; endocrine – chemical/using hormones (1)
Note that both need to be correct for (1) mark

Speed of response: nervous – fast (milliseconds); endocrine – slower, taking minutes, hours or years (with the exception of adrenaline) (1)
Note that both need to be correct for (1) mark

7 Follicle stimulating hormone (FSH) (1)

8

a. Accept either for 1 mark:

- The eyeball is too long for the strength of the lens
- The cornea is too sharply curved

b. A person with myopia wears a concave lense.(1) This causes the light rays to diverge slightly before they reach the eye and the image is then correctly focused on the retina. (1)

c. c. 34/100x1456 = 495 people

9 Four correct answers from:

slight increase in risk of Type 2 diabetes when intake is less than 1 per month to 1-4 per month (1)
more marked/larger increase in risk when intake of sugar-sweetened soft drinks increased (1)
suggests that intake of sugar is linked with increased risk of diabetes (1)
but this is a correlation and not a proven cause/other ingredients in the soft drinks could be involved (1)
further investigation is needed (1)

10 the hormones inhibit the release of the pituitary hormones that control egg maturation and release (1)

11 Two correct answers from:
(promote growth of plantlets) in tissue culture media (1)
used in hormone-rooting powders to promote root growth (1)
selective weed killers (1)

12 Two correct answers from:
growth rate greater in plants grown in 12 h light: 12 h dark, than in 24 h light (1)

suggests that constant light restricts growth (1)
constant/linear growth rate for plants grown in 12 h light: 12 h dark (1)
slow initial growth rate for plants grown in 24 h light, then more rapid growth, with growth levelling off after 14 days (1)

13 Two correct answers from:
when body temperature falls, detected by thermoregulatory centre (1)
nervous signal leads to release of adrenaline (by adrenal medulla) (1)
thyroid-stimulating hormone (TSH) released by pituitary gland leads to secretion of thyroxine (by thyroid gland) (1)
metabolic rate increases and temperature rises (1)
when body temperature becomes too high, secretion of adrenaline and thyroxine stopped/reduced (1)

14 Four correct answers from:
mother's finger on baby's palm detected by receptors (in baby's skin) (1)
message/nerve impulse sent to spinal cord (1)
along sensory neurone (1)
impulse travels through relay neurone (1)
to motor neurone (1)
impulse causes contraction of muscles in hand (1)
so that baby grasps finger (1)

15 Four correct answers from:
most of cones around visual axis/within around 4° of the horizontal axis of the eye (1)
means that colour vision most acute around axis of the eye (1)
few rods around the visual axis/largest density around 20° from horizontal axis/significant number around the periphery of eye/70° of horizontal axis (1)
means that peripheral vision is the result of rods being stimulated (1)
and is (almost) monochrome (1)
no receptors around 13-18° of horizontal axis (1)
This is where the optic nerve leaves the eye and is known as the 'blind spot'.

Chapter 6: Genetics

Lesson 6.1 DNA and genes

1 DNA/deoxyribonucleic acid.
2 A short section of DNA that codes for the production of a particular protein.
3 At the same position on each of the pair.
4 Two (one on each chromosome of the pair).
5 Nucleus, mitochondrion.
6 Can give antibiotic resistance.
7 They are external to the chromosome of the bacterium.
8 Genes exist on chromosomes and chromosomes make up DNA.

Lesson 6.2 The human genome

1 Increase understanding of how genes are involved in a disease and other conditions/change when they become defective; so develop effective treatments, or correct defective gene/allele.
2 Cystic fibrosis (or any other).
3 recommend better preventative medicine; improve treatment of disease by identifying targets.
identify reasons for effective and ineffective medicines or side effects for specific people/personalised medicine.
4 Around 1.5%
5 Switching genes on and off/gene regulation/controlling gene expression.
6 Genes for making the cell function as a muscle cell, e.g. for synthesis of muscle filaments, enzymes of respiration, etc., are switched on; others, e.g. genes for digestive enzymes, etc., etc., are switched off.
7 Arguments for yes it was worth the money might include the medical and therapeutical benefits of knowing the genetic code such as testing for genetic diseases. We also understand more about how genes can be switched on or off (gene expression).
Arguments for no might include money could have been better spent on other medical research or treatment. Only a small percentage of the genome was found to code for protein.

Lesson 6.3 Tracing human migration

1 Too much information to analyse; too costly to analyse it all; better to focus on regions known to show most variation.
2 The Y-chromosome;
DNA found in mitochondria.
The cytoplasm of the fertilised egg, from our mothers, contained mitochondria. All our other cells are produced from this first cell (fertilised egg or zygote). The mitochondria are replicated before each cell divides. (the mitochondria from the sperm do not survive)
3 Better to obtain analyses by sampling across all populations. If populations were left out, incorrect conclusions would be drawn.
4 By 2015, data from 1000 populations and 130 countries has been sampled. This involved data from 700 000 people.
5 Genomic data could find its way into the hands of medical or pharmaceutical companies.
6 Analysis was carried out of regions of DNA that did not code for proteins and had no known function.
7 Understanding disease, single gene disorders, tracing migration.
8 Accept any plausible answer such as change in climate or natural disaster.

Lesson 6.4 The structure of DNA

1 Double helix.
2 Like a twisted ladder.
3 A, T, C and G.
4 4 – because there are 4 different bases.
5 Diagram to show A paired with T and C paired with G.
6 By the sequence of bases in a gene.

Lesson 6.5 Proteins

1 antibody; collagen; enzyme; keratin; muscle protein/actin or myosin; some hormones, e.g. insulin.
2 20.
3 Three.
4 On a ribosome.
5 Attractive and repulsive forces between amino acids (and with the medium).
6 Intermolecular forces.
7 Flow diagram to show steps:
1. DNA unzips and bases act as template for messenger molecule
2. Messenger molecule leaves nucleus and attaches to ribosome
3. carrier proteins match amino acids to complementary base pairs
4. Peptide bonds form between amino acids
5. 5.A protein chain is formed as the messenger molecule moves along the ribosome

Lesson 6.6 Mutations

1 A change in an organism's DNA (these can be on a chromosome level or changes to the base sequence of the DNA).
2 If they bring about something that helps with the organism's survival, e.g. a melanic form, that improves camouflage.
3 Chromosome and gene.
4 The genetic code allows for some errors; when a protein may be changed, but is still able to carry out its function.
5 If an extra base is inserted; if a sequence of bases is copied and repeated; if one base is substituted with another; if bases are deleted.
6 Accept diagram showing how change in amino acid changes the intermolecular forces between the amino acids. Diagram should also show the enzyme shape is changed so the enzyme and substrate are no longer complementary.

Lesson 6.7 Meiosis

1 Four.
2 Half the parent number/23 in humans.
3 One.
4 Two.
5 Each gamete is unique; the is some exchange of genetic material between chromosomes during meiosis (as [homologous] chromosomes pair up and the separate).
6 Fertilisation is random; some unknown environmental effect.

Lesson 6.8 Asexual and sexual reproduction

1 It preserves desirable characteristics; if the chances of meeting another individual are rare; it produces many offspring quickly; requires less energy.
2 Introduces variation, increasing the chances of survival (particularly if the environment changes); can be manipulated to produce new varieties/breeds.
3 Daffodil; strawberry; fungus, e.g. bread mould.
4 Produces many identical plants quickly/without the need for sowing seed.
5 Asexual in human; sexual in mosquito.
6 If conditions change (e.g. an immune response or anti-malarial drug), it increases the chances of survival.
7 The temperature and plentiful food supply

Lesson 6.9 Genetics

1 Allele.
2 A person with two normal alleles will (clearly) be normal.

A person with two defective genes will be unable to produce the CFTR protein and therefore have cystic fibrosis.

A person with one normal allele will be able to produce the CFTR protein and therefore not suffer from cystic fibrosis.
3 The allele is dominant if it expressed, i.e. the characteristics associated with it are always visible.
4 The genotype refers to the alleles for a particular gene that an organism has, e.g. AA, Aa or aa. The phenotype refers to the visible appearance or characteristic.
5 No this is not true. Describing a trait as dominant does not mean it is the most common in the population; it means that it is expressed over the recessive trait. How frequently a trait is observed in a population is not related to whether or not it is dominant or recessive. Instead, it is a reflection of how frequently the gene responsible for causing a trait is found in people

Lesson 6.10 Genetics

1 BB.
2 B.
3

		Mother - tt gametes	
		t	t
Father - Tt gametes	T	Tt	Tt
	t	tt	tt

4

		Female - Bb gametes	
		B	b
Male - Bb gametes	B	BB brown eyes	Bb brown eyes
	b	Bb brown eyes	Bb red eyes

Lesson 6.11 Tracking gene disorders

1 PP – polydactyly

Pp – polydactyly

pp – normal.
2

		Mother - pp gametes	
		p	p
Father - Pp gametes	P	Pp polydactyly	Pp polydactyly
	p	pp normal	pp normal

3 He must have at least one copy of the dominant allele, C. But we do not know whether he has one or two.
4 Charles, Sarah, David, Stephen, Anthony, Daniel, Linda and Christian.
5 Medical – the opportunity for a couple not pass on the defective allele to their child; (the possible opportunity for research).

Social – (possible) requirement of social support for parents and/or child; selecting the gender of a child.

Economic – cost of any medical support required; insurance premiums; employment possibilities.
6 Student answer. The possibility of destroying an embryo should be weighed against the possibility of developing, or not developing breast cancer. Students may also appreciate that many breast cancer sufferers do not develop cancer until later life.

Lesson 6.12 Gregor Mendel

1 Wide range of varieties available; fertilisation easy to control; easy to cultivate; grow, flower and set seed in one growing season.

2 Mendel was investigating inheritance of traits in peas. If the peas were not of pure lines, and other characteristics appeared, then the investigation would not do what it set out to do, investigating one trait at a time.

3

		Plant 1 - PP gametes	
		P	P
Plant 2 - pp gametes	p	Pp purple	Pp purple
	p	Pp purple	Pp purple

You do not really need a Punnett square to show this cross.

For the self-fertilisation/cross of the purple-flowered plants:

		Plant 1 - Pp gametes	
		P	p
Plant 2 - Pp gametes	P	Pp purple	Pp purple
	p	Pp purple	pp white

4 Yes, there's a 1:1 ratio of yellow seeds to green seeds. (but there would have been even more confidence with a greater number of seeds).

5

Scientist	Ratio, yellow:green seeds
Gregor Mendel (1866)	3.01
Carl Correns (1900)	3.08
Erich von Tschermak-Seysenegg (1900)	3.01
William Bateson (1909)	3.05
Charles Hurst (1902)	2.94
Robert Lock (1908)	2.80

6 Sample sizes were large so conclusions on the reproducibility of results can be drawn. The ratios show a high degree of reproducibility across the six scientists.

Lesson 6.13 Key concept: Genetics is simple – or is it?

1 A genetic disorder whereby sufferers mix up all colours that have shades of red or shades of green in them

2 On the X-chromosome

3 Cystic fibrosis; Huntington's disease, haemophilia; muscular dystrophy.

4 Heart disease, diabetes, obesity.

5 Mutation in *Staphylococcus aureus* giving rise to antibiotic resistance. The antibiotic-resistant genes have spread throughout the population, giving rise to MRSA.

6 The genome analysis would indicate the strain that was involved in the outbreak so this could be matched with the most appropriate antimicrobial agent.

7 Analysing the genomes of pathogens helps us to understand and control infectious disease as it allows us to select the best treatment according to the genomics of the strain of the pathogen. It is a worthwhile investment as it could save lives and money wasted on other less effective treatments

Lesson 6.14 Maths skills: Fractions, ratio, proportion and probability

1 2/4 or 1/2.

(the offspring would be made up of 2 heterozygous individuals: 2 homozygous recessive.

As the condition is recessive, there will be two with the condition to two without

2 5 715 000

Calculation: $\frac{9}{100} \times 63\ 500\ 000 = 5\ 715\ 000$

3 Huntington's sufferers: 1 normal

(the offspring would be made up of 1 homozygous dominant: 2 heterozygous individuals: 1 homozygous recessive.

As the Huntington's allele is dominant, there will be three with the condition to one without)

4 600.

Calculation: $\frac{40}{100\ 000} \times 1\ 500\ 000 = 600$

5 Fertilisation is random. The probability of this couple, or any other with identical genotypes, producing a child with cystic fibrosis, will be 1 in 4 for every child they have.

6 Fertilisation is random, so in a family, there are likely to be insufficient children for the expected ratio to be observed. There may be hundreds or thousands of offspring from a plant, sufficient to give numbers close to the expected ratio.

End of chapter questions

1 Identical positions; each chromosome (1)

Note that both are required for 1 mark.

DNA; bases (1)

Note that both are required for 1 mark.

2 Two from:

- Understand/treat (complex) disease
- Understand/treat gene disorders
- Trace human migration patterns from the past (2)

3 a Two from:

- Dominant
- The phenotype for the tongue-roller (from the annotation of the diagram)
- Is shown in upper case letters (2)

b Tt, tt (1)

Note that both are required for 1 mark.

c Tongue-roller; non-tongue-roller (1)

Note that both are required for 1 mark.

4 Two from:

- The cell is undergoing meiosis (1)

1 mark for either explanation:

- Four daughter cells are produced
- Each with half the number of chromosomes as the parent cell (1)

5 Meiosis – sperm production

Mitosis – growth of a human embryo

Note that both are required for 1 mark.

6 Malarial parasite, daffodil, strawberry, mushroom (1)

Note that all four are required for 1 mark.

7 Two from:

The mother is XX/has two X-chromosomes/produces X gametes (1)

The father is XY/produces X and Y gametes (1)

So offspring will be 50% male:50% female (1)

OR,

correctly drawn Punnett square (2)

		Mother – XX gametes	
		X	X
Father – XY gametes	X	XX female	XX female
	Y	XY male	XY male

8

a. Gametes of parent mice (1)

Correct genotypes of offspring (2)

2 marks all correct; 1 mark 2-3 correct

Correct phenotypes of offspring (1)

		Female – Bb gametes	
		B	b
Male – Bb gametes	B	BB black	Bb black
	b	Bb black	bb white

b. 1/4

c. 2.2 ×108

9 The larger the genome size, the greater the rate of mutation (1)

Accept answers relating to complexity of the organism and mutation rate.

Collect more data/there are only two values for mammals/one value for plants, etc. (1)

10 A mutation changes the base sequence of DNA (that codes for the protein) (1)

If this is changed, the correct amino acid may not (accept will not) be assembled into the protein (1)

11 (Through a) dominant (1)

Allele (of a gene) (1)

12 No genetic variation/if exposed to a disease, they are all likely to die (1)

13 (B) complementary (1)

14

Male – Pp gametes	P	PP purple	Pp purple
	p	Pp purple	Pp Blue

Purple-flowered plant may be PP or Pp (1)
Cross with blue-flowered plant (1)
If PP, all the offspring will have purple flowers (1)
If Pp, offspring will be 50% purple-flowered, 50% blue-flowered (1)

15

a. Some gametes receive chromosomes of maternal origin, some paternal (1)
Some exchange of genetic material between chromosomes during meiosis/correct diagram of crossing over (1)

16 There must have been 7 binary divisions therefore 175/7 (1)

= 25minutes (1)

17 12600/14000×100= 90% (1)

90% of 20,000 = 18,000 (1)

18 Adenine (A) is <u>directly proportional</u> to thymine (T) (1) The answer must say directly proportional for the mark.

Data point for proportion of adenine (A) at 0.35 and thymine (T)at 0.130/0.125 is anomalous (1)

The proportion of bases that are adenine (A) and thymine (T) do not add up to 1 because guanine (G) and cytosine (C) are not included in the data. (1)

The scattergraph is based on data from just 32 species of bacteria (1)

Analyses of more bacterial genomes should be carried out to increase confidence in the conclusion (1)

Chapter 7: Variation and evolution

Lesson 7.1 Variation

1 genetic variation: blood group; eye colour; natural colour of hair; shape of earlobe; other reasonable answers

environmental variation: dyed hair colour; hair length; scars; tattoos; other reasonable answers.

2 height; skin colour; weight; sporting achievements.

3 it is adapted/has adaptations that enable it to survive.

4 fur colour/camouflaged, so better able to catch prey; thick fur for insulation; fat for insulation; large paws assist with walking on snow.

5 mutation; sexual reproduction; meiosis.

6 (melanic) peppered moth in polluted areas; antibiotic resistance in bacteria; warfarin (and other poison) resistance in rats.

Lesson 7.2 The theory of evolution

1 characteristics, e.g. a longer neck in giraffes, acquired during an organism's lifetime are passed on to the next generation.

2 inherited characteristics are dependent on genes; the characteristics acquired during an organism's lifetime do not modify genes (though we now know that the environment can affect the genome, though not in the way Lamarck thought).

3 ability to survive harsh environments, e.g. very cold, very hot, low light intensities, flooding, seawater; resistance to disease; attractive flowers to attract insect pollinators; other reasonable answers.

4 in a struggle for existence when organisms best-suited to the environment survive and pass on their beneficial genes, while those least-suited will not survive to mate.

People often use the phrase 'survival of the fittest'. It's worth noting that it was used by biologist Herbert Spencer in 1864; don't attribute it to Charles Darwin!

5 change in climate; geographical isolation, e.g. by mountains, glaciers, valley or water.

6 when the new species is no longer able to produce fertile offspring if it mates with its ancestral species.

Lesson 7.3 The origin of species by natural selection

1 in particular, his visit to the Galápagos Islands (though his fossil finds elsewhere were also important).

2 Darwin found different, though similar animals on different islands. As these must have come from mainland South America, originally, it was evidence that showed that species could change, and were not fixed.

3 artificial selection (/selective breeding).

4 rock dove.

5 organisms produced larger numbers of offspring than they need to sustain their numbers.

6 separation and isolation of organisms that had originated from the mainland (birds may have flown or been blown over; animals may have swum, with others being carried over by currents or have floated on rafts of vegetation); different selection pressures on different islands.

7 Darwin and Lamarck both thought that life had changed gradually over time and was still changing, that living things change to be better suited and adapted to their environments, and that all organisms are related. Darwin and Lamarck also agreed that life evolved from fewer, simpler organisms to many, more complex organisms.

Differences:

Lemarck believed theory centred on two ideas:

- the law of use and disuse
- the law of inheritance of acquired characteristics

His theory stipulated that a characteristic which is used more and more by an organism becomes bigger and stronger. One that is not used disappears eventually. Any characteristic of an organism that is improved through use is passed to its offspring. He also believed that one organism developed into another rather than becoming extinct.

In contrast Darwin believed that:

- individuals compete for limited resources
- individuals in a population show natural variation
- individuals with characteristics best suited to their environment are more likely to survive to reproduce
- 'successful' characteristics are inherited

Species unable to compete successfully eventually become extinct.

Lesson 7.4 Fossil evidence

1 an organism dies and sinks to the bottom of the water; the organism becomes covered in sediment; soft parts of the body decay; the sediment begins to turn to rock; the sediments are compressed by further layers; an exchange of minerals occurs between the skeleton and the water; the skeleton is turned to rock, or mineralised.

2 hard parts, i.e. the bones, shell or teeth of animals; woody parts of plants.

3 footprints, burrows or spaces left by plant roots.

4 if preserved in amber; when buried in mud or volcanic ash.

5 by their presence in rock sediments/layers. Older species are found in deeper rock sediments.

6 radiometric dating.

Note that radioactive carbon dating is not appropriate because the half-life of ^{14}C is too short. Potassium-argon dating is suitable.

Lesson 7.5 How much have organisms changed?

1 early life soft-bodied; conditions must be appropriate for fossil formation; most fossils

destroyed by geological activity, and the older a fossil is, the more likely this is to have happened.
2 some Archaea and Bacteria.
3 *Hyracotherium*.
4 loss of toes and development of hooves; animals had become larger.
5 bacteria and archaeans.
6 unique chemicals from their cell membranes.

Lesson 7.6 Darwin and Wallace

1 he was accumulating evidence for the theory.
2 Charles Darwin and Alfred Russell Wallace.
3 warning coloration warns predators that an organism is distasteful. Predators learn to avoid it, so the genes giving warning coloration are passed on. Organisms not having the warning coloration are likely to be eaten.
4 wasp, cinnabar moth larva/caterpillar; other named examples.
5 predators will avoid the mimic as well as the distasteful organism; forms of certain organism, e.g. certain butterflies; have developed (/evolved) to resemble distasteful butterflies.
6 the predator will 'take a chance' and eat an organism (the mimic or the 'model') if there is a low probability of finding that it is distasteful.

Lesson 7.7 A new species

1 an organism that is unable to interbreed successfully and produce fertile offspring.
2 because of the odd number of chromosomes, chromosomes cannot pair up for meiosis to occur (to produce gametes).
3 evolution is slow, occurring over millennia.
4 a short life cycle/rapid reproductive rate.

Lesson 7.8 Evidence of natural selection and evolution?

1 mutation.
Note that pollution is NOT the answer.
2 pollution was blackening bark and killing lichens on trees. The pale moths were easily spotted by birds and eaten; the dark/melanic form was camouflaged. The dark form therefore had a selective advantage.
Note that Tutt would not have known about genes at this time.
3 released identical numbers of pale and melanic moths. Carrying out the experiment in additional areas.
4 collecting/breeding sufficient numbers of moths for the investigation; recapturing the moths.
In fact, Kettlewell had used a combination of moth trap caught and laboratory-bred moths. Their behaviours could have been different from wild moths. It would have been better to have obtained them from a single source. Recent studies by Professor Mike Majerus sourced all the moths from within 5 km of the test site and analysed the data from moths sourced in different ways separately.
With small proportions being recaptured, it is difficult to know the extent to which the recapture technique and the numbers of moths being eaten have each contributed.
5 anti-pollution measures (including the Clean Air Act, 1956).
6 melanism was found in other animals, such as ladybirds.
Many other studies confirmed Kettlewell's findings. These included, in 1972, a series of investigations at seven sites along a transect from urban Liverpool to rural North Wales.
As anti-pollution measures of the 1950s and 1960s took effect, the dark form also decreased in abundance in the USA and in The Netherlands.
Experiments by Mike Majerus over a 7 year period in the early 2000s supported Kettlewell's findings.

Lesson 7.9 Key concept: Evolution: fitting the pieces of the jigsaw

1 Mendel's paper was not widely read. It was published in a local journal and in German.
2 no one yet knew.
It was thought to be some kind of 'blending' of characteristics from parents.
3 geneticists had only worked with characteristics that showed either/or traits, and could not explain the variation in populations that led to natural selection.
4 scientists (J.B.S. Haldane, R.A. Fisher and S. Wright) who produced mathematical models of natural selection and mutations.
5 Thomas Hunter Morgan found that it was genes, located on chromosomes that controlled inheritance. His work confirmed the chromosomal theory of inheritance that we use today.
6 It has taken years to develop the technology to be able to investigate the genetic code in detail. The fact that only a very small percentage of genes code for proteins (the rest code for switching on and off of genes) has made the job of deciphering the genetic code more complex and time consuming than first thought

Lesson 7.10 Antimicrobial resistance

1 in a population of bacteria exposed to the antibiotic, some are resistant because of mutations; bacteria with the resistant genes survive and pass these on to offspring as they divide; the whole population becomes resistant to the antibiotic.
2 owing to genetic variation, some bacteria are better able to survive in this (changed) environment; characteristic/antibiotic resistance spreads throughout the population.
Note that new species of bacteria are not produced, simply resistant strains.
3 'last-resort' antibiotics.
4 we won't be able to treat common infectious diseases with common antibiotics, and some of these may be fatal. Organ transplants would not

be possible because of the risk of infection (which normally would be controlled by antibiotics) and certain cancer treatments that weaken or destroy the immune system would not be possible as they are likely to lead to the death of the patient owing to infection.

5 over-prescription of antibiotics.

6 resistance is proportional to the use of penicillin; it is highest in countries where prescription is highest, e.g. France, and lowest where penicillin use is lowest; there are some anomalies, e.g. Spain and Hungary, where resistance is perhaps higher than expected from the trend.

Lesson 7.11 Combatting antimicrobial resistance

1 any bacteria left may have a slightly higher resistance that those killed first.

2 limit the prescription of antibiotics; not prescribe antibiotics inappropriately, i.e. when not absolutely necessary, or for diseases caused by viruses.

3 currently lower than cancer, diabetes, diarrhoeal disease and road traffic accidents/higher than measles, cholera, tetanus;

By 2050, will be the biggest killer (assuming other death rates do not increase markedly).

4 deaths continue at the same rate; rate of developing resistance continues at the same rate; no way of making antimicrobials more effective discovered; no new antimicrobials/other method of treatment discovered that reduce the rate.

Note that this is just *one* mathematical model/other mathematical models may suggest a (slightly) different trend.

It's also appropriate to say that some other catastrophic event may occur which could affect the world's population.

5 cost of treatment/more medical staff required; research and development/search for new antimicrobials/treatments; health effects on the workforce.

6 research, development and trialling of new antibiotics is costly and takes a long time; drugs companies may be reluctant to invest money as an antibiotic developed is likely to be used only as a last resort and therefore not generate a profit.

7 student answer. Sponsorship by government or entrepreneurs would provide an incentive for pharmaceutical companies and offset potential losses. It's a global problem that needs a global solution which perhaps should not be down to individual governments/pharmaceutical companies/research institutes.

Lesson 7.12 Selective breeding

1 amount of muscle/meat; amount/length/quality of wool; amount and quality of milk.

2 animals/plants with a desirable characteristic(s) bred; others prevented from breeding; the offspring with the best of the characteristic selected and bred; the process is repeated over many generations.

3 hunting; temperament; named features relating to appearance.

4 capable of interbreeding and producing fertile offspring.

5 the dogs become inbred; have limited gene pool/the incidence of rare disease alleles will increase with no outbreeding.

6 hip dysplasia; heart valve problems; deafness; puppies may need to be born by Caesarean section.

7 Benefits:

Desirable traits can be bred, distinctive features may be aesthetically pleasing for owners, more predicable temperament and behaviour of dogs

Health risks:

Conditions such as hip dysplasia, deafness, hips to narrow for birth can occur from inbreeding to maintain pedigree lines.

Lesson 7.13 Producing new plant varieties

1 grow and mature quickly; taste, aroma, colour; long shelf-life, store well or can be frozen.

2 if susceptible/unable to tolerate the change, they may all die.

3 physically transfer pollen to flower to be fertilised; place a bag over the fertilised flower to prevent further pollen reaching the flower.

Note that some plants, e.g. hops, have male or female flowers. For plants having flowers with male and female parts, the male parts of the flower to be fertilised would have to be removed to prevent self-pollination.

4 because they may have genes/alleles, bred out of recent varieties/cultivars, that will be an advantage for new varieties to have, e.g. for disease resistance.

5 many plants are grown and trialled over many generations; the offspring each go through a selection process, e.g. exposure to fungal diseases, testing for desirable properties/qualities.

6 bitterness; yield; resistance/tolerance to fungal disease.

Lesson 7.14 Genetic engineering

1 taking specific genes from one organism and introducing them into the genome of another.

2 *Escherichia coli* (*E. coli*); yeast (*Saccharomyces cerevisiae*).

3 increase resistance to disease; increase yields; bigger and better fruit.

4 fungi; viruses.

Note that bacteria cause far fewer diseases in plants; they are more likely to cause disease in animals.

5 an enzyme.

6 using a vector, which is often a plasmid (or a gene gun).

Lesson 7.15 Genetically modified crops: the science

1 crops could grow in poor soils and other harsh environments.

2 crops could be fortified with extra nutrients.

3 weeds will compete with crops for light, water and nutrients (and can also encourage fungal growth); spraying a field of a herbicide-tolerant crop with

the herbicide will kill weeds but not the crop; there will be no loss of yield resulting from weed growth .

4 because of mutation, genes/alleles making the insect resistant to insecticide will arise; other insects killed; because of reproductive rate, resistant gene/allele will spread quickly throughout the population.

5 seed carried by wind/other agents; may cross-pollinate with a wild or a cultivated relative of the GM crop.

6 possible effects on non-target organisms/studies have suggested that pollen *may* be toxic to non-target organisms, e.g. the monarch butterfly.

7 Issues with safety may include:

- Cross pollination of GM crops with wild or cultivated relative of the crop
- Studies have shown GM crops can swap man-made genes.
- Pollen of GM crops can be carried by wind and may be toxic to other insects that are essential pollinators of crops and other plants (although evidence suggests pollen levels of GM crops are not high enough to be toxic)

Lesson 7.16 Is genetic modification safe?

1 *E. coli* lives in the human gut, so an escaped GM version could colonise it; yeast lives on the surface of fruit, so the GM version could be transferred to humans through fruit.

2 deactivate the genes involved in infection.

3 the marker is needed to so that any cells not having taken up the gene are killed/isolate cells having taken up the gene successfully.

4 kanamycin is an antibiotic that is no longer used, so the effectiveness of antibiotics will not be affected by an organism developing antibiotic resistance to this antibiotic.

5 most scientists have no found no evidence that GM crops are harmful; no evidence that they cause allergies; some studies have suggested that they may cause cancer, but these are controversial.

6 evidence of correlation between glyphosate use and an increase in a number of conditions ranging from autism to cancer (though this does not mean cause).

Lesson 7.17 Ethically wrong, or essential?

1 increasing food supply/security/quality for world's populations.

2 it cannot be justified to change the genome of another organism/other reasonable answer.

3 (if cross-pollinated/cross-fertilised,) it will not breed true/have different characteristics.

4 so that the public know that it would be a GM food that they're eating and can make a decision as to whether to buy/eat it, or not.

5 DNA analysis.

6 they caused leukaemia, probably because of the viral vector used.

7 if successful, would improve the life of the patient, but the defective gene/allele would still be passed on to offspring (unless the germline was modified).

8

- Patients have died from massive immune responses or died from leukaemia in early trials of gene therapy. It could be argued that it is unethical to continue conducting gene therapy trials on humans due to this risk.
- Some people might also argue that altering genes goes against nature or god.
- It might save lives but we don't know the long term outcomes.

Lesson 7.18 Cloning

1 tissue culture.

2 (taking) cuttings.

3 so that animals identical genetically to those having desirable characteristics, e.g. producing high milk yields or quality beef, can be produced.

4 a developing embryo is removed from a pregnant female early in pregnancy; the unspecialised cells are separated, grown in culture, then transplanted into a host mother.

5 the nucleus is removed from a body cell and inserted into an egg cell that has had its nucleus removed; the egg cell is stimulated to divide; the embryo, once it has divided to produce a ball of cells, is transplanted into the animal.

6 the embryo is implanted into a 'mother' in adult cell cloning.

Lesson 7.19 The tree of life

1 Carl Linnaeus.

2 binomial system.

3 microscopical examination has improved our understanding of the physical characteristics, and therefore relationships of organisms; understanding of the biochemistry and biochemical processes of an organism has helped us to understand relationships.

4 Archaea. Bacteria, Eukaryota (see pages 10-11).

5 comparing the base sequences of DNA/structures of other chemicals, e.g. proteins, from organisms.

6 1 and 3: distance value =

$$\frac{\text{number of changes}}{\text{length}} = \frac{4}{20} = 0.2$$

2 and 4: distance value =

$$\frac{\text{number of changes}}{\text{length}} = \frac{4}{20} = 0.2$$

1 and 4: distance value =

$$\frac{\text{number of changes}}{\text{length}} = \frac{3}{20} = 0.15$$

Lesson 7.20 Extinction… or survival?

1 value from graph/443 million years ago (Ordovician-Silurian);

value from graph/359 million years ago (late Devonian);

value from graph/248 million years ago (Permian);

value from graph/200 million years ago (Triassic-Jurassic);

value from graph/66 million years ago (Cretaceous-Tertiary).

2 we can only estimate the number of species alive today; the fossil record is very incomplete.
3 because of the introduction of new species, which attack organisms and destroy nests; introduce disease.
4 eradicating new species.
5 transformation of the landscape/habitat loss; overexploitation of species; pollution; introduction of alien species.
6 they pollinate over 70% of the crop species that provide 90% of food supplies; with the loss of bees, the world will lose most of its food supply.
7 Points may include:

- Humans have transformed many landscapes and destroyed the natural habitat and food sources for many species
- Humans have overexploited some species through hunting or fishing
- Pollution has destroyed habitat for many species
- Humans have introduced alien species which have affected numbers of native species

Lesson 7.21 Maths skills: Using charts and graphs to display data

1 student plot.
2 25%
Calculation:
100 – (14 + 33 + 3 + 22 + 3)
3 student chart. Two bars should be drawn, side by side, to show numbers of species threatened with extinction, and number of species.
4 student chart. This should be a stacked bar chart with a single bar.
5 student chart. Note that there are no gaps between the bars of the histogram.
6 student data and chart. Note that there are no gaps between the bars of the histogram.

End of chapter questions

1 Charles Darwin – Wrote *On the Origin of Species*.
Gregor Mendel – Carried out genetics experiments on peas.
Alfred Wallace – Described warning coloration. (1)
Note that all need to be correct for (1) mark.
2 Two correct answers from:
Animal/plant/organism dies and sinks in/becomes covered by water (1)
becomes covered in sediments (1)
soft parts of the body decay (1)
the sediments are compressed by further layers (1)
the skeleton is turned to rock/mineralised as an exchange of minerals occurs between the skeleton and the water (1).
3 The differences between individuals due to our genes. (1)
4 Genetic causes – one correct answer from:
Natural hair colour (1)
Eye colour (1)
Other characteristic relating to inherited facial feature, e.g. shape of nose (1)
Environmental causes – one correct answer from:
Tattoos (1)
Spectacles (though note that eye defects could have a genetic cause) (1)
Hair length/style (1)
Colour of dyed hair (1).
5 Barchart drawn correctly (1)
6 30% (1)
7 One mark for the correct answer:
Trilobites are the oldest;
Scallops are the youngest. (1)
Note that both need to be correct for (1) mark.
One mark for the explanation:
Trilobite is in the lowest/scallop is in the uppermost rock sediment. (1)
8 Three correct answers from:
it could be:

- A predator (1)
- Compete with native organisms for food (1)
- Introduce new diseases (1).

9

a. mutation (1)
b. Plotting (3)
All six points plotted correctly (3); 4-5 points plotted correctly (2); 1-3 points plotted correctly (1)
Points connected using appropriate line of best fit (a smooth curve) (1)
c. One correct answer from:
Reduction in smoke (1)
Means that the dark form of the ladybird is less camouflaged on surfaces with less soot (1).
So eaten/predated on (1).
One mark for prediction:
Further reduction in number (as less pollution) (1).

10 Two correct answers from:
Required gene cut from genome of human cells using enzyme (1)
Plasmid is cut open (using the same enzyme) (1)
Inserted in plasmid of bacterium/E. coli/yeast (1)
Using same enzyme (1).
11 Risk:
Safety related to new food or medical product (1)
Safety related to new organism (containing it/environmental risk to other organisms) (1)
Ethical:
Some groups object to interfering with nature/humans have no right to manipulate the genome of another organism/other suitable answer (1).
Note that there needs to be one appropriate risk and one appropriate ethical consideration for (1) mark.
12 Two from:
Transformation of the landscape (1)
Overexploitation of species (1)
Pollution (1)

Introduction of alien species (1)

13 Four correct answers from:

Choose plants from a wild population/existing cultivars (1)

That are most resistant to drought (1)

Transfer pollen from one to female parts/stigma of the other (1)

Prevent further pollination of plant (1)

From offspring, select those most resistant to drought and cross (1)

Repeated over many generations (1).

14

a. Two correct answers from:

Herbicides intended to kill weeds (1)

Weeds will compete with crop plants/encourage disease and reduce yields (1)

Spraying kills weeds but not crop (1)

Crop yields not reduced by weeds (1).

b. Four correct answers from:

Graph shows a (steady) increase in reports of herbicide-resistant weeds (1)

Resistance develops because mutation in weeds (1)

Means that some weeds will be herbicide-resistant (1)

These weeds will survive spraying with herbicide (1)

And pass on genes/alleles for resistance (1)

Resistance develops rapidly because of rapid reproductive rate of weeds (1)

And possible cross-breeding (1).

c. Four correct answers from:

There is a <u>correlation</u> between the increase in the number diagnosed with diabetes and mass of glyphosate applied/as glyphosate use increases, so does percentage of those with diabetes (1)

Increase in diabetes is far greater than that predicted by trend modelled on 1980s'/1990s' trend (1)

Suggests use of glyphosate is linked with/leads to/causes diabetes (1)

But correlation does not mean cause/humans exposed to other factors that could increase diabetes (1)

Scientists need to look for mechanism of possible link (1).

Chapter 8: Ecology in action

Lesson 8.1 Key concept: Learning about ecosystems

1 The interaction of a community (of living organisms) with the non-living parts of their environment.

2 Large fish, small fish, coral, sea plants

3 a A habitat is where an organism lives but an ecosystem is the interaction of a community with the non-living parts of their environment.

b A population is the total number of one organism in an ecosystem but a community is all the different species of plants and animals in an ecosystem

4 Sea water, dissolved oxygen cold, large and small fish, coral, sea plants etc.
High altitude mountainous terrain with low oxygen levels little plant life and water running through etc
Deciduous forest with trees and bluebells etc
Freshwater pond, water plants, stones etc

5 Plants use carbon dioxide to grow and produce food (and oxygen) for animals. Some animals eat other animals. Animals use oxygen and produce carbon dioxide. Dead animals are broken down by fungi, bacteria etc to provide nutrients for plants etc

6 All animals depend on plants for oxygen and food. Plants depend on animals for carbon dioxide, pollination and seed dispersal.

7 Areas of high biodiversity are important because they;
Allow a wide variation of food sources to support a species survive.
Provide us with food and other products such as timber.
Play a role in maintaining the atmosphere and water cycle.

Lesson 8.2 Changing abiotic factors

1 For example:
Abiotic – light intensity – no photosynthesis so plants would not grow and animals would have less food and oxygen
Biotic – new predator will kill many of one species of animal, this could result in plants not being used for food and growing more

2 Use more of the water and nutrients in the soil, take up more space etc

3 Amount of sea water, temperature, light availability, tides etc
Some sea weeds may not like being submerged, need more light for photosynthesis etc.

4 They can withstand not being submerged in water because of their shell, more food available etc.

5 where the biotic and abiotic factors are in balance so that population sizes remain fairly constant

6 Little grazing allows a few plants to out-compete others. As grazing increases more plant species grow because dominant plants are controlled by the animals, allowing weaker species to grow. Only specially adapted plants can resist the effect of intensive grazing and survive.

Lesson 8.3 Investigation predator-prey relationships

1 Humans eat producers (fruit and vegetables) and other consumers (meat)

2 a Grow and reproduce more

b Grow and reproduce less, may die from starvation and population decreases or dies out

3 Tapeworms feed on food digested by humans and the person suffers from malnutrition

4 In parasitism one organism benefits but the other is harmed by the relationship. If parasites kill their host, they die too.

5 Large predator population causes number of prey to fall as less available to reproduce, or may cause them to die out. Small predator population causes prey numbers to increase as more can reproduce

6 Prey numbers have to increase so more food is available to sustain a larger population of predators before predator numbers can increase
ORA

Lesson 8.4 Looking at trophic levels

1 Feeding positions in food chains

2 a Mackerel – level 4

b Zooplankton – level 2

c Producer – level 1

3 Wet mass can vary

4

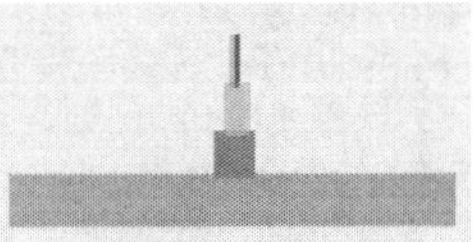

5 Trout – bar with 10 squares
Frogs – bar with 25 squares
Insects – bar with 50 squares
Pondweed – 2500 squares

6 Would have to kill and dry krill, whales and humans

Lesson 8.5 Transferring biomass

1 photosynthesis

2

- some plant material, which cannot be digested, leaves the body as faeces
- some animal material cannot be digested, e.g., bone, horn, hooves, claws and teeth
- biomass eaten by animals is also used in respiration to release energy, and leaves the animal as carbon dioxide and water.

3 There is much more indigestible food in plant material than meat

4 Some is used for respiration, movement, growth and some is waste (urine, faeces and methane)

5 need to eat lots of plant material to convert sufficient amounts into new biomass and much plant material is indigestible

6 a 4%

b Not as much energy is used to keep warm and move around so more can be used for growth

Lesson 8.6 Competing for resources

1 Animals and plants that get more of the resources are more successful than those that get less. Successful organisms are more likely to survive and reproduce so the size of their population is more likely to increase.
Animals will travel to where food is available.

2 For a mate
For water

For water, nutrients and sunlight
For food

3 If they are not matched, one will become more successful than the other

4 Resources in any habitat are limited

5 Any correct examples e.g. seals competing for a mate is intraspecific, desert plants competing for water is interspecific

6 Animals try to avoid competition with other species if they can but need to compete with members of their population for food and to reproduce.

Lesson 8.7 Required practical: Measure the population size of a common species in a habitat

1 a Reduces aeration (amount of air) and irrigation (amount of water) soil is able to hold

b More difficult for roots to penetrate, less water in soil, as well as physical damages of trampling

2 Meristems – daisies unable to grow
Leaves – daisies unable to photosynthesize
Flowers – daisies unable to reproduce

3 Daisies may dehydrate as lack of surrounding grass to protect them. Excessive water loss, even death

4 The areas that are un-trampled will have less daisies due to competition with other plants for resources. Trampled areas will have more daisies because they are tougher than the other plants

5 Make a transect line across each of the investigation areas with the tape, one at a time and place the quadrant every 5m along the line, starting at 0m.
In each placement of the quadrat, count the number of whole daisy plants completely within the quadrant. Count the number of daisy plants partially in the quadrant along 2 adjacent sides only. Add these numbers together and record them in a table.

6 The plant may not be in flower; the flower may be in the quadrat, but not the whole plant; to get a more accurate record of plants growing, not just those in flower.
12 (accept 11)

7 By adding all the daisy plants they found altogether and divide by the number of quadrats – gives a plant density per m^2

8 The other student may have carried out a similar investigation, using a different transect. Pooling the results will make them more valid, and minimise the effect of any anomalies

9 There are more daisies in the trampled area (converse: there are less daisies in the un-trampled area). The daisies in both areas were approximately equally spread along the transect/in the area

10 Other plants (grass) out-competed the daisies for resources in the un-trampled areas.
Trampling had more effect on the other plants, but daises are tougher – leaves not delicate, tough cuticle to withstand extreme temperatures and long fibrous roots (compacted soil less impact than for plants with shallow roots, able to reach deeper water reserves).
When other plants died off due to trampling, daisies had more resources available, and so there were more of them.

11 Proved/upheld

Lesson 8.8 Adapting for survival in animals

1 They allow them to survive and be successful in their habitat.

2 Correct description of any 3 adaptations, e.g.
- Folded skin to increase the surface area for heat loss in elephants
- Thick fur and blubber to reduce heat loss in polar bears etc.

3 Stay under rocks or go underground during hot days and cold nights, where temperatures are more stable.

4 Stay in burrows during hot days and cold nights to try and maintain constant temperature, brown colouring for camouflage, eyes on side of head to look for predators etc.

5 having a large surface area to volume ratio maximises heat loss through the skin; a small surface area to volume ratio reduces heat loss through the skin

6 structural adaptations form part of the body whereas functional adaptations refer to how the body functions or operates as an adaptation to the environment.

Lesson 8.9 Adapting for survival in plants

1 They grow close to the ground (avoids damage by wind) and have small leaves (conserves water).

2 Extremophiles are organisms that can survive in very extreme environments

3 Their enzymes do not denature in extreme temperatures so chemical reactions can still take place in cells

4 Pine trees are found in colder climates where there is ice and snow in winter. Needle leaves have a smaller surface area to reduce evaporation so the trees can save water.

5 Plants in hot climates adapt by reducing their surface area to volume ratio by being for example barrel shaped (so there is a smaller area for water to be lost from to reduce water loss)
Plants in cold climates adapt by having small leaves i.e. small SA:V so smaller area with fewer stomata to lose water from (conserves water).

6 Roses do not have fleshy stems to store water, have a shape with a large SA:V e.g. have large flat leaves so water loss will be very high, have a thin cuticle that allows water loss etc. but cacti have fleshy stems to store water, have a shape with a small SA:V etc.

7 Tropical plants have been found to have a higher density of stomata than desert plants. This is because desert plants need to conserve water more than tropical plants due to the nature of the environment. Water is lost thorugh the stomata during transpiration so a lower density of stomata means less water is lost, helping the plant survive.

Lesson 8.10 Cycling materials

1 Decomposers that break down carbon include worms, bacteria and fungi

2 Decomposers release carbon (and other nutrients) in dead plants and animals

3
- Precipitation: water droplets in clouds get bigger and heavier. They fall as rain, snow or sleet.

- Evaporation: water evaporates as it is heated by the Sun's energy. Water vapour is carried upwards in convection currents.
- Transpiration: water vapour is released into the air through stomata in leaves.
- Condensation: water vapour rises, cools and condenses back into water droplets that form clouds.

4 a When it is water vapour
b When it falls as snow or hail
5 break down the smaller pieces of dead material. Decomposers release waste carbon dioxide, water, heat and nutrients (that plants use).
6 Maintains a supply of freshwater for plants and animals

Lesson 8.11 Cycling carbon

1 Producers use carbon dioxide to photosynthesise. The carbon is used to make carbohydrates, proteins, fats and DNA that form new biomass, which is eaten by consumers.
All organisms respire to release energy for cellular processes. They release waste carbon dioxide back into the environment, which is then used by plants.
2 Because most of the carbon dioxide is combined with other elements, such as in coal and other fossils fuels or in carbonate rocks. Only carbon dioxide in the air can be used by plants for photosynthesis
3 Respiration and combustion
4 To include: animals eat plants and absorb carbon from them. Carbon passes along food chains. Energy is transferred along the food chain and to the environment at each trophic level. Carbon dioxide is returned to the atmosphere by:
- Plants, animals and decomposers respiring:
- glucose + oxygen → carbon dioxide + water
- burning (combustion) of fossil fuels and wood:
- fossil fuel/wood + oxygen → carbon dioxide + water

5 Plants need supplies of carbon dioxide to photosynthesise and carbon is built into glucose which is needed for respiration to release the energy living cells need for chemical reactions.
6 Respiration – reacting oxygen and glucose (containing carbon) to release energy
Photosynthesis – reacting carbon dioxide with water to produce glucose needed for respiration
Combustion – burning wood and fossil fuels which releases carbon dioxide
Feeding – digestion of food containing carbon
Egestion – waste food eliminated from the body
Death – dead plants and animals are broken down by decomposers releasing carbon (and other nutrients)
Dead plants and animals may also form fossil fuels over millions of years

Lesson 8.12 Investigating decay

1 Warmth, oxygen and water (and microorganisms)
2 Compost is made by placing the compost bin in the sun (warmth), mixing the contents to let oxygen in and adding moisture.
3 generators hold rotting organic material (dung, farm and garden waste) which is broken down by anaerobic bacteria
4 As temperature increases biogas production increases
5 Aerobic bacteria cause most decay in compost heaps. They respire using oxygen to release energy for growth and reproduction. The more oxygen present, the more decay happens. Bacteria feed on the waste using extracellular digestion
6 The hotter it is the quicker the biogas is produced until it reaches 40°C because the enzymes have more energy. At 40°C the enzymes are denatured and the process slows down quickly and stops
The more oxygen is available, the faster the rate of respiration. This releases more heat for the enzymes
7 If decomposers did not exist, decomposition would not occur so the remains of dead plants and animals and all their waste products (there would be a lot of excrement lying around!). A lack of decomposers would also affect the recycling of materials such as the carbon cycle.

Lesson 8.13 Required practical: Investigate the effect of temperature on the rate of decay of fresh milk by measuring pH change

1 10cm^3 measuring cylinder or a 10cm^3 calibrated dropping pipette
2 The water baths give the more accurate measurements, because you are able to keep the temperature constant. Warm water in a beaker will quickly cool down, possibly slowing/effecting/changing the reaction rate
3 A pH probe connected to a data logger will give more accurate results, as you will have readings of the exact pH. The phenolphthalein indicator tells you more approximately the pH, depending on comparing the colour in the test tube against a colour card, it is more subjective.
4 Care in measuring and transferring the chemical; wear safety goggles; clean up any spillages appropriately
5 So that he could work out the best intervals for his variable (minutes) – not too long, not too short, around the time when the indicator changed colour
6 Time, in minutes
7 The alkalinity of the solution/ colour change, pH
8 Independent variable values in left hand column 1 – 5
2nd column - test 1 results
3rd column – test 2 results
4th column – test 3 results
5th column – average for each independent variable value
9 At 0°C the reaction only happened very, very slowly and the pH dropped only very slightly.
At 20°C the reaction started after about 2 mins and the pH steadily and quite slowly dropped, eventually reaching to about pH 8.6
At 37°C (human body temperature) the pH initially went up fractionally (anomalous/ inaccurateness of indicator) after 3 minutes, but then dropped quickly and steadily to reach pH just above 8 (8.08)
At 80°C the reaction had finished after little more than 2 minutes, the pH reached after 4 minutes did not dip any lower, about pH 8 (8.04)
Zak's hypothesis was mainly upheld – the warmer the solutions, the faster the pH fell, although the

investigation carried out at 37°C reached a final pH similar to the investigation at 80°C

10 Use a water bath rather than a beaker of warm water and use a pH data logger rather than phenolphthalein indicator

Lesson 8.14 Changing the environment

1 Natural changes e.g seasonal changes, geographic location etc
Changes caused by humans.

2 Their numbers and distribution may change e.g. they may migrate, stay dormant (seeds) etc.

3 global warming the seas are becoming more acidic (and warmer), which means that the shells of calcareous organisms (such as bivalve molluscs) are dissolving and becoming thinner.
Erosion by rivers and tides may gradually destroy habitats.
Volcanic eruptions/tsunamis destroying environments etc.

4 Pollution e.g. pesticides entering waterways and building up in food chains etc.
Dredging sea beds for building projects increasing sand in the water so less sunlight can reach marine plants for photosynthesis

5 Very small rises in sea temperature can have a massive effect on biodiversity e.g. coral reefs, with 20% being destroyed in just 50 years. If this continues, 200 years all coral reefs will have disappeared, but if sea temperatures continue to rise they will disappear even faster, so it is really important that we act quickly to try and stop the rise.

Lesson 8.15 Learning about land use

1

- farming
- building
- quarrying
- dumping waste.

2 To produce as much crop as possible as quickly as possible and in a smaller space

3 Biodiversity will be reduced further. More land is needed for homes, shops, factories and roads. Building sites destroy habitats. Roads divide habitats making it harder for organisms to find food and mates.

4

- New quarries are mined to provide stone, slate and metal ores for building materials. Habitats are destroyed.
- More farmland is needed and fertilisers are used. Many farms grow one crop over huge areas. This affects food availability for insect pollinators. There are fewer available nesting sites for birds.
- More waste is sent to landfill, and more sewage and industrial waste are produced. This can pollute the land.

5 Eutrophication will happen:nitrates and phosphates in the water increase and so then algal growth increases. Algae cover the water surface and prevent light from reaching water plants. The plants and algae die. Bacteria respire as they break down dead plants and use up oxygen in the water. The other living organisms in the water die.

6 e.g. grassland can be changed into landfill sites. Waste is sent to landfill, chemicals can leak into the soil and pollute it. There is very little biodiversity as few organisms can live there. However, when the sites are full, they are re-developed into parks etc. and biodiversity increases again.

7 We are dependent on a rich biodiversity to provide the resources that we need (food, natural materials for building, textiles etc.) and for our survival

Lesson 8.16 Changing the landscape

1 provide land for cattle and rice fields to produce more food and to grow crops

2 Peat is plant material that is not completely broken down because of acidic conditions with little oxygen. It is used as a fuel and compost

3

- increases the release of carbon dioxide into the atmosphere
- reduces the rate that carbon dioxide is removed from the atmosphere (by photosynthesis)
- reduces biodiversity;
- Increases methane in the atmosphere

4 Plant trees to replace those cut down

5 Peat is a really useful fuel and compost but peatlands are also very unique ecosystems with a large biodiversity. Destroying the peatlands can result in many species becoming extinct as well as the loss of an important carbon source. Other fuels are available that could replace the use of peat and compost can be made by other means. The use f peat needs to be controlled to ensure the survival of the ecosystem

Lesson 8.17 Thinking about global warming

1 A rise in the average global temperature of the Earth

2 Increasing levels of carbon dioxide and methane

3

- increases the release of carbon dioxide into the atmosphere because less photosynthesis takes place and therefore less carbon dioxide absorbed. Burning of the trees also releases carbon dioxide as does respiration of the microorganisms that decompose the remaining plant material.
- reduces biodiversity as fewer trees and less habitat and diversity of habitat to support other species
- Increases methane in the atmosphere because land is cleared and used to grow rice in swamp-like fields or to raise cattle both of which produce methane

4

- loss of habitat due to polar ice and glaciers melting and rise in sea levels causing flooding of low-lying land
- changes in the distribution of species in areas where temperature or rainfall has changed
- changes to migration patterns of animals as climates get hotter or cooler
- destruction of coral reefs due to change in sea temperature.

5

- loss of habitat when low-lying areas are flooded by rising sea levels that result from polar ice and glaciers melting.
- changes in the distribution of species in areas where temperature or rainfall has changed.
- changes to migration patterns of animals as seasonal temperatures and times change
- destruction of coral reefs which are sensitive to very small rises in temperature

Lesson 8.18 Looking at waste management

1 As the human population grows and living standards increase, we are using more resources: the demand for agriculture, manufacturing and industry increase, producing more waste.

2
- sewage
- smoke and toxic gases
- herbicides, pesticides and fertilisers
- lead
- paper and cardboard
- plastic products.

3 Fertilisers entering waterways, cause eutrophication.
Toxic chemicals from household and industrial waste can spread into soil at landfill sites and enter waterways. Pesticides and herbicides are also washed into waterways and pollute them. Untreated sewage, chemicals and parasites can enter waterways. Microorganisms that decompose sewage use dissolved oxygen, causing aquatic organisms to die.

4 Soot will cover the leaves and block he stomata so that gas exchange is reduced and photosynthesis is reduced. This will result in insufficient glucose for respiration in the cells, so cells die etc.

5
1. Small amounts of toxic substances - often from human activity - are taken up by plants.
2. These plants are eaten by primary consumers.
3. The primary consumers are eaten by secondary consumers, and the secondary consumers are eaten by higher level consumers.
4. At each stage harmless substances are excreted but the toxins remain in the tissues so the concentration of toxin becomes most concentrated in the body tissues of the animals at the top of the food chain.

Lesson 8.19 Investigating pollution

1 Changes the distribution and numbers of living organisms

2 An organism that is used to measure environmental change e.g. lichens

3 Yes. Stonefly larvae are sensitive to pollution and cannot survive but sludgeworms can tolerate pollution

4 Use pH paper or use a pH probe and meter

5 Fast running stream – A
Stagnant pool – C
Polluted pond - B

6 Sample B contained the least dissolved oxygen. It contained the indicator species which are able to tolerate pollution and low dissolved oxygen levels (rat-tailed maggot, sludgeworm, bloodworm)

7 Advantage - just need a water sample and do not need to survey the habitat
Disadvantage – it is just a short term, spot test

Lesson 8.20 Maintaining biodiversity

1
- Introducing breeding programmes for endangered species.
- The protection and regeneration (restoring) of rare habitats,
- Re-introducing field margins and hedgerows on farmland
- Reducing deforestation and carbon dioxide emissions.
- replanting trees.
- Recycling resources
- Cloning endangered plant species.

2 Hedgerows provide nesting sites for many birds, habitats for many insect species, small mammals etc.

3 We have a moral responsibility to protect species that have become endangered because of our actions

4 The animals are kept in a protected ecosystem, where the animals can breed and the young can be reared without danger from humans or predators.

5 To include:
- ensuring long-term funding
- having qualified scientists who understand the issues
- animals and plants do not recognise boundaries
- many organisations and governments may be involved
- lack of 'policing' of protected areas.

6 planting programmes are introduced and local environment agencies work to maintain healthy, litter-free sustainable forests

7 The protection of biodiversity is important because;
- It allows a wide variation of food sources to support a species survive.
- It can provide us with potential material and economic benefits
- Plays a role in recycling of nutrients
- Morally we should encourage a culture that respects a range of habitats

Lesson 8.21 Learning about food security

1 when all people have access to consistent supplies of safe and nutritious food that is available, affordable and useable to meet their needs for an active healthy life.

2
- increasing birth rate
- changing diets in developed countries
- new pests and pathogens affecting farming
- environmental changes
- cost of agricultural inputs
- conflicts and war.

3 Increasing birth rates mean that populations are growing faster and in some countries demand for food is increasing faster than

the increased food production, which threatens food security.
Decreasing rainfall means that some regions of the world suffer from famine and food insecurity continually

4 Changing diets in developed countries mean that foods are transported from countries where it is scarce, to other countries around the world. These crops are often grown at the expense of staple foods, for example, coffee.

5 Although world population and total crop yield show an increasing trend and world population is rising at a faster rate which will cause more food insecurity in the future

6 Cash crops for export are grown at the expense of food crops for local people
Some areas in developing countries suffer recurring droughts and cannot grow sufficient food crops for their needs
Some regions are more prone to natural disasters e.g. volcanic eruptions, violent storms and tsunamis etc.

Lesson 8.22 Maintaining food security

1 Farming using machines and chemicals to produce large yields very quickly.

2
- using fertilisers to provide essential nutrients to increase yield
- using insecticides and fungicides to kill pests, and herbicides to kill unwanted plants (weeds)
- growing plants in glasshouses where environmental conditions can be controlled e.g. temperature, CO_2 levels etc.
- using hydroponics to grow plants
- factory farming and fish grown in cages or tanks where they are protected from predators, expend little energy so more is used for growth and fed the correct diet.

3
- Mineral supplies can be controlled and unused minerals are recycled. Production costs are lower with no pollution risk to waterways.
- It is under cover so can control conditions and diseases better

4
- Using fishing quotas to conserve stocks at sustainable levels
- controlling net size to ensure that young fish mature and breed

5 Advantages:
- Fish grow quickly so high yield of fish to meet demand from human population
- Obtaining fish from fish farms helps allow wild fish stocks to recover

Disadvantages:
- Disease spreads quickly as fish in very close proximity
- Harm to local environment as chemicals and antibiotics used on fish farms can leach into local environment
- High cost of running fish farm.
- Ethics – is it moral?

6
- restricted movement of animals so that energy used for movement is minimal and more can be used for growth
- controlled temperatures, to reduce the energy that an animal uses to keep cool or warm. The energy saved is used for growth.

7 More food production in smaller spaces and more quickly through intensive farming of plants and animals
but
monocultures can be susceptible to new pests/diseases
Large fields with no hedgerows reduces biodiversity
Pesticides/herbicides can enter waterways and reduce biodiversity
Fertilisers can enter waterways and cause eutrophication etc.
Pens/cages for animals can be cramped and injure livestock
Diseases can spread quickly in cramped conditions
Transportation can be cramped and over huge distances
Slaughtering techniques can be inhumane etc.

Lesson 8.23 Using biotechnology

1 Biotechnology:
1. produces GM crops which can:
- be resistant to insect attack or to herbicides.
- produce increased yields because of the characteristics chosen

2. produce mycoprotein

2 It has been developed using a gene from carrots, which produces the β-carotene (which we say as 'beta-carotene') that makes vitamin A.

3 Advantages: produces large quantities of insulin very quickly; It is less likely to cause an adverse reaction; It overcomes ethical concerns about using insulin from pigs
Disadvantages: ethical concerns about the process and inserted genes having unexpected harmful effects.

4 Diabetes is becoming more common due to obesity rising

5
- constant oxygen supplies for Fusarium to respire
- water jackets to remove heat produced during respiration
- pH and temperature probes to monitor conditions and
- allow adjustments to be made quickly
- stirrers to spread heat, oxygen and syrup evenly through
- the vats and keep the fungus in suspension.

6 Large amounts of high protein foods can be produced very quickly
Foods can be modified to provide essential vitamins
There may be unknown health risks of genetically modified food.
Ethical concerns about creating new life forms, or moving genes between different species, especially if this may have side effects

Lesson 8.24 Maths skills: Using graphs to show relationships

1 At y = 2
2 At 0 (y = 0)
3 a About 15kg
b 15/4 = 3.75kg/month or answer from 3a divided by 4
c About 10kg
d 10/4 = 2.5kg/month or answer from 3c divided by 4
4 To check that their puppies are growing normally and the=at they are not over/under feeding them
5 So they know how big a dog they are thinking of buying, will eventually be
6 20°C
b 2 minutes
c 20/2 = 10°C per minute
7 Because the water has reached 100°C – its boiling point (extra heating will turn more of the water into gas/water vapour)

End of chapter questions

1 More food is produced/food is produced quicker/smaller areas of land needed
2 They eat plants and small animals/insects
3 Moisture/oxygen/warmth
4 So that they can get sufficient light to grow and flower
5 Biogas is flammable and needs to be removed if it escapes into the house
6 If it is sealed there will be a lack of oxygen for the micro-organisms to respire so decomposition will be slower. (1)
If the bag is left in the shade the temperature may be below the optimum temperature for the enzymes that control respiration so decomposition will be slower. (1)
7 C because it has the greatest yield with no fertilizer
8 So that it is a representative sample
9 Trophic levels are feeding positions in food chains. (1)
10

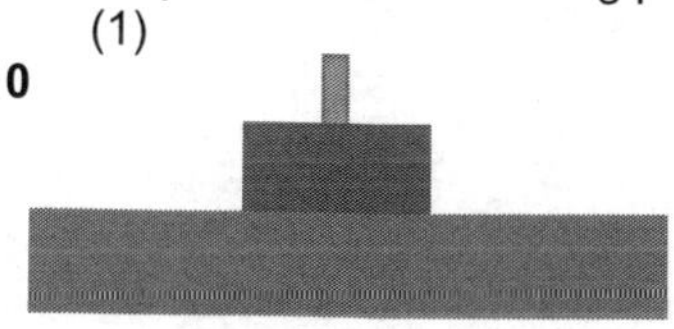

11
a. Total number of dandelions = 18, therefore mean is 18/6=3
b. A description including the following:
  - Divide that field area by the quadrat size. (1)
  - Multiply by the number of dandelions. (1)
c. Accept any one of the following:
  - the trees would reduce the light levels for photosynthesis
  - less water available for photosynthesis
  - less minerals available for growth.

12 The data shows that as the populations of rabbits increases the population of foxes increases (1)
T The cycles are out of phase with each other. The number of rabbits increases and then the number of foxes increases.
13 Impacts of deforestation may include:
  - Increased release of carbon dioxide into the atmosphere (due to burning of trees in slash and burn and respiration of micro-organisms that decompose remainders of the trees.
  - Reduced rate of removal of carbon dioxide from the atmosphere due to less photosynthesis
  - Reduced biodiversity/loss of habitat for species living in/around trees and those dependent on the trees as food source
- Many areas deforested in developing countries are used for rice fields or grazing meat cattle, both of these land uses result in increased methane released into atmosphere.

14
a. 0.675 m^3/day
b. Any 3 from:
  - As the temperature increases enzymes in the bacteria get closer to optimum temperature for respiration and reproduction and so the amount of biogas produced increases (1)
  - Decay continues to speed up and as bacteria respire even more, releasing heat making the detritus hotter and speeding up the decay.(1)
  - Once the temperature reaches 40°C the reaction slows down because the enzymes are denatured (1)

15 From 1800-1960 the human population rose slowly /from 1000 – 2000 millions but then rapidly increased /to 8000 million by 2010 (1)
From 1800 – 1920 there were few extinctions, but these increased slowly /to 5000 in 1980 and then rose sharply/to over 50000 by 2010 (1)
Extinctions show the same trend as human population increase (1)
Reasons include loss of habitat, hunting, poaching, pollution etc.
16 Excessive nitrates and phosphates from fertilisers are washed from the land into rivers/lakes (1)
  - Algal growth increases preventing sunlight reaching other plants. The plants die. (1)
  - Oxygen in the water is depleted as bacteria respire as they decompose the algae (1)
- Other living things in the ecosystem die because oxygen levels reach a point where no life is possible. (1)

17 The insulating blubber and fur (1) would prevent them losing heat so their body temperature could not be controlled (1)
Their white colour (1) means they are not camouflaged (1) etc.
18 As distance from the sewage outlet increases, to 7km downstream: decomposers in the water multiply as they breakdown the sewage / the more decomposers in the water the more dissolved oxygen they use (1)
less dissolved oxygen is in the water for other organisms and biodiversity will decrease (1)

after 7km downstream, dissolved oxygen levels
increase (1)
biodiversity will increase again (1)

Lesson number	Lesson title	Lesson objectives	AQA specification reference	Lesson resources (on CD ROM)	Collins Connect resources
Chapter 1: Cell biology					
1.1	Looking at cells	• Describe the structure of eukaryotic cells. • Recognise the order of magnitude of cells. • Explain how the main sub-cellular structures are related to their functions.	4.1.1.1; 4.1.1.2	Worksheets 1.1.1, 1.1.2 and 1.1.3; PowerPoint presentation	Quick starter Homework worksheet Homework quiz Slideshow
1.2	The light microscope	• Describe how to use a microscope. • Observe plant and animal cells with a light microscope. • Understand the limitations of light microscopy.	4.1.1.5	Worksheet 1.2; Practical sheet 1.2; Technician's notes 1.2; PowerPoint presentation	Quick starter Homework worksheet Homework quiz Slideshow
1.3	Required practical: Using a light microscope to observe and record animal and plant cells	• Apply knowledge to select techniques, instruments, apparatus and materials to observe cells. • Make and record observations and measurements. • Present observations and other data using appropriate methods.	4.1.1.2	Range of materials to look through the microscope, e.g. human hair, newspaper, string, prepared range of animal and plant cells. Worksheets 1.3.1 and 1.3.2; Practical sheets 1.3.1 and 1.3.2; Technician's notes 1.3; PowerPoint presentation	Quick starter Homework worksheet Homework quiz Video
1.4	Looking at cells in more detail	• Identify the differences in the magnification and resolving power of light and electron microscopes. • Describe simply how electron microscopes work in comparison to light microscopes. • Explain how electron	4.1.1.5	Worksheet 1.4	Quick starter Homework worksheet Homework quiz

Lesson number	Lesson title	Lesson objectives	AQA specification reference	Lesson resources (on CD ROM)	Collins Connect resources
		microscopy has increased our understanding of sub-cellular structures.			
1.5	Primitive cells	• Describe and explain the differences between prokaryotic cells and eukaryotic cells. • Explain how prokaryotic and eukaryotic cells evolved over time.	4.1.1.1; 4.1.1.2	Worksheet 1.5; PowerPoint presentation	Quick starter Homework worksheet Homework quiz
1.6	Cell division	• Describe the process of mitosis in growth, and mitosis as part of the cell cycle. • Describe how the process of mitosis produces cells that are identical genetically to the parent cell.	4.1.2.1; 4.1.2.2	Worksheets 1.6.1 and 1.6.2; Technician's notes 1.6; PowerPoint presentation	Quick starter Homework worksheet Homework quiz
1.7	Cell differentiation	• Explain the importance of cell differentiation. • Understand size and scale in relation to cells, tissues, organs and body systems. • Describe how cells, tissues, organs and organ systems are organised to make up an organism.	4.1.1.3; 4.1.1.4	Worksheet 1.7; PowerPoint presentation	Quick starter Homework worksheet Homework quiz
1.8	Cancer	• Describe cancer as a condition resulting from changes in cells that lead to their uncontrolled growth, division and spread. • Understand some of the risk factors that trigger cells to become cancerous.	4.2.2.7	Worksheets 1.8.1, 1.8.2 and 1.8.3; PowerPoint presentation	Quick starter Homework worksheet Homework quiz Slideshow Video

Lesson number	Lesson title	Lesson objectives	AQA specification reference	Lesson resources (on CD ROM)	Collins Connect resources
		• Use data to analyse and evaluate the impact of cancer.			
1.9	Stem cells	• Identify cells as being differentiated, or as stem cells. • Describe the function of stem cells in embryonic and adult animals. • Explain how stem cells can be useful.	4.2.1.3	Plants that have been grown from cuttings; or a plant from which cuttings may be taken, rooting hormone; Worksheets 1.9.1 and 1.9.2; PowerPoint presentation	Quick starter Homework worksheet Homework quiz
1.10	Stem cell banks	• Explore the use of stem cells in medicine. • Identify the risks in using stem cells. • Evaluate the benefits and disadvantages of using stem cells.	4.1.2.3	Worksheet 1.10; PowerPoint presentation	Quick starter Homework worksheet Homework quiz Video
1.11	Key concept: Cell development	• Revise ideas about cell structure, cell division and stem cells. • Apply ideas about cells to unfamiliar contexts. • Describe how plant meristems can be used in cloning.	4.1.2.3	Worksheets 1.11.1 and 1.11.2; PowerPoint presentation	Quick starter Homework worksheet Homework quiz Slideshow Video
1.12	Cells at work	• Recognise that all organisms respire. • Explain respiration as the process of making energy. • Describe aerobic respiration as an exothermic reaction.	4.2.2.1	Worksheet 1.12; Practical sheet 1.12; Technician's notes 1.12; PowerPoint presentation	Quick starter Homework worksheet Homework quiz
1.13	Living without oxygen	• Describe the process of anaerobic respiration. • Explain when anaerobic	4.2.2.1	Worksheet 1.13; PowerPoint presentation	Quick starter Homework worksheet Homework quiz

Lesson number	Lesson title	Lesson objectives	AQA specification reference	Lesson resources (on CD ROM)	Collins Connect resources
		processes occur. • Compare the processes of aerobic and anaerobic respiration.			Slideshow Video
1.14	Growing microorganisms	• Describe the techniques used to produce uncontaminated cultures of microorganisms. • Describe how bacteria reproduce by binary fission. • Calculate the number of bacteria in a population.	4.1.1.6	Worksheet 1.14; Practical sheet 1.14; Technician's notes 1.14; PowerPoint presentation	Quick starter Homework worksheet Homework quiz
1.15	Testing new antibiotics	• Use appropriate apparatus to investigate the effect of antibiotics on bacterial growth. • Use microorganisms safely. • Apply sampling techniques to ensure that samples are representative.	4.3.1.8; 4.3.1.9	Video 'Antibiotics – the beginning or the end?'; Worksheets 1.15.1, 1.15.2 and 1.15.3	Quick starter Homework worksheet Homework quiz
1.16	Required practical: Investigating disinfectants	• Carry out experiments with due regard to health and safety. • Present and process data, identifying anomalous results. • Evaluate methods and suggest further investigations.	4.1.1.6	Worksheet 1.16; Practical sheet 1.16; Technician's notes 1.16	Quick starter Homework worksheet Homework quiz Slideshow
1.17	Maths skills: Size and number	• To make estimates of the results of simple calculations, without using a calculator. • To use ratio and proportion to calibrate a microscope. • To recognise and use numbers in decimal and standard form.		Worksheets 1.17.1 and 1.17.2	Quick starter Homework worksheet Homework quiz Slideshow Video

Lesson number	Lesson title	Lesson objectives	AQA specification reference	Lesson resources (on CD ROM)	Collins Connect resources
Assessments		End of chapter test Student Book End of chapter test Collins Connect			
Chapter 2: Photosynthesis					
2.1	Explaining photosynthesis	• Identify the raw materials and products of photosynthesis. • Describe photosynthesis by an equation. • Explain gas exchange in leaves.	4.4.1.1	Images of unusual plants; Sort cards for photosynthesis equations. Worksheet 2.1; Practical sheet 2.1; Technician's notes 2.1	Quick starter Homework worksheet Homework quiz Slideshow
2.2	Looking at photosynthesis	• Explain the importance of photosynthesis. • Explain how plants use the glucose they produce.	4.4.1.1; 4.2.3.2	Time-lapse photography of growing plants; Sort cards for photosynthesis equations. Image of a woodland in autumn. Worksheet 2.2; Technician's notes 2.2	Quick starter Homework worksheet Homework quiz
2.3	Investigating leaves	• Identify the internal structures of a leaf. • Explain how the structure of a leaf is adapted for photosynthesis. • Recall that chlorophyll pigments in chloroplasts absorb light energy for photosynthesis.	4.2.3.1	Videos showing leaf structure using an electron microscope. Worksheets 2.3.1, 2.3.2 and Technician's notes 2.3	Quick starter Homework worksheet Homework quiz Slideshow
2.4	Required practical: Investigate the effect of light intensity on the rate of photosynthesis using an aquatic organism such as pondweed	• Identify and manage variables. • Process data and identify outliers. • Evaluate an experimental process.	4.4.1.2	Worksheet 2.4; Practical sheets 2.4.1, 2.4.2 and 2.4.3; Technician's notes 2.4	Quick starter Homework worksheet Homework quiz Slideshow

Lesson number	Lesson title	Lesson objectives	AQA specification reference	Lesson resources (on CD ROM)	Collins Connect resources
2.5	Increasing photosynthesis	• Identify factors that affect the rate of photosynthesis. • Interpret data about the rate of photosynthesis. • Explain the interaction of factors in limiting the rate of photosynthesis.	4.4.1.2	Worksheet 2.5; Technician's notes 2.5	Quick starter Homework worksheet Homework quiz
2.6	Increasing food production	• Identify the factors that increase food production. • Explain how these factors can be controlled. • Evaluate the benefits of manipulating the environment to increase food production.	4.4.1.2	Images showing how cold it is in the area by having snow on the ground and if possible evidence of geothermal activity. Worksheet 2.6; Technician's notes 2.6	Quick starter Homework worksheet Homework quiz Slideshow Video
2.7	Key concept: Diffusion in living systems	• Use concentration gradients to explain the direction of diffusion. • Apply the principles of diffusion to movement of different substances in plants.	4.1.3.1	Worksheets 2.7.1, 2.7.2 and 2.7.3; Practical sheet 2.7; Technician's notes 2.7	Quick starter Homework worksheet Homework quiz Slideshow Video
2.8	Looking at stomata	• Describe transpiration in plants. • Explain the structure and function of stomata. • Explain the relationship between transpiration and leaf structure.	4.2.3.1; 4.2.3.2	Celery that has been sitting in a beaker of red dye for 2 days; image of a water lily; Worksheet 2.8; Practical sheet 2.8; Technician's notes 2.8	Quick starter Homework worksheet Homework quiz Video
2.9	Moving water	• Describe the structure and function of xylem and roots. • Describe how xylem and roots are adapted to absorb water. • Explain why plants in flooded or waterlogged soil die.	4.2.3.1; 4.2.3.2	Worksheets 2.9.1, 2.9.2 and 2.9.3; Practical sheet 2.9; Technician's notes 2.9	Quick starter Homework worksheet Homework quiz Video

Lesson number	Lesson title	Lesson objectives	AQA specification reference	Lesson resources (on CD ROM)	Collins Connect resources
		• Explain how wilting occurs.			
2.10	Investigating transpiration	• Describe how transpiration is affected by different factors. • Explain the movement of water in the xylem.	4.2.3.2	Images of plants that are living in hostile environments. Worksheet 2.10; Practical sheet 2.10; Technician's notes 2.10	Quick starter Homework worksheet Homework quiz Slideshow
2.11	Moving sugar	• Describe the movement of sugar in a plant as translocation. • Explain how the structure of phloem is adapted to its function in the plant. • Explain the movement of sugars around the plant.	4.2.3.2	Scanning electron microscope image of a greenfly feeding. Worksheet 2.11; Technician's notes 2.11	Quick starter Homework worksheet Homework quiz
2.12	Maths skills: Surface area to volume ratio	• Be able to calculate surface area and volume. • Be able to calculate surface area to volume ratio. • Know how to apply ideas about surface area and volume.	4.2.3.2	Pieces of paper shaped like leaves or rectangles, material blocks, supply of 30 cm rulers and squared paper. Worksheet 2.12	Quick starter Homework worksheet Homework quiz Slideshow Video
Assessments		End of chapter test Student Book End of chapter test Collins Connect End of teaching block test Collins Connect			
		Chapter 3: Moving and changing materials			
3.1	Explaining water movement	• Describe how water moves by osmosis in living tissues. • Identify factors that affect the rate of osmosis. • Explain what the term 'partially permeable membrane' means.	4.1.3.2	Worksheet 3.1; Practical sheet 3.1; Technician's notes 3.1	Quick starter Homework worksheet Homework quiz Slideshow
3.2	Required practical: Investigate the	• Use scientific ideas to develop a hypothesis.	4.1.3.2	Worksheet 3.2; Practical sheets 3.2.1 and 3.2.2;	Quick starter Homework worksheet

Lesson number	Lesson title	Lesson objectives	AQA specification reference	Lesson resources (on CD ROM)	Collins Connect resources
	effect of a range of concentrations of salt or sugar solutions on the mass of plant tissue	• Plan experiments to test a hypothesis. • Draw conclusions from data and compare these with hypotheses made.		Technician's notes 3.2	Homework quiz Slideshow
3.3	Learning about active transport	• Describe active transport. • Explain how active transport is different from diffusion and osmosis. • Explain why active transport is important.	4.1.3.3	Ball and a slope – the slope could just be a textbook leaning against other books; Worksheet 3.3; Technician's notes 3.3	Quick starter Homework worksheet Homework quiz
3.4	Key concept: Investigating the need for transport systems	• Describe the need for transport systems. Describe how the effectiveness of an exchange surface can be increased. Explain in terms of surface area to volume ratios the need for transport systems	4.1.3.1	Worksheet 3.4; Practical sheet 3.4; Technician's notes 3.4	Quick starter Homework worksheet Homework quiz Slideshow Video
3.5	Explaining enzymes	• Describe what enzymes are and how they work. • Explain the lock-and-key theory. • Use the collision theory to explain enzyme action.	4.2.2.1	Worksheet 3.5; Practical sheet 3.5; Technician's notes 3.5	Quick starter Homework worksheet Homework quiz Slideshow
3.6	Required practical: Investigate the effect of pH on the rate of reaction of amylase enzyme	• Describe how safety is managed, apparatus is used and accurate measurements are made. • Explain how representative samples are taken. • Make and record accurate observations.	4.2.2.1	Worksheets 3.6.1, 3.6.2 and 3.6.3; Practical sheet 3.6; Technician's notes 3.6	Quick starter Homework worksheet Homework quiz Slideshow

Lesson number	Lesson title	Lesson objectives	AQA specification reference	Lesson resources (on CD ROM)	Collins Connect resources
		• Draw and interpret a graph from secondary data using knowledge and observations.			
3.7	Learning about the digestive system	• Identify and locate the organs in the digestive system, and describe their functions. • Describe how the products of digestion are absorbed into the body. • Explain why the small intestine is an efficient exchange surface.	4.2.2.1	Worksheet 3.7; Practical sheet 3.7 (teacher demonstration); Technician's notes 3.7	Quick starter Homework worksheet Homework quiz Slideshow
3.8	Explaining digestion	• Describe how physical digestion helps to increase the rate of chemical digestion. • Name the sites of production and action of specific enzymes. • Interpret data about digestive enzymes.	4.2.2.1	Internet access; materials for making posters. Worksheet 3.8	Quick starter Homework worksheet Homework quiz
3.9	Required practical: Use qualitative reagents to test for a range of carbohydrates, lipids and proteins	• Suggest appropriate apparatus for the procedures. • Describe how safety is managed and apparatus is used. • Describe how accurate measurements are made. • Interpret observations and make conclusions.	4.2.2.1	Image of some food. Practical sheets 3.9.1 and 3.9.2; Technician's notes 3.9	Quick starter Homework worksheet Homework quiz
3.10	Looking at more exchange surfaces	• Identify the structures responsible for gas exchange in fish, amphibians and insects. • Describe the adaptations of	4.1.3.1	Internet access, materials for making posters. Worksheets 3.10.1, 3.10 2 and 3.10.3	Quick starter Homework worksheet Homework quiz Video

Lesson number	Lesson title	Lesson objectives	AQA specification reference	Lesson resources (on CD ROM)	Collins Connect resources
		different gas exchange surfaces. • Explain the gas exchange surfaces in amphibians.			
3.11	Learning about plants and minerals	• Describe how mineral ions from the soil help plants to grow. • Explain how root hair cells are adapted for efficient osmosis. • Describe the function of different mineral ions in a plant.	4.1.3.3; 4.2.3.2	Worksheet 3.11	Quick starter Homework worksheet Homework quiz
3.12	Investigating how plants use minerals	• Describe why plants need different mineral ions. • Explain the effects of mineral deficiencies on plant growth. • Explain the importance of fertilisers.	4.1.3.3	Internet access. Worksheet 3.12; Practical sheet 3.12; Technician's notes 3.12	Quick starter Homework worksheet Homework quiz Video
3.13	Learning about the circulatory system	• Identify the parts of the circulatory system. • Describe the functions of the parts of the circulatory system. • Explain how the structure of each part of the circulatory system relates to its function.	4.2.2.2; 4.2.2.3	Worksheets 3.13.1, 3.13.2 and 3.13.3; Practical sheets 3.13.1 and 3.13.2; Technician's notes 3.13	Quick starter Homework worksheet Homework quiz
3.14	Exploring the heart	• Describe the structure and functions of the heart. • Identify the functions and adaptations of the parts of the heart. • Explain the movement of blood around the heart.	4.2.2.2	Images or video clips of blood passing through the heart; access to the Internet. Worksheet 3.14; Practical sheet 3.14; Technician's notes 3.14	Quick starter Homework worksheet Homework quiz Slideshow Videos
3.15	Studying blood	• Identify the parts of the blood and their functions.	4.2.2.3	Animation showing blood moving through vessels,	Quick starter Homework worksheet

Lesson number	Lesson title	Lesson objectives	AQA specification reference	Lesson resources (on CD ROM)	Collins Connect resources
		• Explain the adaptations of red blood cells. • Explain how red blood cells and haemoglobin transport oxygen efficiently.		microscopic image of blood cells showing red and white blood cells, platelets and plasma; access to the Internet. Worksheets 3.15.1 and 3.15.2	Homework quiz Video
3.16	Investigating gas exchange	• Identify the parts of the human gas exchange system and know their functions. • Explain how gas exchange occurs in humans. • Explain the adaptations of the gas exchange surfaces.	4.1.3.1; 4.2.2.2	Worksheet 3.16; Practical sheet 3.16; Technician's notes 3.16	Quick starter Homework worksheet Homework quiz Video
3.17	Learning about coronary heart disease	• Identify the causes and symptoms of coronary heart disease and heart failure. • Describe possible treatments of coronary heart disease and heart failure. • Evaluate the possible treatments of coronary heart disease and heart failure.	4.2.2.4	Animation of coronary heart disease, video clip of blood moving through a heart valve. Worksheets 3.17.1, 3.17.2 and 3.17.3	Quick starter Homework worksheet Homework quiz
3.18	Maths skills: Extracting and interpreting information	• To extract and interpret information from tables, charts and graphs.		Worksheets 3.18.1, 3.18.2 and 3.18.3	Quick starter Homework worksheet Homework quiz Slideshow Video
Assessments		End of chapter test Student Book End of chapter test Collins Connect			
		Chapter 4: Health matters			
4.1	Learning about health	• Recall the difference between health and disease. • Explain how some diseases	4.2.2.6; 4.2.2.7	Worksheets 4.1.1, 4.1.2 and 4.1.3; Practical sheet 4.1; Technician's notes 4.1	Quick starter Homework worksheet Homework quiz

Lesson number	Lesson title	Lesson objectives	AQA specification reference	Lesson resources (on CD ROM)	Collins Connect resources
		interact. • Evaluate data about lifestyle and health.			
4.2	Key concept: Looking at risk factors	• Recall the causes of some non-communicable diseases. • Describe the impact of lifestyle on non-communicable diseases. • Explain the impact of lifestyle on non-communicable diseases.	4.2.2.6	Internet access; poster material. Worksheets 4.2.1 and 4.2.2	Quick starter Homework worksheet Homework quiz Slideshow Video
4.3	Exploring non-communicable diseases	• Identify risk factors for cancer. • Explain the differences between types of tumours. • Explain the impact of non-communicable diseases	4.2.2.6; 4.2.2.7	Internet access. Worksheet 4.3; Practical sheet 4.3; Technician's notes 4.3	Quick starter Homework worksheet Homework quiz
4.4	Analysing and evaluating data	• Translate information between graphical and numerical forms. • Use scatter diagrams to identify correlations. • Evaluate the strength of evidence.	4.2.2.5; 4.2.2.6	Graph paper; Worksheets 4.4.1, 4.4.2 and 4.4.3; Practical sheet 4.4; Technician's notes 4.4	Quick starter Homework worksheet Homework quiz
4.5	Studying pathogens	• Recall the definition of a pathogen. • Explain how communicable diseases can be controlled. • Distinguish between epidemics and pandemics.	4.3.1.1	Poster material; Internet access. Worksheets 4.5.1 and 4.5.2	Quick starter Homework worksheet Homework quiz Videos
4.6	Learning about viral diseases	• Describe the symptoms of some viral diseases. • Describe the transmission and control of some viral diseases.	4.3.1.2	Images of viruses including HIV; poster material; access to Internet. Worksheets 4.6.1, 4.6.2 and 4.6.3	Quick starter Homework worksheet Homework quiz Slideshow

Lesson number	Lesson title	Lesson objectives	AQA specification reference	Lesson resources (on CD ROM)	Collins Connect resources
		• Explain how some viral diseases are spread.			
4.7	Studying bacterial diseases	• Describe the symptoms of some bacterial diseases. • Explain how some bacterial diseases can be controlled. • Compare and contrast bacterial and viral diseases.	4.3.1.3	Video of bacteria reproducing without the sound on; poster material; access to the Internet. Worksheets 4.7.1, 4.7.2 and 4.7.3	Quick starter Homework worksheet Homework quiz
4.8	Looking at fungal diseases	• Recall the name and symptoms of a fungal disease. • Describe the transmission and treatment of rose black spot. • Explain how rose black spot affects the growth of the plant.	4.3.1.4	Video about black spot; Worksheet 4.8	Quick starter Homework worksheet Homework quiz Slideshow
4.9	Learning about malaria	• Recall that malaria is a protist disease. • Describe the lifecycle of the malarial vector. • Evaluate control methods for the spread of malaria.	4.3.1.5	Videos about malaria; Internet access or a fact sheet on malaria. Worksheets 4.9.1 and 4.9.2	Quick starter Homework worksheet Homework quiz
4.10	Protecting the body	• Describe how the body protects itself from pathogens. • Explain how the body protects itself from pathogens. • Explain how communicable diseases can be spread.	4.3.1.6	A video clip about how the blood clots; Internet access or factsheet printed from the NHS website. Worksheets 4.10.1, 4.10.2 and 4.10.3	Quick starter Homework worksheet Homework quiz
4.11	Exploring white blood cells	• Describe phagocytosis. • Explain how antibody production can lead to immunity. • Explain the specificity of immune system responses	4.3.1.6	A video of white blood cells in action without the sound; a video clip about the production of antibodies; image of a graph that shows primary and secondary	Quick starter Homework worksheet Homework quiz

Lesson number	Lesson title	Lesson objectives	AQA specification reference	Lesson resources (on CD ROM)	Collins Connect resources
				infection. Worksheet 4.11	
4.12	Using antibiotics and painkillers	• Describe the uses of antibiotics and painkillers. • Explain how antibiotics and painkillers can be used to treat diseases. • Explain the limitations of antibiotics.	4.3.1.8	Worksheet 4.12; Practical sheet 4.12; Technician's notes 4.12	Quick starter Homework worksheet Homework quiz
4.13	Building immunity	• Recall how vaccinations prevent infection. • Explain how mass vaccination programmes reduce the spread of a disease. • Evaluate the global use of vaccination.	4.3.1.7	Videos about the first vaccinations and the polio eradication programme; timeline for smallpox; Worksheets 4.13.1, 4.13.2 and 4.13.3	Quick starter Homework worksheet Homework quiz Slideshow Video
4.14	Making new drugs	• Recall some traditional drugs and their origins. • Describe how new drugs are developed. • Explain why 'double-blind' trials are conducted.	4.3.1.9	A video of drugs being manufactured using a tablet press. Worksheets 4.14.1 and 4.14.2; Practical sheet 4.14; Technician's notes 4.14	Quick starter Homework worksheet Homework quiz Slideshow Video
4.15	Investigating monoclonal antibodies	• Describe uses of monoclonal antibodies. • Explain how monoclonal antibodies are produced. • Evaluate the use of monoclonal antibodies.	4.3.2.1; 4.3.2.2	A video that illustrates how monoclonal antibodies are made; poster material; access to the Internet. Worksheet 4.15	Quick starter Homework worksheet Homework quiz
4.16	Looking at plant diseases	• Recall the causes of plant diseases. • Describe the symptoms and identification methods of some plant diseases.	4.1.2.3; 4.3.3.1	Image of plant with rose black spot; image of a plant with tobacco mosaic virus; video of aphids feeding; poster material; access to	Quick starter Homework worksheet Homework quiz

Lesson number	Lesson title	Lesson objectives	AQA specification reference	Lesson resources (on CD ROM)	Collins Connect resources
		• Explain the use of monoclonal antibodies in identifying plant pathogens.		Internet. Worksheets 4.16.1, 4.16.2 and 4.16.3	
4.17	Learning about plant defences	• Recall some physical plant defence responses. • Explain how mechanical plant defence systems help them survive. • Explain how chemical plant defence systems help them survive.	4.3.3.2	Poster material, access to Internet, *Mimosa pudica* plant or a video showing the plant folding its leaves; image of leaf structure; a video about the uses of witch hazel; Worksheet 4.17	Quick starter Homework worksheet Homework quiz
4.18	Maths skills: Sampling and scientific data	• Understand why sampling is used in science. • Be able to explain different sampling techniques.		Worksheet 4.18; Practical sheet 4.18; Technician's notes 4.18	Quick starter Homework worksheet Homework quiz Slideshow Video
Assessments		End of chapter test Student Book End of chapter test Collins Connect End of teaching block test Collins Connect End of year test Collins Connect			
Chapter 5: Coordination and control					
5.1	Homeostasis	• Explain the importance of homeostasis in regulating internal conditions in the body. • Recall that these control systems involve nervous or chemical responses. • Describe how control systems involve receptors, coordination centres and effectors.	4.5.1	Worksheets 5.1.1 and 5.1.2; PowerPoint presentation	Quick starter Homework worksheet Homework quiz
5.2	The nervous system	• Describe the structure and function of the nervous system. • Explain how the nervous	4.5.2.1	List of head injury symptoms; video of what happens at the synapse; materials for	Quick starter Homework worksheet Homework quiz

Lesson number	Lesson title	Lesson objectives	AQA specification reference	Lesson resources (on CD ROM)	Collins Connect resources
		system is adapted to its functions. • Describe the structure of sensory, motor and relay neurones.		making models of neurones, e.g. cardboard, rubber tubing, cling film, jelly, string, sticky tape, modelling clay. Worksheets 5.2.1 and 5.2.2; PowerPoint presentation	
5.3	Reflex actions	• Explain the importance of reflex actions. • Describe the path of a reflex arc. • Explain how the structures in the reflex arc relate to their function.	4.5.2.1	Simulation from www.sumanasinc.com. Worksheets 5.3.1, 5.3.2 and 5.3.3; Practical sheet 5.3; Technician's notes 5.3; PowerPoint presentation	Quick starter Homework worksheet Homework quiz Slideshow
5.4	The brain	• Recall that the brain controls complex behaviour using billions of interconnected neurones. • Identify the three main regions of the brain and describe their functions. • Describe how the regions of the brain are mapped.	4.5.2.2	Videos about brain wiring and brain surgery; Internet access. Worksheets 5.4.1 and 5.4.2	Quick starter Homework worksheet Homework quiz Video
5.5	Required practical: Investigating reaction time	• Select appropriate apparatus and techniques for the measurement of biological processes. • Carry out physiological experiments safely. • Use appropriate techniques in problem-solving contexts.	4.5.2.1	Video 'How synapses work' from BBC Bitesize; information about caffeine from www.sciencemuseum.org.uk. Worksheets 5.5.1,5.5.2 and 5.5.3; Practical sheet 5.5; Technician's notes 5.5	Quick starter Homework worksheet Homework quiz Slideshow
5.6	The eye	• Relate the structures of the eye to their functions.	4.5.2.3	Internet access; websites with optical illusions for students to test; video on	Quick starter Homework worksheet Homework quiz

Lesson number	Lesson title	Lesson objectives	AQA specification reference	Lesson resources (on CD ROM)	Collins Connect resources
		• Explain how the eye is adapted to seeing in colour and in dim light.		how the eye works; strong elastic bands, sticky tape, card, scissors, string. Worksheets 5.6.1, 5.6.2 and 5.6.3; Practical sheet 5.6; PowerPoint presentation	
5.7	Seeing in focus	• Relate the structures of the eye to their functions. • Understand how the eye is able to focus on near or distant objects.	4.5.2.3	Simulation 'Lens Accommodation in the Human Eye'; string, elastic bands, balloon, ray box, fat convex lens, thin convex lens and screen. Worksheets 5.7.1, 5.7.2 and 5.7.3; PowerPoint presentation	Quick starter Homework worksheet Homework quiz Video
5.8	Eye defects	• Describe and understand why short-sightedness (myopia) occurs. • Describe and understand why long-sightedness (hyperopia) occurs. • Demonstrate how techniques are used to correct eye defects.	4.5.2.3	Vision simulations. Worksheets 5.8.1, 5.8.2 and 5.8.3; Practical sheet 5.8	Quick starter Homework worksheet Homework quiz Slideshow
5.9	Controlling body temperature	• Describe the mechanisms by which body temperature is controlled when too hot or cold. • Explain how body temperature can be controlled in a specific context.	4.5.2.4	Video clip of vasodilation. Worksheet 5.9; Practical sheet 5.9; Technician's notes 5.9; PowerPoint presentation	Quick starter Homework worksheet Homework quiz Slideshow
5.10	The endocrine system	• Recall that the endocrine system is made up of glands that secrete hormones into the blood. • Know the location of the major	4.5.3.1	Video of endocrine glands; Internet access. Worksheets 5.10.1 and 5.10.2	Quick starter Homework worksheet Homework quiz

Lesson number	Lesson title	Lesson objectives	AQA specification reference	Lesson resources (on CD ROM)	Collins Connect resources
		endocrine glands. • Understand why the pituitary gland is the 'master gland'.			
5.11	Controlling blood glucose	• Recall that blood glucose is monitored and controlled by the pancreas. • Understand how insulin controls blood glucose levels. • Understand how insulin works with another hormone – glucagon – to control blood sugar levels.	4.5.3.2; 4.5.3.7	Simulation 'What happens when you have diabetes'; coloured card, scissors, pens, rulers. Worksheets 5.11.1, 5.11.2 and 5.11.3	Quick starter Homework worksheet Homework quiz Video
5.12	Diabetes	• Understand the causes of Type 1 and Type 2 diabetes. • Compare Type 1 and Type 2 diabetes. • Evaluate information on the relationship between obesity and diabetes, and make appropriate recommendations.	4.5.3.2; 4.5.3.7	Videos 'What happens when you have diabetes' and 'Type 2 diabetes; props for role play – coloured card and scissors. Worksheets 5.12.1 and 5.12.2; PowerPoint presentation	Quick starter Homework worksheet Homework quiz Video
5.13	Diabetes recommendations	• Understand the causes of Type 1 and Type 2 diabetes. • Compare Type 1 and Type 2 diabetes. • Evaluate information on the relationship between obesity and diabetes, and make appropriate recommendations.	4.5.3.2; 4.5.3.7	Worksheet 5.13	Quick starter Homework worksheet Homework quiz Slideshow
5.14	Water balance	• Recall the ways in which the body loses water. • Explain why cells do not function efficiently if they lose	4.5.3.3; 4.5.3.7	Small mirror, a piece of cobalt chloride paper. Worksheets 5.14.1, 5.14.2 and 5.14.3; PowerPoint presentation	Quick starter Homework worksheet Homework quiz

Lesson number	Lesson title	Lesson objectives	AQA specification reference	Lesson resources (on CD ROM)	Collins Connect resources
		or gain too much water. • Explain how excess protein is converted to urea for excretion.			
5.15	The kidneys	• Recall that excess water, ions and urea are removed from the body by the kidneys in urine. • Describe how the kidneys produce urine. • Explain how the hormone ADH regulates the amount of water in the urine, and therefore, in the body.	4.5.3.3	A real kidney; paper, coloured card, pipe cleaners to make a filter, sticky tape, cardboard boxes, rubber tubing and glue. Worksheets 5.15.1 and 4.15.2; PowerPoint presentation	Quick starter Homework worksheet Homework quiz Video
5.16	Negative feedback (Higher tier only)	• Explain the role of thyroxine in the body. • Understand the principles of negative feedback, as applied to thyroxine.	4.5.3.7	Internet access; website about thyroxine; Worksheet 5.16; PowerPoint presentation	Quick starter Homework worksheet Homework quiz
5.17	Kidney failure	• Recall that people who suffer from kidney failure can be treated by dialysis or kidney transplant. • Understand the principles of dialysis. • Evaluate the advantages and disadvantages of treating organ failure using a mechanical device or transplant.	4.5.3.3	Materials to make dialysis models – tubing, plastic box with holes at either end for tubing to pass through, fluid. Worksheet 5.17	Quick starter Homework worksheet Homework quiz Slideshow
5.18	Dialysis or transplant?	• Describe different forms of dialysis. • Use data to draw conclusions. • Use data to evaluate the advantages and disadvantages	4.5.3.3	Worksheets 5.18.1, 5.18.2 and 5.18.3	Quick starter Homework worksheet Homework quiz Video

Lesson number	Lesson title	Lesson objectives	AQA specification reference	Lesson resources (on CD ROM)	Collins Connect resources
		of dialysis and kidney donation.			
5.19	Human reproduction	• Describe the roles of hormones in sexual reproduction. • Explain how hormones interact in the menstrual cycle.	4.5.3.4	Videos of the menstrual cycle and hormone levels; large sheet of paper to draw a timeline of the menstrual cycle – poster paper and four different colours. Worksheets 5.19.1 and 5.19.2	Quick starter Homework worksheet Homework quiz
5.20	IVF (Higher tier only)	• Explain the use of hormones in technologies to treat infertility. • Describe the technique of in-vitro fertilisation. • Evaluate the scientific, emotional, social and ethical issues of in-vitro fertilisation.	4.5.3.6	Worksheet 5.20; PowerPoint presentation	Quick starter Homework worksheet Homework quiz
5.21	IVF evaluation (Higher tier only)	• Evaluate data regarding in-vitro fertilisation and use this to draw conclusions. • Evaluate the scientific, emotional, social and ethical issues of in-vitro fertilisation.	4.5.3.6	Article from BBC News Magazine 'I wish IVF had never been invented' (www.bbc.co.uk). Worksheet 5.21	Quick starter Homework worksheet Homework quiz
5.22	Systems working together (Higher tier only)	• Describe the effects of adrenaline. • Understand that automatic control systems may involve nervous responses and chemical responses. • Understand that combinations of hormones work to produce a response.	4.5.3.7	Animation 'More hormones – adrenaline and ADH' from www.abpischools.org.uk. Worksheet 5.22	Quick starter Homework worksheet Homework quiz
5.23	Contraception	• Define the purpose of contraception. • Describe hormonal methods	4.5.3.5	Worksheets 5.23.1 and 5.23.2	Quick starter Homework worksheet Homework quiz

Lesson number	Lesson title	Lesson objectives	AQA specification reference	Lesson resources (on CD ROM)	Collins Connect resources
		and non-hormonal methods of contraception. • Explain how these methods are effective at preventing pregnancy.			
5.24	Which contraceptive?	• Describe the advantages and disadvantages of different contraceptive methods. • Use data to evaluate the effectiveness of different contraceptive methods.	4.5.3.5	Worksheet 5.24	Quick starter Homework worksheet Homework quiz
5.25	Auxins	• Recall that plants produce hormones to coordinate and control growth, and responses to light and gravity. • Describe how unequal distributions of auxins cause unequal growth rates in plant shoots and roots.	4.5.4.1; 4.5.4.2	Plant which has been growing on its side for at least a week; normal plant; Venus fly trap; mimosa plant. Worksheets 5.25.1, 5.25.2 and 5.25.3; Practical sheet 5.25; Technician's notes 5.25	Quick starter Homework worksheet Homework quiz
5.26	Applications of auxins	• Explain how auxins coordinate and control responses to light and gravity. • Explain that auxins act on 'stem cells' in plants called meristems. • Describe some applications of auxins.	4.5.4.2	Auxins animation. Worksheets 5.26.1, 5.26.2 and 5.26.3; Practical sheet 5.26; Technician's notes 5.26	Quick starter Homework worksheet Homework quiz
5.27	Required practical: The effect of light and gravity on the growth of germinating seeds	• Describe how an experiment is planned for a specific purpose. • Make and record observations and translate data from one form to another.	4.5.4.1	Recently germinating cress or radish or white mustard plants; one group of seedlings already prepared where the light intensity has been applied at three	Quick starter Homework worksheet Homework quiz Slideshow

Lesson number	Lesson title	Lesson objectives	AQA specification reference	Lesson resources (on CD ROM)	Collins Connect resources
		• Interpret observations and other data, identifying patterns and trends, make inferences and draw conclusions.		different distances the week before the lesson. Worksheet 5.27; Practical sheet 5.27; Technician's notes 5.27	
5.28	Other plant hormones (Higher tier only)	• Recall that gibberellins are important in seed germination, and ethene in cell division and ripening of fruit. • Explain the application of the plant hormones ethane and gibberellins.	4.5.4.2	Some real plants, fruits, germinating seeds and larger fruits. Worksheet 5.28	Quick starter Homework worksheet Homework quiz
5.29	Maths skills: The spread of scientific data	• Be able to use range bars on graphs. • Understand how box and whisker plots can be used to show the spread of data. • Understand how to use percentiles.		Graph paper, calculators, pencils and rulers. Worksheets 5.29.1, 5.29.2 and 5.29.3	Quick starter Homework worksheet Homework quiz Slideshow Video
Assessments		End of chapter test Student Book End of chapter test Collins Connect			
		Chapter 6: Genetics			
6.1	DNA and genes	• Describe the structure of DNA. • Describe a gene as a small section of DNA that codes for a protein.	4.6.1.4; 4.6.1.5	Worksheet 6.1; Practical sheet 6.1; Technician's notes 6.1	Quick starter Homework worksheet Homework quiz Slideshow Video
6.2	The human genome	• Describe a gene as a small section of DNA that codes for a protein. • Explain the importance of understanding the human genome.	4.6.1.4; 4.6.1.5	Videos about the Human Genome Project; poster-making equipment; Internet access. Worksheet 6.2	Quick starter Homework worksheet Homework quiz

Lesson number	Lesson title	Lesson objectives	AQA specification reference	Lesson resources (on CD ROM)	Collins Connect resources
6.3	Tracing human migration	• Explain the importance of understanding the human genome. • Discuss the use of the human genome in understanding human migration patterns.	4.6.1.4; 4.6.1.5	Videos about the Genographic Project and 'The journey of your past'; Internet access; poster-making equipment. Worksheets 6.3.1 and 6.3.2	Quick starter Homework worksheet Homework quiz
6.4	The structure of DNA	• Describe the structure of DNA as repeating nucleotide units. • Identify the four bases in DNA. • Explain that the bases A and T, and C and G, are complementary.	4.6.1.5	Video of DNA structure. Worksheets 6.4.1 and 6.4.2 Practical sheet 6.4; Technician's notes 6.4	Quick starter Homework worksheet Homework quiz
6.5	Proteins	• Describe how proteins are synthesised according to the DNA template of a gene. • Explain that the genetic code of a gene specifies the protein to be made.	4.6.1.5	Video of protein synthesis; poster material. Worksheets 6.5.1 and 6.5.2	Quick starter Homework worksheet Homework quiz Slideshow Video
6.6	Mutations	• Model changes to the base sequences of DNA to illustrate mutations. • Describe the negative and, sometimes, positive effects of mutations. • Describe how mutations can affect protein function.	4.6.1.5	Images of mutations, such as sickle cells; images of sickle-cell haemoglobin. Worksheet 6.6	Quick starter Homework worksheet Homework quiz Slideshow Video
6.7	Meiosis	• Explain how meiosis halves the number of chromosomes for gamete production. • Explain how fertilisation restores the chromosome	4.6.1.1; 4.6.1.2; 4.6.1.8	Video clip of sperm swimming to an egg; video or images of meiosis; string or wool. Worksheets 6.7.1 and 6.7.2	Quick starter Homework worksheet Homework quiz Video

Lesson number	Lesson title	Lesson objectives	AQA specification reference	Lesson resources (on CD ROM)	Collins Connect resources
		number. • Understand that the four gametes produced by meiosis are genetically different.			
6.8	Asexual and sexual reproduction	• Understand that asexual reproduction involves just one parent and produces genetically identical offspring. • Understand that sexual reproduction leads to variety in the offspring.	4.6.1.1; 4.6.1.3	Video clip of aphids reproducing asexually and sexually. Worksheets 6.8.1 and 6.8.2	Quick starter Homework worksheet Homework quiz Slideshow
6.9	Genetics	• Understand and be able to use genetics terms, such as dominant, recessive, genotype, phenotype, homozygous and heterozygous. • Know that some human conditions are caused by a recessive allele.	4.6.1.6; 4.6.1.7	Image of mice of different colours; Internet access. Worksheets 6.9.1, 6.9.2 and 6.9.3	Quick starter Homework worksheet Homework quiz Video
6.10	Genetic crosses	• Use the terms dominant, recessive, genotype, phenotype, homozygous and heterozygous. • Know that some human conditions, such as cystic fibrosis, are caused by a recessive allele. • Complete or construct a Punnett square to predict the outcome of a genetic cross.	4.6.1.6; 4.6.1.7	Worksheets 6.10.1, 6.10.2 and 6.10.3	Quick starter Homework worksheet Homework quiz Video
6.11	Tracking gene disorders	• Understand the use of a family tree to show the inheritance of a characteristic.	4.6.1.6; 4.6.1.7	Worksheets 6.11.1 and 6.11.2	Quick starter Homework worksheet Homework quiz

Lesson number	Lesson title	Lesson objectives	AQA specification reference	Lesson resources (on CD ROM)	Collins Connect resources
		• Explain economic, social and ethical issues concerned with embryo screening.			Video
6.12	Gregor Mendel	• Plan experiments to explore phenomena and test hypotheses. • Draw conclusions from given observations. • Evaluate data in terms of reproducibility.	4.6.3.3	Worksheet 6.12.1; Worksheet 6.12.2; Worksheet 6.12.3	Quick starter Homework worksheet Homework quiz
6.13	Key concept: Genetics is simple – or is it?	• Explain how certain characteristics are controlled by a single gene. • Understand that many characteristics are the result of multiple genes which interact. • Describe the search for genes that are linked to disease.	4.6.1.6; 4.6.1.7	Ishihara cards; sheets of A3 paper, coloured pens. Worksheet 6.13	Quick starter Homework worksheet Homework quiz Slideshow Video
6.14	Maths skills: Fractions, ratio, proportion and probability	• Understand and use fractions and percentages. • Understand and use ratio and proportion. • Understand and use probability when predicting the outcomes of genetic crosses.		Practical sheet 6.14; Technician's notes 6.14	Quick starter Homework worksheet Homework quiz Slideshow Video
Assessments		End of chapter test Student Book End of chapter test Collins Connect End of teaching block test Collins Connect			
Chapter 7: Variation and evolution					
7.1	Variation	• Recall that differences in the characteristics of individuals in a population is called variation.	4.6.2.1	Worksheets 7.1.1, 7.1.2 and 7.1.3; Practical sheet 7.1; Technician's notes 7.1;	Quick starter Homework worksheet Homework quiz

Lesson number	Lesson title	Lesson objectives	AQA specification reference	Lesson resources (on CD ROM)	Collins Connect resources
		• Understand the genetic and environmental differences leading to variation.		PowerPoint presentation	
7.2	The theory of evolution	• Recall that all species of living things have evolved from simple life forms. • Explain how evolution occurs through natural selection,	4.6.2.2	Worksheets 7.2.1, 7.2.2 and 7.2.3; PowerPoint presentation	Quick starter Homework worksheet Homework quiz Slideshow
7.3	The origin of species by natural selection	• Explain the evidence that led Darwin to propose the theory of evolution by natural selection. • Describe the process of natural selection.	4.6.3.1	Animation from. Worksheets 7.3.1 and 7.3.2; Practical sheet 7.3; Technician's notes 7.3; PowerPoint presentation	Quick starter Homework worksheet Homework quiz Video
7.4	Fossil evidence	• Understand how, and the situations in which, fossils are formed. • Understand how fossils are used as evidence for evolution of species from simpler life forms.	4.3.6.4; 4.3.6.5	Video of how fossils are made. Worksheets 7.4.1 and 7.4.2 ; PowerPoint presentation	Quick starter Homework worksheet Homework quiz
7.5	How much have organisms changed?	• Understand why the fossil record is incomplete. • Use the fossil record to understand how much, or how little, organisms have changed as life developed on Earth.	4.6.3.5	Useful websites with 'missing links'; Worksheets 7.5.1, 7.5.2 and 7.5.3; PowerPoint presentation	Quick starter Homework worksheet Homework quiz
7.6	Darwin and Wallace	• Recognise how Darwin and Wallace proposed, independently, the theory of evolution. • Describe how Alfred Wallace gathered evidence for	4.6.3.2	Internet access. Worksheets 7.6.1, 7.6.2 and 7.6.3; PowerPoint presentation	Quick starter Homework worksheet Homework quiz

Lesson number	Lesson title	Lesson objectives	AQA specification reference	Lesson resources (on CD ROM)	Collins Connect resources
		evolution, including warning coloration and mimicry.			
7.7	A new species	• Understand that when natural selection operates differently on populations, a new species is produced. • Understand that during evolution, new species are formed when populations become so different that they can no longer interbreed.	4.6.3.2	Worksheets 7.7.1, 7.7.1 and 7.7.3; PowerPoint presentation	Quick starter Homework worksheet Homework quiz
7.8	Evidence of natural selection and evolution?	• Understand how scientific theories develop over time. • Plan experiments to test hypotheses	4.6.3.4	Worksheet 7.8; Practical sheets 7.8.1 and 7.8.2	Quick starter Homework worksheet Homework quiz Slideshow
7.9	Evolution: fitting the pieces of the jigsaw	• Describe the work of Mendel, Darwin and Wallace. • Explain how they contributed to the theory of evolution. • Appreciate that many scientists have contributed to the gene theory.	4.6.3.1; 4.6.3.2; 4.6.3.3	Videos on the life of Mendel; and the work of William Bateson. Worksheets 7.9.1, 7.9.2 and 7.9.3	Quick starter Homework worksheet Homework quiz
7.10	Antimicrobial resistance	• Recall that bacteria develop that are resistant to antibiotics, which is evidence of evolution. • Understand the mechanism by which antibiotic resistance develops. • Understand the effects of the development of antibiotic resistance on the treatment of disease.	4.6.3.7	Modelling clay; poster paper; sticky notes for each pair. Video about the discovery of penicillin; two short clips: one from the Chief Medical Adviser and one about the threat from antibiotic-resistant bacteria. Worksheets 7.10.1, 7.10.2 and 7.10.3; PowerPoint presentation	Quick starter Homework worksheet Homework quiz

Lesson number	Lesson title	Lesson objectives	AQA specification reference	Lesson resources (on CD ROM)	Collins Connect resources
7.11	Combatting antimicrobial resistance	• Describe how to reduce the rate of development of antibiotic resistance. • Understand the requirement for, and the impact of, new antibiotics. • Recognise the difficulties associated with developing new antibiotics.	4.6.3.7	Internet access. Worksheets 7.11.1 and 7.11.2	Quick starter Homework worksheet Homework quiz
7.12	Selective breeding	• Describe the process of selective breeding. • Recall how selective breeding enables humans to choose desirable characteristics in animals and plants. • Explain how selective breeding can lead to inbreeding.	4.6.2.3	Videos on selective breeding. Worksheets 7.12.1 and 7.12.2; PowerPoint presentation	Quick starter Homework worksheet Homework quiz Slideshow
7.13	Producing new plant varieties	• Describe the process of selective breeding. • Recall how selective breeding enables humans to choose desirable characteristics in plants. • Evaluate the benefits and risks of selective breeding in plants.	4.6.2.3	Selection of flowers, fruits and vegetables. Worksheets 7.13.1 and 7.13.2; PowerPoint presentation	Quick starter Homework worksheet Homework quiz
7.14	Genetic engineering	• Explain what is meant by the term genetic engineering. • Give examples of how plant crops have been genetically engineered to improve products. • Describe how fungus cells are	4.6.2.4	Pipe cleaners of different colours, scissors, small boxes and tubs to represent cells. Worksheets 7.14.1 and 7.14.2; PowerPoint presentation	Quick starter Homework worksheet Homework quiz Video

Lesson number	Lesson title	Lesson objectives	AQA specification reference	Lesson resources (on CD ROM)	Collins Connect resources
		engineered to produce human insulin			
7.15	Genetically modified crops: the science	• Explain the benefits of genetic modification in a range of crops. • Explain the concerns about genetic modification. • Explain the ethical concerns about genetic engineering.	4.6.2.4	Video on the development of genetically modified papaya plants. Worksheet 7.15	Quick starter Homework worksheet Homework quiz
7.16	Is genetic modification safe?	• Explore the benefits of genetic modification in medicine. • Explain the concerns that people have about genetic modification. • Explain the possible safety issues of genetic engineering in agriculture and medicine.	4.6.2.4	Worksheets 7.16.1 and 7.16.2; PowerPoint presentation	Quick starter Homework worksheet Homework quiz
7.17	Ethically wrong, or essential?	• Explain the benefits of, and concerns about, genetic modification. • Explain the ethical issues of genetic engineering in agriculture and medicine.	4.6.2.4	Infographic from http://us.naturespath.com; video from http://news.discovery.com. Worksheets 7.17.1, 7.17.2 and 7.17.3; PowerPoint presentation	Quick starter Homework worksheet Homework quiz Slideshow
7.18	Cloning	• Describe how cuttings and tissue culture are used to produce new plants. • Describe the use of embryo transplants and adult cell cloning in animals.	4.6.2.5	Videos on how to clone a cauliflower and how jellyfish reproduce. Worksheets 7.18.1, 7.18.2 and 7.18.3; Practical sheet 7.18; Technician's notes 7.18; PowerPoint presentation	Quick starter Homework worksheet Homework quiz Video
7.19	The tree of life	• Describe how living things have been classified into groups	4.6.4	Videos of Linnaean classification and	Quick starter Homework worksheet

Lesson number	Lesson title	Lesson objectives	AQA specification reference	Lesson resources (on CD ROM)	Collins Connect resources
		using a system devised by Linnaeus. • Describe how new models of classification have developed.		evolutionary tree. Worksheets 7.19.1, 7.19.2 and 7.19.3	Homework quiz Video
7.20	Extinction…or survival?	• List the causes of extinction. • Explain how new predators, competitors and diseases can lead to extinctions.	4.6.3.6	Worksheets 7.20.1, 7.20.2 and 7.20.3; PowerPoint presentation	Quick starter Homework worksheet Homework quiz Slideshow Video
7.21	Maths skills: Using charts and graphs to display data	• Understand when and how to use bar charts. • Understand how to show sub-groups on bar charts. • Understand how to plot histograms.		Worksheets 7.21.1 and 7.21.2	Quick starter Homework worksheet Homework quiz Slideshow Video
Assessments		End of chapter test Student Book End of chapter test Collins Connect			
		Chapter 8: Ecology in action			
8.1	Key concept: Learning about ecosystems	• Describe what an ecosystem is. • Explain the importance of high biodiversity. • Explain what is meant by a self-supporting ecosystem	4.7.1.1; 4.7.3.1	Video of an ecosystem. Worksheets 8.1.1, 8.1.2 and 8.1.3	Quick starter Homework worksheet Homework quiz Slideshow Video
8.2	Changing abiotic factors	• Identify abiotic factors that affect ecosystems. • Explain changes in the distribution of species in an ecosystem. • Describe stable and unstable populations.	4.7.1.2; 4.7.2.4	Video of a rocky shore; image of a kite diagram from a rocky shore. Worksheets 8.2.1, 8.2.2 and 8.2.3; Practical sheets 8.2.1 and 8.2.2; Technician's notes 8.2	Quick starter Homework worksheet Homework quiz Slideshow
8.3	Investigating	• Describe how changes in one	4.7.2.1	Image of a food web.	Quick starter

Lesson number	Lesson title	Lesson objectives	AQA specification reference	Lesson resources (on CD ROM)	Collins Connect resources
	predator–prey relationships	population affect another. • Explain interdependent relationships. • Explain how predator–prey population cycles have cyclical changes.		Worksheets 8.3.1 and 8.3.2; Practical sheet 8.3; Technician's notes 8.3	Homework worksheet Homework quiz
8.4	Looking at trophic levels	• Explain trophic levels. • Explain and construct pyramids of biomass. • Explain the difficulties in constructing pyramids.	4.7.2.1; 4.7.4.1; 4.7.4.2	Video clip showing food chain in action. Worksheets 8.4.1, 8.4.2 and 8.4.3	Quick starter Homework worksheet Homework quiz
8.5	Transferring biomass	• Identify how biomass is lost. • Calculate the efficiency of biomass transfers. • Explain the impact of biomass loss on the numbers of organisms.	4.7.4.3	Video of an ecosystem. Worksheets 8.5.1, 8.5.2 and 8.5.3	Quick starter Homework worksheet Homework quiz Video
8.6	Competing for resources	• Describe how competition impacts on populations. • Explain why animals in the same habitat are in competition. • Explain interspecific and intraspecific competition.	4.7.1.1	Animals competing for a mate video. Worksheets 8.6.1, 8.6.2 and 8.6.3	Quick starter Homework worksheet Homework quiz Slideshow
8.7	Required practical: Measure the population size of a common species in a habitat	• Describe a suitable method to investigate a population. • Estimate the size of a population. • Explain the effect of sample size.	4.7.2.1	Worksheet 8.7; Practical sheet 8.7; Technician's notes 8.7	Quick starter Homework worksheet Homework quiz Slideshow
8.8	Adapting for survival in animals	• Recall why animals have adaptations.	4.7.1.4	A video clip of huddling penguins. Worksheets 8.8.1;	Quick starter Homework worksheet

Lesson number	Lesson title	Lesson objectives	AQA specification reference	Lesson resources (on CD ROM)	Collins Connect resources
		• Explain some adaptations. • Use surface area to volume ratios to explain some adaptations.		8.8.2 and 8.8.3; Practical sheet 8.8; Technician's notes 8.8	Homework quiz Slideshow
8.9	Adapting for survival in plants	• Identify some adaptations of plants and bacteria. • Explain the importance of plant adaptations. • Explain a range of plant adaptations.	4.7.1.4	Image of a cactus; image of a cross section of a marram grass leaf; video clip of bacteria living in deep sea vents; image of a pine tree in the snow. Worksheets 8.9.1, 8.9.2 and 8.9.3; Practical sheet 8.9; Technician's notes 8.9	Quick starter Homework worksheet Homework quiz
8.10	Cycling materials	• Recall that many materials are recycled in nature. • Explain the stages in the water and decay cycles. • Explain the importance of recycling materials.	4.7.2.2; 4.7.4.1	Time-lapse photography of a decaying animal. Worksheets 8.10.1 and 8.10.2; Practical sheet 8.10 (demonstration); Technician's notes 8.10	Quick starter Homework worksheet Homework quiz Video
8.11	Cycling carbon	• Recall that plants take in carbon as carbon dioxide. • Explain how carbon is recycled. • Interpret a diagram of the carbon cycle.	4.7.2.2	Worksheets 8.11.1 and 8.11.2; Practical sheet 8.11; Technician's notes 8.11	Quick starter Homework worksheet Homework quiz Video
8.12	Investigating decay	• Recall the factors needed for decay. • Describe how different factors affect decay. • Explain extracellular digestion.	4.7.2.2	An image of Otzi the iceman or a frozen woolly mammoth; a video or pictures of worm compost bins. Worksheets 8.12.1, 8.12.2 and 8.12.3; Practical sheet 8.12; Technician's notes 8.12	Quick starter Homework worksheet Homework quiz
8.13	Required practical: Investigate the	• Describe how safety is managed, apparatus is used	4.7.2.3	Milk and sour milk; Worksheets 8.13.1, 8.13.2	Quick starter Homework worksheet

Lesson number	Lesson title	Lesson objectives	AQA specification reference	Lesson resources (on CD ROM)	Collins Connect resources
	effect of temperature on the rate of decay of fresh milk by measuring pH change	and accurate measurements are made. • Make and record observations and make accurate measurements. • Evaluate methods and suggest possible improvements and further investigations.		and 8.13.3; Practical sheet 8.13; Technician's notes 8.13	Homework quiz Slideshow
8.14	Changing the environment	• Recall causes of environmental change. • Describe the impact of environmental change. • Explain the impact of an environmental change.	4.7.2.4; 4.7.3.1; 4.7.3.2	Video of migrating animals, such as caribou. Worksheet 8.14	Quick starter Homework worksheet Homework quiz Videos
8.15	Learning about land use	• Identify why land use has changed. • Describe the effects of changing land use. • Evaluate a change in land use.	4.7.3.3	A video news clip about building houses on greenbelt land; Internet access. Worksheets 8.15.1 and 8.15.2	Quick starter Homework worksheet Homework quiz Video
8.16	Changing the landscape	• Identify the reasons for deforestation. • Describe the impact of peat bog destruction and deforestation. • Evaluate the destruction of peat bogs and forests.	4.7.3.3; 4.7.3.4	A video clip showing deforestation; a video showing the effects of deforestation on orangutans or another species; Internet access; sheets of A3 plain paper. Worksheets 8.16.1 and 8.16.2	Quick starter Homework worksheet Homework quiz
8.17	Thinking about global warming	• Recall what global warming is. • Describe the causes of global warming. • Explain how global warming impacts on biodiversity.	4.7.3.5	Video clip about global warming; Internet access. Worksheet 8.17	Quick starter Homework worksheet Homework quiz Slideshow Videos

Lesson number	Lesson title	Lesson objectives	AQA specification reference	Lesson resources (on CD ROM)	Collins Connect resources
8.18	Looking at waste management	• Describe how waste production is linked to human population growth. • Describe the impact of waste on ecosystems. • Explain how waste impacts on biodiversity.	4.7.3.2	Video clip about bioaccumulation; Internet access. Worksheet 8.18	Quick starter Homework worksheet Homework quiz Video
8.19	Investigating pollution	• Identify pollution levels using indicator species. • Explain how indicator species measure pollution. • Compare different methods of measuring pollution.	4.7.3.2	Short video clip about indicator species. Worksheets 8.19.1 and 8.19.2; Practical sheet 8.19; Technician's notes 8.19	Quick starter Homework worksheet Homework quiz
8.20	Maintaining biodiversity	• Describe some conservation measures. • Describe the impact of breeding programmes. • Explain how habitats are regenerated.	4.7.3.6	Video clip about coral reefs; Internet access. Worksheets 8.20.1, 8.20.2 and 8.20.3	Quick starter Homework worksheet Homework quiz Videos
8.21	Learning about food security	• Identify factors affecting food security. • Describe how different factors affect food security. • Interpret data to evaluate food security.	4.7.5.1	Short video news clip about food shortages; news clip about food wastage; paper and pens for poster. Worksheets 8.21.1, 8.21.2 and 8.21.3	Quick starter Homework worksheet Homework quiz
8.22	Maintaining food security	• Describe some intensive farming methods. • Explain ethical issues related to intensive farming. • Evaluate modern farming techniques. • Describe methods to maintain	4.7.5.2	Images of battery chickens and free range chickens; video about one example of biological control. Worksheet 8.22	Quick starter Homework worksheet Homework quiz

Lesson number	Lesson title	Lesson objectives	AQA specification reference	Lesson resources (on CD ROM)	Collins Connect resources
		sustainable fisheries.			
8.23	Using biotechnology	• Describe some uses of biotechnology. • Explain the advantages of some uses of biotechnology. • Evaluate some uses of biotechnology.	4.7.5.4	Images of unusual genetically engineered organisms and a fermentation vat. Worksheets 8.23.1 and 8.23.2	Quick starter Homework worksheet Homework quiz Videos
8.24	Maths skills: Using graphs to show relationships	• To recognise direct proportionality in a graph. • To calculate reaction rates in linear graphs. • To use the gradient of a graph to calculate the rate.		Image of the World population growth curve. Worksheet 8.24	Quick starter Homework worksheet Homework quiz Slideshow Video
Assessments		End of chapter test Student Book End of chapter test Collins Connect End of teaching block test Collins Connect End of course test Collins Connect			

Using Collins AQA GCSE (9-1) Biology Student Book and Teacher Pack to teach AQA GCSE (9-1) Combined Science: Trilogy

The following table gives an overview of the Student Book pages and Lesson Plans that can be used to teach Combined Science: Trilogy.

Student Book and Lesson Plan Reference	**Chapter**	**Notes on how to use the resources for Combined Science: Trilogy**
Chapter 1	**Cell Biology**	
Student Book only	Chapter Introduction	Combined Science: Trilogy students do not need to know the content in the following boxes: Microorganisms can help to keep us healthy and provide us with food. Why is it important to study microorganisms, and how do we grow them in the lab and commercially?
1.1	Looking at cells	Combined Science: Trilogy students need to know and understand all content.
1.2	The light microscope	Combined Science: Trilogy students need to know and understand all content.
1.3	Looking at cells in more detail	Combined Science: Trilogy students need to know and understand all content.
1.4	Required practical: Using a light microscope to observe and record animal and plant cells	Combined Science: Trilogy students need to know and understand all content.
1.5	Primitive cells	Combined Science: Trilogy students need to know and understand all content.
1.6	Cell division	Combined Science: Trilogy students need to know and understand all content.
1.7	Cell differentiation	Combined Science: Trilogy students need to know and understand all content.
1.8	Cancer	Combined Science: Trilogy students need to know and understand all content.
1.9	Stem cells	Combined Science: Trilogy students need to know and understand all content.
1.10	Stem cell banks	Combined Science: Trilogy students need to know and understand all content.
1.11	Key concept: Cell development	Combined Science: Trilogy students need to know and understand all content.
1.12	Cells at work	Combined Science: Trilogy students need to know and understand all content.
1.13	Living without oxygen	Combined Science: Trilogy students need to know and understand all content.
1.14	Growing microorganisms	Combined Science: Trilogy students do not need to know any of this content.
1.15	Testing new antibiotics	Combined Science: Trilogy students do not need to know any of this content.
1.16	Required practical: Investigating disinfectants	Combined Science: Trilogy students do not need to know any of this content.

1.17	Maths skills: Size and number	Combined Science: Trilogy students need to know and understand all content.
	Check your progress/Worked example	Combined Science: Trilogy students do not need to know the content in the following boxes: [green] describe equipment, materials and procedures required to work with microorganisms [blue] describe the process of binary fission [pink] be able to calculate numbers of microorganisms produced given the mean generation time
	End of chapter questions	All questions are appropriate for Combined Science: Trilogy students
Chapter 2	**Photosynthesis**	
2.1	Explaining photosynthesis	Combined Science: Trilogy students need to know and understand all content.
2.2	Looking at photosynthesis	Combined Science: Trilogy students need to know and understand all content.
2.3	Investigating leaves	Combined Science: Trilogy students need to know and understand all content.
2.4	Required practical: Investigate the effect of light intensity on the rate of photosynthesis using an aquatic organism such as pondweed	Combined Science: Trilogy students need to know and understand all content.
2.5	Increasing photosynthesis	Combined Science: Trilogy students need to know and understand all content.
2.6	Increasing food production	Combined Science: Trilogy students need to know and understand all content.
2.7	Key concept: Diffusion in living systems	Combined Science: Trilogy students need to know and understand all content.
2.8	Looking at stomata	Combined Science: Trilogy students need to know and understand all content.
2.9	Moving water	Combined Science: Trilogy students need to know and understand all content.
2.10	Investigating transpiration	Combined Science: Trilogy students need to know and understand all content.
2.11	Moving sugar	Combined Science: Trilogy students need to know and understand all content.
2.12	Maths skills: Surface area to volume ratio	Combined Science: Trilogy students need to know and understand all content.
	Check your progress/Worked example	Combined Science: Trilogy students need to know and understand all content.
	End of chapter questions	All questions are appropriate for Combined Science: Trilogy students
Chapter 3	**Moving and changing materials**	
3.1	Explaining water movement	Combined Science: Trilogy students need to know and understand all content.
3.2	Required practical: Investigate the effect of a range of salt or sugar solutions on the mass of plant tissue	Combined Science: Trilogy students need to know and understand all content.

3.3	Learning about active transport	Combined Science: Trilogy students need to know and understand all content.
3.4	Key concept: Investigating the need for transport systems	Combined Science: Trilogy students need to know and understand all content.
3.5	Explaining enzymes	Combined Science: Trilogy students need to know and understand all content.
3.6	Required practical: Investigate the effect of pH on the rate of reaction of amylase enzyme	Combined Science: Trilogy students need to know and understand all content.
3.7	Learning about the digestive system	Combined Science: Trilogy students need to know and understand all content.
3.8	Explaining digestion	Combined Science: Trilogy students need to know and understand all content.
3.9	Required practical: Use qualitative reagents to test for a range of carbohydrates, lipids and proteins	Combined Science: Trilogy students need to know and understand all content.
3.10	Looking at more exchange surfaces	Combined Science: Trilogy students need to know and understand all content.
3.11	Learning about plants and minerals	Combined Science: Trilogy students need to know and understand all content.
3.12	Investigating how plants use minerals	Combined Science: Trilogy students do not need to know the content of: [blue section] Investigating minerals
3.13	Learning about the circulatory system	Combined Science: Trilogy students need to know and understand all content.
3.14	Exploring the heart	Combined Science: Trilogy students need to know and understand all content.
3.15	Studying blood	Combined Science: Trilogy students need to know and understand all content.
3.16	Investigating gas exchange	Combined Science: Trilogy students need to know and understand all content.
3.17	Learning about coronary heart disease	Combined Science: Trilogy students need to know and understand all content.
3.18	Maths skills: Extracting and interpreting information	Combined Science: Trilogy students need to know and understand all content.
	Check your progress/Worked example	Combined Science: Trilogy students need to know and understand all content.
	End of chapter questions	All questions are appropriate for Combined Science: Trilogy students.
Chapter 4	**Health matters**	
Student Book only	Chapter Introduction	Combined Science: Trilogy students do not need to know the content in the box: How are plants affected by disease and protected from attack?
4.1	Learning about health	Combined Science: Trilogy students need to know and understand all content.

4.2	Key concept: Looking at risk factors	Combined Science: Trilogy students need to know and understand all content.
4.3	Exploring non-communicable diseases	Combined Science: Trilogy students need to know and understand all content.
4.4	Analysing and evaluating data	Combined Science: Trilogy students need to know and understand all content.
4.5	Studying pathogens	Combined Science: Trilogy students need to know and understand all content.
4.6	Learning about viral diseases	Combined Science: Trilogy students need to know and understand all content.
4.7	Studying bacterial diseases	Combined Science: Trilogy students need to know and understand all content.
4.8	Looking at fungal diseases	Combined Science: Trilogy students need to know and understand all content.
4.9	Learning about malaria	Combined Science: Trilogy students need to know and understand all content.
4.10	Protecting the body	Combined Science: Trilogy students need to know and understand all content.
4.11	Exploring white blood cells	Combined Science: Trilogy students need to know and understand all content.
4.12	Using antibiotics and painkillers	Combined Science: Trilogy students need to know and understand all content.
4.13	Building immunity	Combined Science: Trilogy students need to know and understand all content.
4.14	Making new drugs	Combined Science: Trilogy students need to know and understand all content.
4.15	Investigating monoclonal antibodies	Combined Science: Trilogy students do not need to know any of this content.
4.16	Looking at plant diseases	Combined Science: Trilogy students do not need to know any of this content.
4.17	Learning about plant defences	Combined Science: Trilogy students do not need to know any of this content.
4.18	Maths skills: Sampling and scientific data	Combined Science: Trilogy students need to know and understand all content.
	Check your progress/Worked example	Combined Science: Trilogy students do not need to know the content in the following Check your progress boxes: [green] recall some plant diseases and plant defences [blue] explain the identification and symptoms of some plant diseases [blue] explain some plant defences [pink] explain the use of monoclonal antibodies in identifying plant pathogens
	End of chapter questions	Combined Science: Trilogy students do not need to answer the following questions: 11, 12 18.
Chapter 5	**Coordination and control**	

Student Book only	Chapter Introduction	Combined Science: Trilogy students do not need to know the content in the following boxes: Conditions in the body, processes and organs are coordinated and controlled: the sub points relating to the brain and eye. Control of metabolism and levels of chemicals in the body: the two sub points relating to the kidneys. Plants respond to stimuli to control important processes: all content.
5.1	Homeostasis	Combined Science: Trilogy students do not need to know about thermoregulation.
5.2	The nervous system	Combined Science: Trilogy students need to know and understand all content.
5.3	Reflex actions	Combined Science: Trilogy students need to know and understand all content.
5.4	The brain	Combined Science: Trilogy students do not need to know any of this content.
5.5	Required practical: Investigating reaction time	Combined Science: Trilogy students need to know and understand all content.
5.6	The eye	Combined Science: Trilogy students do not need to know any of this content.
5.7	Seeing in focus	Combined Science: Trilogy students do not need to know any of this content.
5.8	Eye defects	Combined Science: Trilogy students do not need to know any of this content.
5.9	Controlling body temperature	Combined Science: Trilogy students do not need to know any of this content.
5.10	The endocrine system	Combined Science: Trilogy students need to know and understand all content.
5.11	Controlling blood glucose	Combined Science: Trilogy students need to know and understand all content.
5.12	Diabetes	Combined Science: Trilogy students need to know and understand all content.
5.13	Diabetes recommendations	Combined Science: Trilogy students need to know and understand all content.
5.14	Water balance	Combined Science: Trilogy students do not need to know any of this content.
5.15	The kidneys	Combined Science: Trilogy students do not need to know any of this content.
5.16	Negative feedback	Combined Science: Trilogy students need to know and understand all content.
5.17	Kidney failure	Combined Science: Trilogy students do not need to know any of this content.
5.18	Dialysis or transplant?	Combined Science: Trilogy students do not need to know any of this content.
5.19	Human reproduction	Combined Science: Trilogy students need to know and understand all content.
5.20	IVF	Combined Science: Trilogy students need to know and understand all content.
5.21	IVF evaluation	Combined Science: Trilogy students need to know and understand all content.

5.22	Systems working together	Combined Science: Trilogy students need to know and understand all content.
5.23	Contraception	Combined Science: Trilogy students need to know and understand all content.
5.24	Which contraceptive?	Combined Science: Trilogy students need to know and understand all content.
5.25	Auxins	Combined Science: Trilogy students need to know and understand all content.
5.26	Applications of auxins	Combined Science: Trilogy students do not need to know any of this content.
5.27	Required practical: The effect of light and gravity on the growth of germinating seedlings	Combined Science: Trilogy students do not need to know any of this content.
5.28	Other plant hormones	Combined Science: Trilogy students do not need to know any of this content.
5.29	Maths skills: The spread of scientific data	Combined Science: Trilogy students need to know and understand all content.
	Check your progress/Worked example	Combined Science: Trilogy students do not need to know the content in the following Check your progress boxes: [green] recall the three regions of the brain and their location [blue] describe the function of the brain regions [pink] explain the processes involved and why investigating the brain and treating disorders is difficult [green] identify the structure and function of parts of the eye [blue] describe how images are focused on the retina and what causes eye defects [pink] explain how eye defects can be corrected [blue] describe how these are controlled by the nervous and/or endocrine system [green] identify types of plant hormone and their role in plant growth [blue] describe the role of redistribution of auxins in tropisms [pink] describe applications of plant hormones The examples in the Worked example are not relevant for Combined Science: Trilogy students.
	End of chapter questions	Combined Science: Trilogy students do not need to answer the following questions: 6, 9, 10, 14, 16, 17.
Chapter 6	**Genetics**	
Student Book only	Chapter Introduction	Combined Science: Trilogy students do not need to know the content in the following boxes: Our understanding of DNA and the way genes work: sub points relating to nucleotides, bases, protein synthesis and mutations. Characteristics are inherited from one generation to the next: sub point relating to Gregor Mendel.
6.1	DNA and genes	Combined Science: Trilogy students do not need to know the following: [green section] Genes work by providing the code for the production of a particular protein. [pink section] Certain other parts of the cell also contain DNA – for example, mitochondria

		Plasmids are not part of the chromosome, so do not contribute to the running of the cell, but they have genes that help the survival of the organism.
6.2	The human genome	Combined Science: Trilogy students do not need to know the following: Gene expression.
6.3	Tracing human migration	Combined Science: Trilogy students need to know and understand all content.
6.4	The structure of DNA	Combined Science: Trilogy students do not need to know any of this content.
6.5	Proteins	Combined Science: Trilogy students do not need to know any of this content.
6.6	Mutations	Combined Science: Trilogy students do not need to know any of this content.
6.7	Meiosis	Combined Science: Trilogy students need to know and understand all content.
6.8	Asexual and sexual reproduction	Combined Science: Trilogy students do not need to know: [green section] the advantages and disadvantages of sexual and asexual reproduction.
6.9	Genetics	Combined Science: Trilogy students need to know and understand all content.
6.10	Genetic crosses	Combined Science: Trilogy students need to know and understand all content.
6.11	Tracking gene disorders	Combined Science: Trilogy students need to know and understand all content.
6.12	Gregor Mendel	Combined Science: Trilogy students do not need to know any of this content.
6.13	Key concept: Genetics is simple – or is it?	Combined Science: Trilogy students need to know and understand all content.
6.14	Maths skills: Fractions, ratio, proportion and probability	Combined Science: Trilogy students need to know and understand all content.
	Check your progress/Worked example	Combined Science: Trilogy students do not need to know the content in the following Check your progress boxes: [pink] understand that genes work by coding for the production of a particular protein, and non-coding genes switch genes on and off [blue] describe the structure of nucleotides as a sugar molecule, phosphate groups and bases [pink] explain that strands of DNA are complementary [green] identify how a gene works in coding for the production of a protein. [blue] describe how a gene controls the order of amino acids in a protein. [pink] describe in outline how a protein is synthesised. [green] know that there are many different types of protein [blue] explain that the structure of a protein is related to its molecular shape [pink] explain that a protein folds into a unique shape that enables it to function [blue] describe advantages and disadvantages of each type (asexual and sexual reproduction)

		[green] identify a mutation as a change in the DNA. [blue] describe how mutations can, on rare occasions, affect how an organism functions. [pink] explain how mutations can affect protein synthesis and structure.
	End of chapter questions	Combined Science: Trilogy students do not need to answer the following questions: 10–14, 18.
Chapter 7	**Variation and evolution**	
7.1	Variation	Combined Science: Trilogy students need to know and understand all content.
7.2	The theory of evolution	Combined Science: Trilogy students do not need to know: [green section] An early theory of evolution: all content.
7.3	The origin of species by natural selection	Combined Science: Trilogy students need to know and understand all content.
7.4	Fossil evidence	Combined Science: Trilogy students need to know and understand all content.
7.5	How much have organisms changed?	Combined Science: Trilogy students need to know and understand all content.
7.6	Darwin and Wallace	Combined Science: Trilogy students do not need to know any of this content.
7.7	A new species	Combined Science: Trilogy students do not need to know any of this content.
7.8	Evidence of natural selection and evolution?	Combined Science: Trilogy students need to know and understand all content.
7.9	Key concept: Evolution: fitting the pieces of the jigsaw	Combined Science: Trilogy students do not need to know any of this content.
7.10	Antimicrobial resistance	Combined Science: Trilogy students need to know and understand all content.
7.11	Combatting antimicrobial resistance	Combined Science: Trilogy students need to know and understand all content.
7.12	Selective breeding	Combined Science: Trilogy students need to know and understand all content.
7.13	Producing new plant varieties	Combined Science: Trilogy students need to know and understand all content.
7.14	Genetic engineering	Combined Science: Trilogy students need to know and understand all content.
7.15	Genetically modified crops: the science	Combined Science: Trilogy students need to know and understand all content.
7.16	Is genetic modification safe?	Combined Science: Trilogy students need to know and understand all content.
7.17	Ethically wrong, or essential?	Combined Science: Trilogy students need to know and understand all content.
7.18	Cloning	Combined Science: Trilogy students need to know and understand all content.
7.19	The tree of life	Combined Science: Trilogy students need to know and understand all content.

7.20	Extinction ... or survival?	Combined Science: Trilogy students need to know and understand all content.
7.21	Maths skills: Using charts and graphs to display data	Combined Science: Trilogy students need to know and understand all content.
	Check your progress/Worked example	Combined Science: Trilogy students do not need to know the content in the following boxes: [blue] describe the evidence that led to the development of the theory of evolution by Darwin and Wallace [pink] explain the evidence for the occurrence of evolution and natural selection
	End of chapter questions	All questions are appropriate for Combined Science: Trilogy students.
Chapter 8	**Ecology in action**	
Student Book only	Chapter introduction	Combined Science: Trilogy students do not need to know the content in the following Check your progress boxes: What factors affect living organisms in a habitat? The following sub points: How energy is lost from each trophic level in a food chain. Why there is a maximum size to food chains. How pyramids of biomass show the mass of organisms in a food chain. How are materials in a community cycled? The following sub point: How we use the process of decay to produce compost and biogas.
8.1	Key concept: Learning about ecosystems	Combined Science: Trilogy students need to know and understand all content.
8.2	Changing abiotic factors	Combined Science: Trilogy students need to know and understand all content.
8.3	Investigating predator–prey relationships	Combined Science: Trilogy students need to know and understand all content.
8.4	Looking at trophic levels	Combined Science: Trilogy students do not need to know the sections relating to trophic levels and pyramids of biomass
8.5	Transferring biomass	Combined Science: Trilogy students do not need to know any of the content on this page.
8.6	Competing for resources	Combined Science: Trilogy students need to know and understand all content.
8.7	Required practical: Measure the population size of a common species in a habitat	Combined Science: Trilogy students need to know and understand all content.
8.8	Adapting for survival in animals	Combined Science: Trilogy students need to know and understand all content.
8.9	Adapting for survival in plants	Combined Science: Trilogy students need to know and understand all content.
8.10	Cycling materials	Combined Science: Trilogy students do not need to know [green section] factors that speed up decay]

8.11	Cycling carbon	Combined Science: Trilogy students need to know and understand all content.
8.12	Investigating decay	Combined Science: Trilogy students do not need to know any of the content on this page.
8.13	Required practical: Investigate the effect of temperature on the rate of decay of fresh milk by measuring pH change	Combined Science: Trilogy students do not need to know any of the content on this page.
8.14	Changing the environment	Combined Science: Trilogy students do not need to know any of the content on this page.
8.15	Learning about land use	Combined Science: Trilogy students need to know and understand all content.
8.16	Changing the landscape	Combined Science: Trilogy students need to know and understand all content.
8.17	Thinking about global warming	Combined Science: Trilogy students need to know and understand all content.
8.18	Looking at waste management	Combined Science: Trilogy students need to know and understand all content.
8.19	Investigating pollution	Combined Science: Trilogy students need to know and understand all content.
8.20	Maintaining biodiversity	Combined Science: Trilogy students need to know and understand all content.
8.21	Learning about food security	Combined Science: Trilogy students do not need to know any of the content on this page.
8.22	Maintaining food security	Combined Science: Trilogy students do not need to know any of the content on this page.
8.23	Using biotechnology	Combined Science: Trilogy students do not need to know any of the content on this page.
8.24	Maths skills: Using graphs to show relationships	Combined Science: Trilogy students need to know and understand all content.
	Check your progress/Worked example	Combined Science: Trilogy students do not need to know the content in the following Check your progress boxes: [green] explain what a trophic level is [pink] construct accurate pyramids of biomass [green] identify factors affecting food security [blue] describe how different factors affect food security [pink] interpret data to evaluate food security [green] recall causes of environmental change [blue] describe the impact of environmental change [pink] explain the impact of environmental change [green] describe some uses of biotechnology [blue] explain the advantages of some uses of biotechnology [pink] evaluate some uses of biotechnology
	End of chapter questions	Combined Science: Trilogy students do not need to answer the following questions: 3, 5, 7, 8, 10, 11, 14, 15, 16, 17, 19.